SEMICONDUCTOR DEVICES

McGraw-Hill Series in Electrical Engineering

Consulting Editor

Stephen W. Director, Carnegie-Mellon University

CIRCUITS AND SYSTEMS

COMMUNICATIONS AND SIGNAL PROCESSING

CONTROL THEORY

ELECTRONICS AND ELECTRONIC CIRCUITS

POWER AND ENERGY

ELECTROMAGNETICS

COMPUTER ENGINEERING

INTRODUCTORY

RADAR AND ANTENNAS

VLSI

Previous Consulting Editors

Ronald N. Bracewell, Colin Cherry, James F. Gibbons,
Willis W. Harman, Hubert Heffner, Edward W. Herold, John G. Linvill,
Simon Ramo, Ronald A. Rohrer, Anthony E. Siegman, Charles Susskind,
Frederick E. Terman, John G. Truxal, Ernst Weber, and John R. Whinnery

Electronics and Electronic Circuits

Consulting Editor

Stephen W. Director: *Carnegie-Mellon University*

SEMICONDUCTOR DEVICES

Mauro Zambuto

Department of Electrical Engineering
New Jersey Institute of Technology

McGraw-Hill Book Company

New York St. Louis San Francisco Auckland Bogotá Caracas
Colorado Springs Hamburg Lisbon London Madrid Mexico Milan
Montreal New Delhi Oklahoma City Panama Paris San Juan
São Paulo Singapore Sydney Tokyo Toronto

This book was set in Times Roman.
The editors were Alar E. Elken and John M. Morriss;
the production supervisor was Denise L. Puryear.
The cover was designed by Joseph Gillians.
Project supervision was done by Santype International Limited.
R. R. Donnelley & Sons Company was printer and binder.

SEMICONDUCTOR DEVICES

234567890 DOCDOC 89432109

ISBN 0-07-072700-7

Library of Congress Cataloging-in-Publication Data

Zambuto, Mauro.
    Semiconductor devices/Mauro Zambuto.
      p. cm.
    Includes bibliographies and index.
    ISBN 0-07-072700-7 (text).
    ISBN 0-07-072701-5 (solutions manual)
    1. Semiconductors. I. Title.
 TK7871.85.Z35 1989   88-16969
 621.36'6--dc19

ABOUT THE AUTHOR

Dr. Mauro Zambuto holds Ph.D.s in physics and electrical engineering from the Universities of Padua and Rome, Italy, where he also conducted early research in quantum and nuclear physics under Professor Enrico Fermi.

As a motion picture engineer he built, equipped, and directed motion picture studios on both sides of the Atlantic. His research contribution to the field includes several novel techniques and specialized equipment and to date he maintains his activity and interest in the audiovisual industry as a researcher and consultant.

Since 1957 his primary activity has been academic. He has been a member of the Electrical Engineering Faculty of the City College of New York, Manhattan College, and is now a Distinguished Professor of Electrical Engineering at the New Jersey Institute of Technology. As a consultant to the United Nations he has contributed to the establishment of universities in Africa and South America and to the development of novel educational techniques.

During the last 20 years his research has been primarily concerned with quantum electronics, coherent optics, optical and acoustical holography, and nuclear physics, both in the United States and abroad. He is also a consultant to industry, scientific, and learning institutions on a wide variety of subjects. A master of ten languages, he frequently presides over international scientific and professional conferences.

CONTENTS

PREFACE

This textbook is intended primarily for a one-semester course for electrical engineering students at the sophomore-junior level. To allow implementation of several possible educational preferences, provision is made for presentation in a number of modes to be discussed later in this preface.

Intuitive visualization of the phenomena involved is emphasized throughout. To this end each subject is first introduced in a logical, but purely qualitative way, then, after qualitative understanding is achieved, the logical sequence is repeated in the language of mathematics, resulting in a more rigorous, quantitative discussion. The underlying educational purpose is manyfold:

(1) To allow the student to concentrate attention on the physically intuitive aspects of the phenomenon, free from the masking effect sometimes exerted by the intricacies of the mathematical process.

(2) To discourage the student's tendency to reduce his effort to the memorization of formulas, to be used as a recipe for problem solving, and instead focus his attention on visualization of the physical concepts.

A formula, as a sequence of mathematical symbols, can be easily retained for a short time, but, after the quiz is over, tends to fade away, while a physical concept tends to stay with us for the rest of our lives.

(3) To show that a valid mathematical formulation can usually be obtained by simply translating into the language of mathematics the same common sense reasoning that led us to the intuitive understanding of the phenomena in the first place. This fact is systematically emphasized in the text.

Training oneself to perform such translation from common sense to formula and vice versa is, in the author's opinion, the most important single asset of the successful engineer and scientist.

(4) To permit limiting the presentation to the qualitative discussion, if so desired, by omitting the mathematical formulation, or postponing its analysis to a subsequent phase, as discussed later in this preface.

Presentation of phenomena at the intuitive level usually leads to the formulation of models and simulations. Validity and rigor of the formulation may suffer, especially when quantum physics concepts are involved. Whenever such a danger occurs, the reader is alerted that the model is used merely as a plausibility argument, rather than valid proof. The facts are then restated in more rigorous form and, if dogmatic assertions are introduced, usually reference is made to more advanced literature.

Model presentation can greatly benefit from the use of modern technological aids. To this end several computer-aided simulations are provided as supporting software. The decision whether to use, not to use, or even mention the availability of the software is left to the exclusive discretion of the instructor. The textbook is designed to be efficiently usable in the conventional way, without this aid.

Extensive class testing under statistically valid sample conditions indicates that use of the software may permit wider coverage of the subject matter (about 35 percent). Significantly higher student performance in conventional tests and improved carryover are also consistently indicated.

As previously stated the book can be used in several modes of presentation:

(1) As a stand-alone, one semester course. If desired, several variations are possible in this mode:

Appendices can be bypassed without prejudice for the continuity, with the exception of Sec. 10.10, which uses material from App. 6.

The discussion of approximate formulas in Secs. 8.3, 8.5, 9.4, and 9.5 and the examples discussing their accuracy can be omitted if the more accurate formulas are consistently used (e.g. if computers are utilized).

Several of the more specialized topics can evidently be omitted or just referred to in passing, at the instructor's discretion.

(2) If the subject matter is to be presented not as a stand-alone course, but incorporated in one or more courses of the usual electronics sequence, the book can be used together with any one of several electronic circuit textbooks.

In this case it is recommended that the material be presented in two parts:

(*a*) Chapters 1 through 4 should be introduced at or near the beginning of the course (possibly omitting the more mathematical portions of Chapters 3 and 4). Then, the introduction of each new device as a circuit element should be followed by a purely qualitative description of its physical principles of operation exclusively to justify the device circuit characteristics. The separate quantitative sections should be omitted. A sufficient number of qualitative problems is provided to support this mode.

(*b*) Much later, after familiarity with the device has been achieved, device physics should again be addressed covering the quantitative sections of the text. The engineering maturity and experience gathered in the meantime will ensure prompter and deeper appreciation of the meaning and implications of the material, with significant saving of class time.

(3) It is sometimes argued that quantitative information about the physics and manufacturing technology of semiconductor devices is not required of all

engineers and that only a qualitative introduction is sufficient for the non-specialists.

If it is desired to implement this educational philosophy, simply omit all quantitative analysis sections. This can be done without loss of continuity.

Similarly, if it has been decided not to include a detailed discussion of the fabrication techniques, the topic of fabrication can be omitted without prejudice for the continuity. Chapter 5 and the fabrication sections of Chapters 6 and 8 through 10 can be bypassed. Conversely a brief (half hour) introduction to the basic fabrication steps (epitaxy, oxidation, photolithography, diffusion, and metallization) is sufficient to support the presentation of the fabrication sequences in Chapters 6 through 10.

(4) The above abridged mode is also appropriate for an engineering technology or introductory junior college course. The special emphasis on intuitive visualization and the computer simulations are particularly effective in this case.

(5) An eight week minicourse, corresponding to 1.5 credits, has been successfully conducted by omitting some of the more advanced topics and abridging the fabrication sections as previously indicated. Computer simulation has significantly contributed in saving class time.

(6) Computer simulation has proved well suited to individual study, especially by engineers who desire to become familiar with, or to expand their understanding of, device physics and fabrication.

The software provided can be used at three different levels:

(1) In the classroom, as a complement to (not a substitute for) the blackboard. Advantages are: quantitatively accurate figures and data, animation (adding the time dimension), saving of class time when complex drawings are involved, and, most importantly, the ease and speed of presentation and analysis of many examples permitted by the interactive feature. By appropriate choice of parameters, specific peculiarities of the phenomena can be vividly pointed out and class questions answered most effectively and expeditiously. Statistical evidence also indicates that computer simulation promotes more immediate and permanent recall and better overall visualization.

(2) Use at home, or in study periods, depending on computer availability (the students can be provided with copies of the software, which is not protected). This provides more than just animation of the book figures. The effect of varying parameters can be analyzed by making quantitative predictions and checking them against the simulation. The ease with which the computer can handle calculations affords the opportunity to develop familiarity with realistic values and to clarify elusive concepts, such as sensitivity, by observing the effects of parameter variations, providing the equivalent of hours of computation and laboratory practice.

(3) Development of personalized CAD-type software for the solution of device-related problems. The set of problems singled out for computer solution at the end of several chapters can be of assistance in reaching this goal, especially if

the instructor provides useful hints about the appropriate use of subroutines, as exemplified in the instructor manual.

A copy of the software will be supplied to all instructors who adopt this text for their course. For those who purchase this book independently, you can receive the disk by contacting McGraw-Hill, College Division at 212/512-2756.

It is not possible to mention individually all the people who have significantly contributed to this work. I wish to thank each and all of them collectively. Special thanks are due to AT&T for their donation of equipment and funds without which class experimentation of the new pedagogical methods would not have been possible. I would like to express my special appreciation to the reviewers of the manuscript for their valued constructive criticism: Franco Cerrina, University of Wisconsin—Madison; David Dumin, Clemson University; Anand K. Kulkarni, Michigan Technological University; Wolfgang Porod, University of Notre Dame; and Patrick Roblin, The Ohio State University. Thanks also go to my colleagues at the New Jersey Institute of Technology, especially Dr. W. Carr for making available the equipment of the MJIT Microelectronics Laboratory, Dr. R. Misra, Dr. A. Meyers, Dr. R. Cornely, and Dr. J. Carpinelli for their advice and suggestions and to the administration of the New Jersey Institute of Technology, without whose support this work could not have been completed.

Mauro Zambuto

INTRODUCTION

Modern semiconductor devices can perform some remarkable tasks. To understand how such behavior is possible it is necessary to visualize the physical mechanisms by which atomic particles in solids interact with energy, storing and releasing it.

As is well known, atomic phenomena cannot be accurately described in terms of classical physics. A rigorous discussion of these mechanisms requires application of the methods of modern quantum theory. Unfortunately, the level of proficiency in both physics and mathematics required to perform this task successfully is not available to the majority of engineering undergraduates.

In the following, however, an effort will be made to generate an intuitive insight into these phenomena (and even some proficiency in handling quantitative results) exclusively on the basis of the student's background in basic classical physics. In addition, whenever necessary, some of the fundamental results and tenets of quantum physics will be introduced in a purely dogmatic way. Previous familiarity with modern quantum theory, with the attendant comparatively sophisticated level of mathematical skills, is therefore not required, making things easier for everybody concerned.

The resultant semiclassical treatment and the "models" introduced will, of course, be somewhat lacking in rigor, but experience shows that a satisfactory level of understanding can be achieved by these means, at least for the purposes of the nonspecialist in the field, and such semiclassical treatment is often adopted in the literature.

The previous considerations should, however, be kept in mind as they limit both the validity of some of the models drawn and the rigor of the presentation offered. The interested reader is referred to the abundant higher-level literature in the field and is definitely encouraged to acquire some background in modern quantum physics.

For some particularly important or interesting background material, specific reference to the literature, often including chapter and page, is offered. Some of this material is very elementary and in many cases the student can be expected to have been exposed to it in previous courses; some is definitely at the graduate level. Furthermore, a few of these references offer a justification in terms of quantum physics of the most important dogmatic statements made in the text, in an effort to accommodate the reader having some familiarity with quantum theory.

None of this material is essential, at least if the reader is willing to accept some basic tenets on faith.

Wherever the presentation requires familiarity with quantum theory, or is above the average undergraduate level, the reader is alerted by an appropriate remark.

Mauro Zambuto

SEMICONDUCTOR DEVICES

CHAPTER

1

PHYSICAL SYSTEMS

1.1 ENERGY AND STATES

A physical *system* consists of a set of physical components and of elements interconnecting them. Because they are interconnected, the components exert actions on one another, so that, as long as they are in the system, they are subject to mutual *constraints*. As a consequence, the components can coexist in the system only in physical configurations consistent with the constraints. Each of these configurations constitutes a *state*.

For instance, define a system composed of a train and its tracks: the only states in which this system can exist are those for which the train is on the tracks and its velocity is tangent to the rail.

To each state there generally corresponds a certain amount of energy stored in the system in some form. Consequently, to undergo a transition from one state to another characterized by a different amount of stored energy, the system must acquire or release some energy, so that some means to impart (or, as commonly stated, *input*) energy into the system must be available. It should be noted that the energy input to the system can be positive or negative. A negative energy input represents a loss of energy (energy released).

Without entering into deeper and more general considerations, the important concepts of system and state will be illustrated by reference to Fig. 1.1.1.

The physical system is represented in the left portion of the figure. The main physical component of this system is the weight. The constraining elements are the rope, the pulley, and the little man, providing the balancing force required to keep the system in equilibrium with the weight at any height. At first it will be assumed that the system is static, i.e., we shall consider it only while at rest. The

1

Mass = 7 kg
Height = 7.2812 m

Energy = 500 J

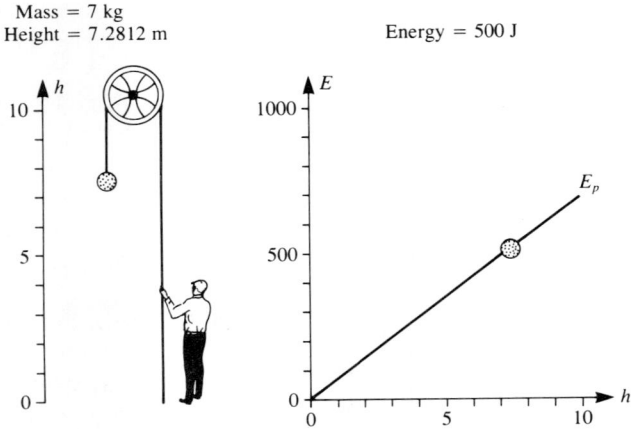

FIGURE 1.1.1
Classical static physical system.

motion necessary for the transition from one position to another will be discounted.

With this limitation, to each possible height between 0 and 10 m of the weight there corresponds a system state, and each state is characterized by an energy:

$$E = 9.81mh \qquad \text{J} \qquad (1.1.1)$$

stored in the system in the form of potential energy E_p.

It is important to realize that each state can be characterized not only by the geometrical configuration of the physical system (the height of the weight), but also by other parameters, such as the potential energy. Therefore the states can be represented in *state diagrams*, such as the one in the right portion of Fig. 1.1.1, in which potential energy is mapped vs. weight position and the current condition is represented by the dot shown in the figure.

Suppose that initially the weight is on the ground. Then the potential energy initially stored in the system (reference value) is zero. The little man can raise or lower the weight by inputting the corresponding amount of energy into the system, thereby forcing a *transition* to another state. The dot representing the current condition in the state diagram will shift accordingly.

Example 1.1.1. Assume the mass and the potential energy of the weight to be 7 kg and 500 J respectively. Compute the height of the weight and draw the energy position diagram indicating the instantaneous condition point.

Solution. Using (1.1.1),

$$h = \frac{500}{9.81 \times 7} = 7.2812 \text{ m}$$

The potential energy line has equation

$$E = 68.67 \, h$$

and therefore is a straight line between point (0, 0) and point (10, 686.7). On it the instantaneous condition point has coordinates (7.2812, 500). The line and point are both shown in Fig. 1.1.1.

Notice that the state diagram provides complete information about all of the above data, whereas additional information and some computation are required to obtain it from the geometric system representation. It often happens that important physical properties of the system under consideration are easier to visualize with the help of an appropriate state diagram than by observing the physical system configuration alone. For instance, if the system is constrained to be a static one (i.e., it can store energy only in potential form) then the only states in which it can exist (the *permissible* states) are the ones represented by points along the line shown in the diagram. All other states are *forbidden* ones. Notice how permissible and forbidden states can be identified with greater ease in the state diagram.

The student is encouraged to consider several transitions, using different values of masses, heights, and/or energies, predicting the outcomes and drawing state diagrams and instantaneous condition points.

In this case, because of the static constraint imposed on the system, the terms *state* and *instantaneous system condition* can be used interchangeably. This, however, is not always the case, as will become apparent in Sec. 1.3, when the static constraint will be released.

In practice the concepts of *physical system, states, permissible and forbidden states, state diagrams*, etc., are all extremely useful in discussing atomic systems and the student is urged to gain familiarity with them with the aid of Fig. 1.1.1. In particular, it should again be emphasized that, in our example, not all points of the state diagram represent permissible conditions. Indeed, this system is constrained to be a static one; consequently, as already pointed out, all points of the diagram outside of the line shown are forbidden states. The student should justify the above statement by appropriate considerations.

In the example given, the state diagram is represented in terms of energy vs. position (or, in the jargon of physicists and mathematicians, in *energy-position space*), but representations in other spaces are also possible. Several examples will be encountered later on.

In the system shown the little man at the rope has a double role: he constitutes the means for permitting the system to exist in any one state (by providing the balancing force required to keep the weight at any given height) and also for introducing energy into (or releasing it from) the system. By thus changing the energy stored in the system, it is possible to force a *transition* from one state to another.

Notice that, in this case, no matter what state the system is in, it is possible to impart to (or take away from) it any amount of energy, even an infinitesimal one. The system will store or release it and a transition to another permissible state will occur. Such a system is said to be *continuous*. This does not happen in all systems.

1.2 CLASSICAL DISCONTINUOUS ENERGY SYSTEM

In the system shown in Fig. 1.2.1 the main physical component is the weight, just as in the previous system, but the constraining element is now the staircase-like structure indicated in the left side of the figure. As in the previous example, in order to draw the energy-position diagram, the mass of the weight must be quantitatively specified. In Fig. 1.2.1 this mass is assumed to be 8.5 kg. The state of the system can be changed by moving the weight up and down along the staircase. This, of course, requires a corresponding energy input. The means for imparting this energy to the system is not indicated, being immaterial.

Let the static constraint be still in effect. If an arbitrary amount of energy E is imparted to the system, when in the reference state, the weight is raised to a height

$$h = \frac{E}{9.81m} \qquad \text{m} \tag{1.2.1}$$

just as in the previous system; however, if this height does not correspond to that of an existing step, the staircase is unable to support the weight at that height. The weight will then fall onto the next lower step and find there the closest permissible condition of static equilibrium.

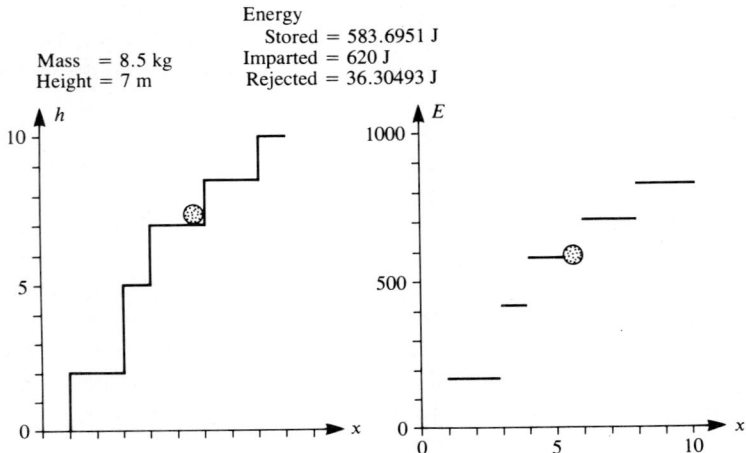

Energy
Stored = 583.6951 J
Mass = 8.5 kg   Imparted = 620 J
Height = 7 m   Rejected = 36.30493 J

FIGURE 1.2.1
Classical static discontinous system.

It may be useful to restate the above considerations in different terms. Under static conditions this system is able to store energy only in amounts corresponding to the available step heights. Such amounts characterize the permissible static states of the system. They constitute a discontinuous (or *discrete*) set. All other static states are forbidden states. If an arbitrary amount of energy is imparted, in most cases the system will not be able to store it all, but will instead undergo a transition to the next lower permissible state, *rejecting the extra energy*.

Example 1.2.1. Assume the system of Fig. 1.2.1 to be originally in the ground (zero energy) state with a mass of 8.5 kg. At $t = 0$, 620 J of energy are imparted to the system. Draw the state diagram and describe quantitatively the sequence of events both in the physical system and from the point of view of the state diagram.

Solution. The energy diagram after the transition is as indicated in Fig. 1.2.1. At $t = 0$ the weight jumps to a height of $620/(9.81 \times 8.5)$ or 7.435 m. As there is no step at this height, the weight will come to rest at the next lower step height, or 7 m, a fall of $7.435 - 7 = 0.435$ m. In the state diagram the instantaneous condition point will first go to $x = 5.5$ m (by inspection) and $E = 620$ J. As this is a forbidden energy level, however, it will store only an energy $7 \times 9.81 \times 8.5 = 583.7$ J, rejecting $620 - 583.7 = 36.3$ J. This quantity agrees with the 0.435-m fall computed above $(0.435 \times 9.81 \times 8.5 = 36.3$ J).

The student should make sure that these simple concepts are crystal clear by assuming several different values of system parameters and quantitatively predicting the sequence of events.

In terms of the energy-position state diagram, the above means that the potential energy curve is discontinuous, indicating that, in any given state, the system is constrained to store energy only in finite amounts (*quantum jumps*).

Notice that, in this classical system, this property is contingent on the static constraint imposed on the system. If this constraint were released, the system would jump to a new, permissible dynamic state, as illustrated in the following section.

1.3 CLASSICAL DYNAMIC SYSTEM

In the system of Fig. 1.3.1, the mass is $m = 7$ kg and the only constraining element is the floor. *Let the static constraint be released*, so that the system is permitted to store energy in both potential and kinetic forms. Permissible conditions are now not limited to points on the potential energy line.

For instance, at time $t = 0$ let the system be in the condition depicted in Fig. 1.3.1. From the state diagram we read an initial weight height of 5 m which, for a mass of 7 kg, represents a potential energy of 343.35 J ($5 \times 9.81 \times 7$) and an initial kinetic energy of 200 J (again notice that the state diagram contains much more quantitative information than the physical system representation). Assume that the weight is made of perfectly elastic material and that the initial velocity is

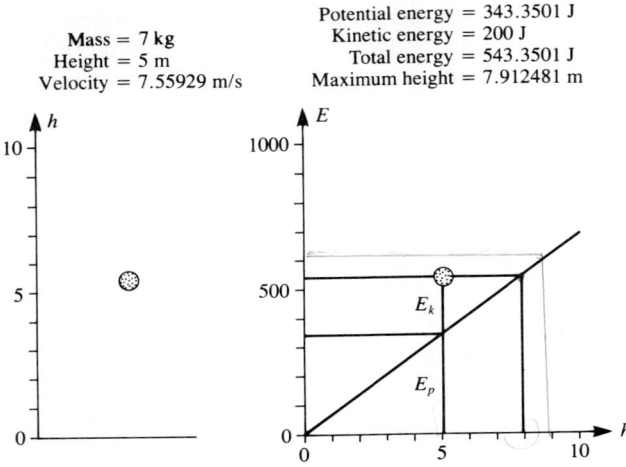

FIGURE 1.3.1
Classical dynamic system.

in the upward direction. Under these conditions, initially the system transforms kinetic energy into potential energy, and the weight moves upwards, decelerating to a position where all the energy is in potential form [at a maximum height of $(343.35 + 200)/(9.81 \times 7) = 7.9$ m]; then it starts moving downwards, transforming potential into kinetic energy and accelerating to the ground level. There all of its acquired kinetic energy is transformed into elastic form and then back into kinetic form, so that the weight rebounds upwards. This cycle is repeated ad infinitum.

On the energy-position diagram, the instantaneous condition point correspondingly moves back and forth along a horizontal line of constant total energy $E = 543.35$ J. At each position, the height above the potential energy line represents the *kinetic energy* E_k; the remaining distance to the position axis is the *potential energy* E_p, as shown in the figure. As the point moves toward the left, the instantaneous potential energy decreases and the kinetic energy correspondingly increases; the mass accelerates, showing that the system is transforming potential into kinetic energy. The reverse occurs when the point moves toward the right. When the point reaches the potential energy line, all the energy in the system is in potential form (the mass comes to rest before inverting the direction of motion), so that the point is now compelled to rebound to the left.

In conclusion, in the energy-position diagram, the potential energy line behaves as a barrier that the instantaneous condition point cannot cross and from which it must consequently rebound. This is not a peculiarity of this particular example, so in all energy-position diagrams we are justified in considering this line as a *potential energy barrier*.

The horizontal trajectory of the instantaneous condition point is the *state line* characterizing this particular state. Notice that the state line extends from the energy axis to the potential energy barrier, never crossing it. The concept of potential energy barrier is so fundamental that a more rigorous discussion is in order.

At any instant, the total system energy E_t equals the sum of the potential and kinetic energies:

$$E_t = E_p + \tfrac{1}{2}mv^2 \tag{1.3.1}$$

from which

$$v = \sqrt{2\,\frac{E_t - E_p}{m}} \tag{1.3.2}$$

However, remembering the previously discussed general properties of the energy-position diagram, it is evident that, in the region below the potential barrier, $E_p > E_t$. Here, therefore, the speed v in (1.3.2) *becomes imaginary*, so that these points cannot represent physically permissible conditions. In conclusion, the region of the energy-position diagram below the potential energy barrier is a *physically forbidden region* and no instantaneous condition point (or state line) can enter it.

If any further energy is input into the system, the instantaneous condition point simply moves onto another constant energy state (at a different energy level). The permissible energy levels constitute a continuous set, indicating that this classical system can store and release energy continuously (i.e., stored energy can be varied in infinitesimal increments).

Again the student is urged to generate a generous number of examples by varying the system parameters, predicting state lines, instantaneous kinetic and potential energy, the law of motion of the instantaneous condition point on the state line, etc.

1.4 QUANTUM SYSTEMS— INTRODUCTORY SURVEY

Most of the material covered in this section should be familiar to the student from previous physics and chemistry courses. It is presented here in capsule form (and some mathematical details are developed at an introductory level and by way of examples in Apps. 1A.1 and 1A.2) for the interested student who may have some background in quantum physics. This material is not strictly necessary for the understanding of the rest of the text, provided the reader is willing to accept a few facts on faith.

The Bohr Atom

The Rutherford model of the atom, as a miniature planetary system in which negatively charged electrons orbit around a heavy, positively charged nucleus, is not compatible with the laws of classical electromagnetic theory: the orbiting electron is a charge moving under acceleration (centripetal) and so must radiate a time-varying electromagnetic field [1]. The rate of energy radiation, computed from electrodynamics, is so high that the available energy would be exhausted in a very short time and the electron would spiral into the nucleus.

To avoid this inconsistency, Bohr, inspired by the discoveries of Planck, postulated that the laws of classical electrodynamics must be modified, that there exist privileged orbits in which the electron can move under acceleration without radiating and that these orbits correspond to the only stable states in which the atom can exist. Each orbit is characterized by a quantity of energy stored by the electron in both potential and kinetic form and the difference of stored energy between any two permissible states is finite. Thus the energy an atom can store is *quantized*.

Bohr also postulated that emission or absorption of radiation may occur only when the electron "jumps" from one such orbit to another. The finite difference between the energies stored in the two orbits may then be emitted, or absorbed, in the form of a photon of frequency proportional to this energy difference, as predicted by Planck's theory.

The Hydrogen Atom

Expressing this hypothesis quantitatively for the case of a one-electron atom (the H atom, cf. App. 1A.1), Bohr calculated the radii of the stable orbits, the corresponding energies stored, and the wavelengths of the photons emitted in the possible transitions from one permissible state to another. These results [Eqs. (1A.1.8) and (1A.1.11)] are in excellent agreement with the experimentally determined radius of the H atom, its ionization energy, and the frequencies of the Balmer series of the H spectral emission. Notice that the accord is not merely qualitative: the value of the Rydberg constant $R_H = 109,677$ cm^{-1} is consistent with Eq. (1A.1.11).

Taking into account additional properties of the H spectral emission, such as other spectral series, the Zeeman effect, etc., Bohr was led to the hypothesis that not only energy but also other physical quantities, such as angular momentum, magnetic moment, and electron spin, are quantized in the atom. The permissible values of these characteristics can be expressed in terms of *quantum numbers* and the quantization properties are indicated by the fact that these numbers can take on only specific values determined by a set of *selection rules*.

The quantum numbers and selection rules for the Bohr atom are:

n (energy); values: integer ≥ 1

l (angular momentum); values: integer $0 \leq l \leq n - 1$

m_l (orbital magnetic moment); values: integer $-l \leq m \leq l$

m_s (electron spin); values $-\frac{1}{2}$ and $\frac{1}{2}$.

The physical property to which each is related is indicated in parentheses. Each permissible orbit is uniquely determined by the four values of the corresponding quantum numbers.

Shell Structure of the Atom

Considering the chemical and physical properties (especially light emission spectra) of the elements in Mendeleyev's table, Bohr was led to the hypothesis that the orbits occupied by electrons in atoms of atomic numbers higher than 1 are grouped in *shells*. All orbits within each shell have the same principal quantum number n; consequently their energies (and so the orbit radii) are essentially the same. The orbits in each shell differ from one another by the values of the other quantum numbers and all orbits having the same value of the angular momentum quantum number constitute a further grouping, called a *subshell*.

In accordance with the above description, each subshell is characterized by the values of its principal and angular momentum quantum numbers. For historical reasons, linked to the importance of spectroscopic data in the development of the Bohr theory, the value of the angular momentum quantum number is designated by a set of letters, corresponding to the traditional classification of spectroscopic lines, as follows:

$$l = \qquad 0, 1, 2, 3, 4, 5$$

$$\text{Symbol} \qquad s, p, d, f, g, h$$

Thus a subshell is identified by a code consisting of a number, indicating the value of n, followed by a letter, indicating the value of l. For instance, $2p$ indicates the subshell in shell 2 having $l = 1$ (notice that a subshell $2d$ cannot exist, because it would have $l = 2 = n$, which contravenes the selection rules).

As the permissible values of the quantum numbers are limited by the selection rules, it follows that each shell and subshell may contain no more than a specific maximum number of orbits. For instance, in a p subshell, m_l is limited to three permissible values: -1, 0, and $+1$, for each of which m_s can assume two values ($-\frac{1}{2}$ and $+\frac{1}{2}$), so that the subshell can contain no more than $2 \times 3 = 6$ orbits. When a subshell is filled, the next electrons generally occupy the next higher subshell.

Each atom of the table of elements is uniquely described by the distribution of its electrons in the shell structure. This distribution is usually indicated by a numerical superscript added to the subshell designation and indicating the number of occupied subshell orbits.

As a useful example, one way to indicate the unique structure of the Si atom is shown in Fig. 1.4.1. Si has atomic number $Z = 14$, so the lowest 14 orbits are filled, corresponding to $1s^2$, $2s^2$, $2p^6$, $3s^2$, $3p^2$, as also shown in the figure. Notice that shells 1 and 2 and subshell s of shell 3 are filled, whereas subshell p of shell 3 contains only two filled orbits, missing four outer shell electrons to fill the subshell.

The electrons in the outer shell determine the chemical properties of the element. For instance, an atom with filled inner shells and one electron in the outer shell behaves as if the nucleus, together with the inner shells tightly surrounding it, were one compact nucleus with a charge of 1 atomic unit. The outer shell electron is in orbit around this quasi-nucleus. Such a structure would

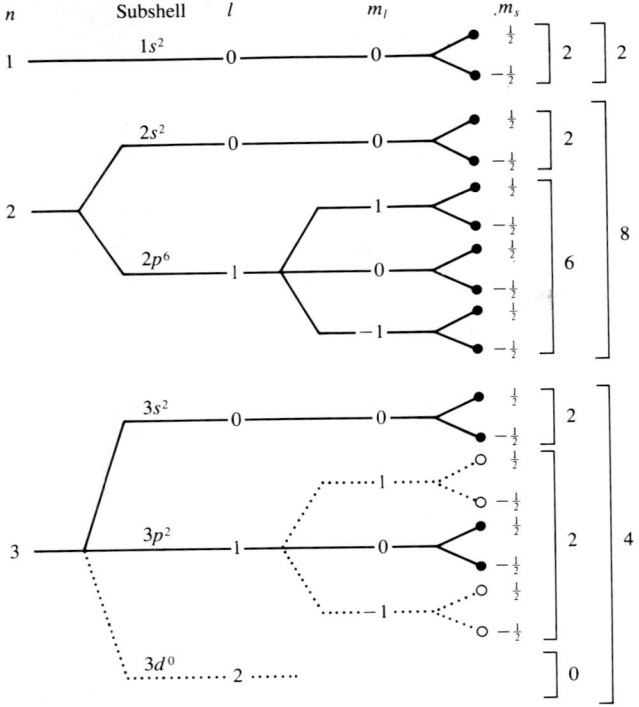

FIGURE 1.4.1
Shell structure of the Si atom; symbolic graphic representation with indication of the number of orbits in each subshell.

behave similarly to the hydrogen atom, and is sometimes called a hydrogenoid atom. A structure with a full outer shell would behave similarly to He ($1s^2$, with shell 1 full) and so elements with full outer subshells are noble elements, as can be observed from the table of elements.

The theory predicts that Ge and Si have analogous characteristics, due to the similarity of their outer shell structure, as shown in Fig. 1.4.2.

The periodicity of the chemical and physical properties of the elements in the table is therefore due to the repetition of the electronic patterns in the outer shell after an inner shell has been filled.

As shown by the previous considerations, this very qualitative system of seemingly arbitrary hypotheses proves amazingly successful in "explaining" a number of atomic phenomena. However, the many successes of the Bohr atomic model are accompanied by several inconsistencies and discrepancies with a number of observed facts.

With the wisdom of hindsight we now realize that any physical model is severely hampered by the fact that, to be acceptable to us, it must satisfy our physical common sense, and this is based on strictly macroscopic experience, as

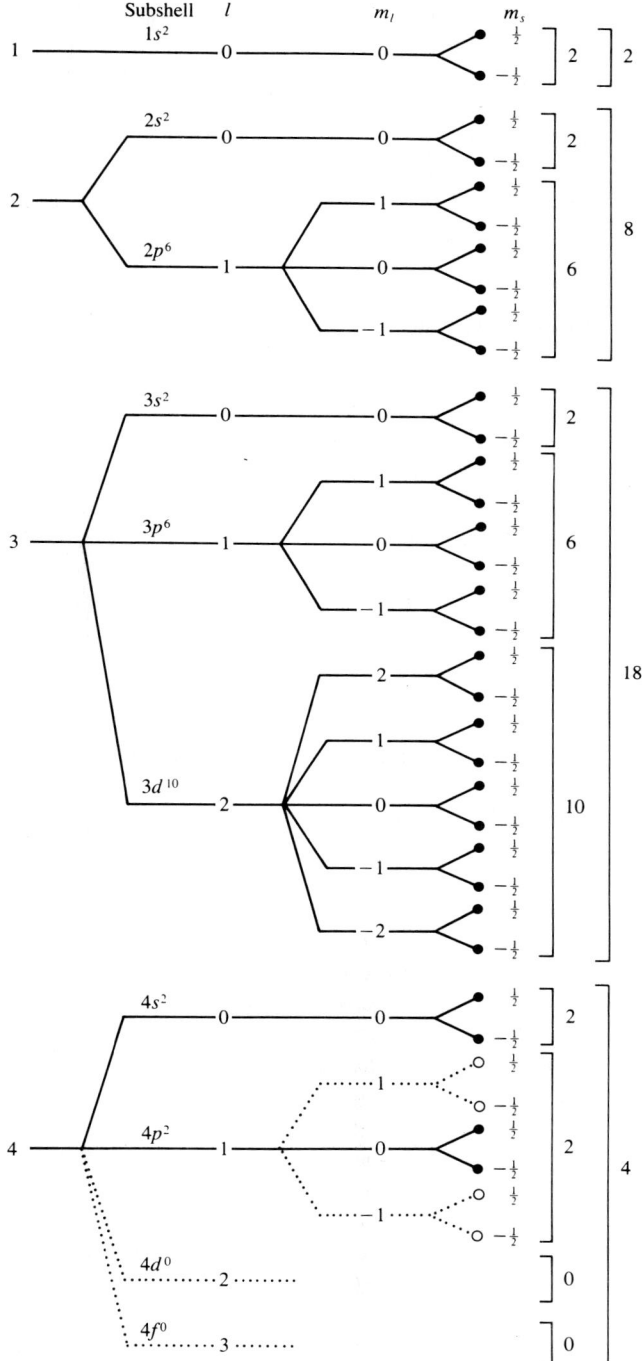

FIGURE 1.4.2
Shell structure of the Ge atom. Notice the similarity of the structure of the outer shell with that of the Si atom of Fig. 1.4.1.

expressed by the experimental laws of classical physics. Such macroscopic experience is sometimes at odds with the reality of microscopic systems, necessarily limiting the ability of a macroscopic physical model to represent accurately such microscopic systems.

The solution to the problem is found in the construction of mathematical models of quantum structures. Such models, following the rigorous laws of mathematical logic, are not subject to the prejudice of our physical common sense, while retaining the logical self-consistency of mathematical thought. Unfortunately, for this very same reason, while these models lend themselves to amazingly accurate predictions of physical fact, provided we can handle the attendant mathematics, they still disclose sometimes bewildering and frustrating horizons, which may lead us to wonder and doubt about some of our most cherished assumptions (consider, for instance, the uncertainty principle [2, p. 86]).

Modern quantum physics has produced two different formulations of such mathematical models: the Schroedinger wave equation and the Heisemberg matrix theory. Appendix 1A.2 offers an introduction to the Schroedinger theory. For more exhaustive treatments the interested student is referred to the eminently legible textbooks on the subject, some of which are listed in the references and suggested additional reading at the end of the chapter.

The Schroedinger Equation

At first, the discovery of the dual nature of the photon (Planck, Einstein) motivated theoretical scientists to search for some physical model of the atom in which the components displayed both particle and wave properties. One such model, the Bohr atom, as we saw, proved successful in predicting several observed facts. However, further observations evidenced several serious limitations of the model.

A more successful approach was taken by Schroedinger. He reasoned that, if all particles possessed wave properties, then some physical quantity related to the particle must be described by a function representing a wave. Not knowing yet what property this function represents, let us call it the *wave function* (or *state function*).

The wave properties of the photon can be imposed on the wave function by assuming that it has the familiar form of the equation of a wave. The corresponding particle properties can also mathematically be imposed on the state function by properly modifying its elements.

Following this mathematical procedure (for the mathematical details cf. App. 1A.2), Schroedinger was led to the formulation of a partial differential equation [(1A.2.11), assumed to describe the state functions of a photon] and to a set of formal rules for obtaining it for each case from a classical mathematical description of the phenomenon under discussion [Eqs. (1A.2.13) and (1A.2.14)]. He then postulated that the state functions of any system can be obtained by solving this equation, provided the rules are formally adhered to.

This theory may leave us a little doubtful and perplexed. Logically we may well ask ourselves: Why should these rather outlandish and formalistic rules be accepted; indeed, why should they work at all? The strength of the theory does not rest on the logic (or lack of logic) that led us to the postulation, but on the fact that, despite its dogmatic and apparently capricious nature, its application leads to conclusions correctly predicting experimentally verifiable facts, as proved by an imposing wealth of supporting experimental evidence gathered over a period of over 60 years.

There remains the important question: What is the physical meaning of this mysterious wave function? What does it tell us about the behavior of the system?

Sommerfeld proved [3, p. 63] that the square magnitude of the (usually complex) wave function can be used to compute the probability of finding the particle in any position in space [Eq. (1A.2.18)]. Actually, by operating on the wave function with appropriate operators, it is possible to compute the most probable value, not only of the particle position in space but of any other physical characteristic of the system in question [3, p. 63].

Notice that the theory permits computation only of the probability of the occurrence of an event, rather than yielding a deterministic description of the outcome of the phenomenon. This is one of the most intriguing and possibly disappointing features of quantum phenomena, quantitatively described by Heisenberg's uncertainty principle [3, p. 24].

When applied to the one-electron atom, the Schroedinger formulation, on the whole, confirms the general structure of the H atom described by the Bohr theory, but in a more rigorous and valid way. The quantum numbers and their selection rules come automatically out of the mathematical process of solving the Schroedinger equation by the method of separation of variables by product in a spherical system of coordinates and imposing the condition that the state function must be well behaved [2, p. 164; 3, p. 110]. The deterministic concept of orbit loses much of its meaning and is substituted by a rather nebulous outline of regions of space in which the electron is most probably to be found for each state. The concept of state and probability of transition from one state to the other statistically describes the interaction of large numbers of atoms with the outside world.

The Schroedinger formulation of multielectron atoms yields an extremely complex partial differential equation with a number of position variables equal to three times the number of electrons present (even assuming that the nucleus remains fixed in space). In practice the mathematical solution is possible only under several simplifying assumptions and often only by numerical computer means [2, p. 246; 3, p. 126]. Again, on the whole, the shell-like structure predicted by the Bohr theory is confirmed and similarly related to the properties of the different elements. Much of the nomenclature of the Bohr theory is retained in the common scientific jargon because of its intuitive appeal, but the limitations of this deterministic representation must always be kept in mind. For the purposes of visualization, a "picture" of several atomic structures is shown in Fig. 1.4.3.

Using the Schroedinger equation it is also possible to introduce some quantum phenomena that have no correspondent in classical physics and some-

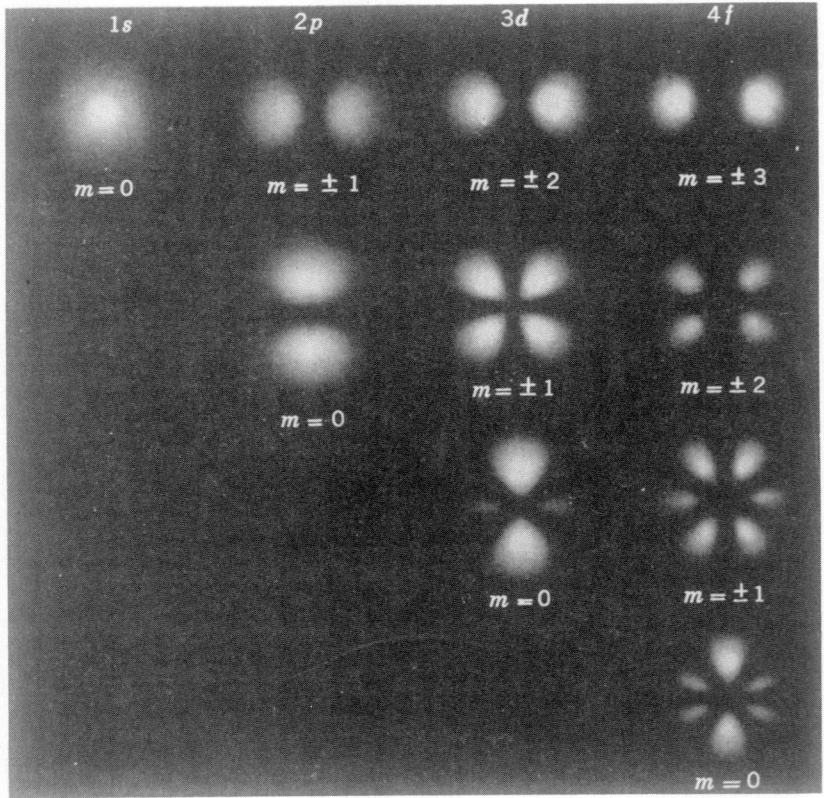

FIGURE 1.4.3
Graphic representation of the structures of several elements. Highlighted regions correspond to locations with high probability of electron occupation [2].

times play an important role in semiconductor physics. One such phenomenon is *tunneling*, the possibility that a particle may cross a potential barrier higher than the particle energy. Classically, as we have seen, this is impossible, but quantistically there is a finite probability of the occurrence of such a transition, provided the barrier is thin, only moderately high, and the initial and final states on the two sides of the barrier are almost at the same energy level. A discussion of tunneling is presented in App. 1A.2 as an example of an application of the Schroedinger equation.

1.5 HYDROGENOID ATOM

The mechanism of interaction between atomic systems and energy plays a very important part in the study of solid state devices and deserves a somewhat detailed discussion. As already stated, any rigorous description of atomic

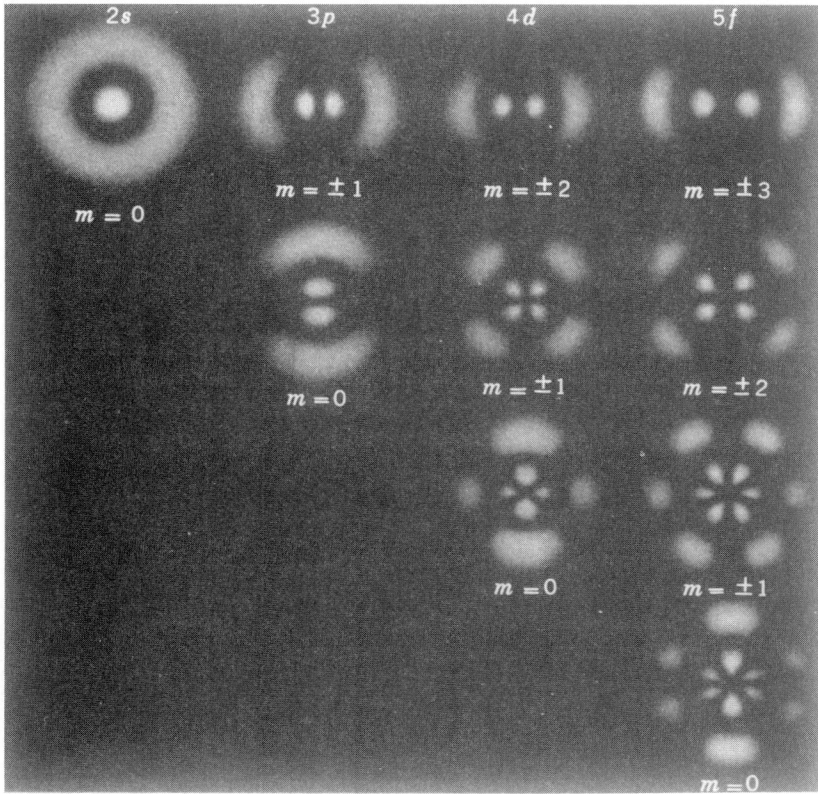

phenomena requires quantum physical considerations. A satisfactory analysis, however, can be based on semiclassical models, such as the energy-position diagram of Fig. 1.5.1, representing a hydrogenoid atom. In the following presentation, whenever any of the tenets of quantum physics are invoked dogmatically, references will be provided to the literature that rigorously justifies it. Most of the quantum concepts involved have already been introduced in Sec. 1.4.

In Fig. 1.5.1, as customary for atomic systems, the vertical axis represents the *total electron energy E* and the horizontal axis, the distance *r* of the electron from the nucleus.

In a three-dimensional system, a point with abscissa *r* in this diagram represents any of the points on the surface of a sphere of radius *r* centered in the nucleus. For the sake of easy visualization, however, the student is advised to think of this as a one-dimensional atom in which the electron moves along a straight line, right through the center of the nucleus and back again, in a kind of

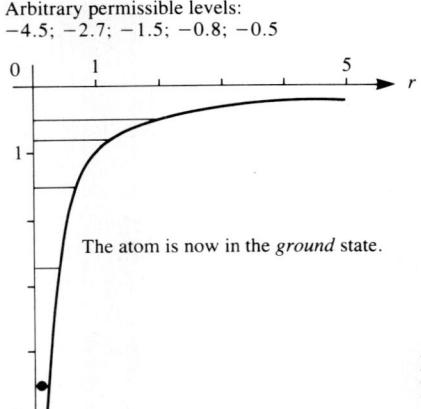

Arbitrary permissible levels:
−4.5; −2.7; −1.5; −0.8; −0.5

The atom is now in the *ground* state.

FIGURE 1.5.1
Hydrogenoid atom energy-position diagram: ground state.

pendular motion. From a more rigorous quantum physical point of view this much simpler model has the same limited validity as the three-dimensional one.

The total electron energy, plotted along the vertical axis, consists essentially of two components: kinetic (depending on the electron velocity and therefore on its orbit) and potential energy arising from the potential distribution of several fields—electric, gravitational, magnetic, etc.—in which the electron is immersed. Of these only electric energy need be considered, all others being negligibly small in comparison.

Suppose that an externally applied field envelops the system and let its electric potential distribution be expressed by $\psi(x)$. The potential energy acquired by an electron at position x in this field is $-q\,\psi(x)$ ($-q$ is the electron charge). Upon establishment of the field, the quantity $-q\,\psi(x)$ is added to the total energy of this electron. Consequently the system's energy-position diagram *undergoes a displacement* $-q\,\psi(x)$ *at point* x. If $\psi(x)$ is a function of position, then this displacement is not uniform and *the energy-position diagram is distorted* by the application of a nonuniform electric field. Notice that, because the charge of the electron is negative, a positive potential results in a downward displacement.

This property of energy-position diagrams will play an important role in future discussions of semiconductor systems and we shall often make use of it in our analysis of semiconductor device behavior.

The potential energy of the electron in the field set up by the nucleus can be computed from classical electric field theory [1]:

$$E_p = -\frac{q^2}{4\pi\varepsilon_0 r} \tag{1.5.1}$$

and is shown in the energy-position diagram as a hyperbola. This is the potential energy barrier and the points below it form a forbidden region, so that the system states are represented by horizontal segments extending from the energy axis to the barrier.[1]

In a classical system, any of these states would be permissible, but, as we saw (cf. App. 1A.1), in atomic systems the laws of quantum physics limit such permissible states to a discrete set, as indicated qualitatively in the figure. An electron in the atom can exist only in one of these permissible states.[2]

Finally, yet another quantum physics law, *Pauli's principle*, limits occupation of any one state to no more than one electron. Therefore each permissible state can be either empty or occupied by no more than one electron.[3]

Notice that, in any of the permissible states indicated in the figure, the maximum distance from the nucleus that the electron can reach is limited by the potential barrier, so that an electron in any of these states is constrained to remain in the atom (it is bound to the atom).

In the energy-position diagram, the region characterized by positive electron energies lies above the potential barrier. Notice that an electron in this region is not bound to the atom any more, but is instead free to move anywhere in space. Quantum physics shows that this region contains a very large number of very closely spaced permissible states. However, to avoid cluttering in the figure, these states are not shown.

Normally the atom as a whole remains in the *ground state*, in which electrons occupy the lowest permissible energy levels. The electron occupying the highest of these levels is known as the *valence* electron and will be of particular

---

[1] Electron energy E is generally expressed in electronvolts (eV). As with all energy levels, choice of the reference zero level is arbitrary. Notice that Eq. (1.5.1) expresses potential energy in joules (divide by $q = 1.6 \times 10^{-19}$ C to convert to electronvolts). The student should prove to his or her satisfaction that (1.5.1) implies a choice of the arbitrary zero energy level as the energy of an electron at rest at an infinite distance from the nucleus (i.e., outside of the atom). With this choice all electrons inside the atom are characterized by negative energy levels.

As q in (1.5.1) is the charge of the proton, it might appear that a hydrogen atom (one electron orbiting a one-proton nucleus) is implied. For other atoms the electron atmosphere is composed of several electrons, as discussed in Refs. 4 (p. 7), 5 (p. 239*), 6 (p. 135) (easy to read), 2 (p. 246*), and 7 (p. 637) (very easy to read) and in Sec. 1.4. However, the formula is still valid to a reasonable approximation when applied to the outermost election of a hydrogenoid atom (cf. Sec. 1.4). (An asterisk next to the page number means that this reference assumes some previous familiarity with quantum theory.)

[2] It is assumed that the student has acquired some familiarity with the Bohr atom theory (cf. Sec. 1.4 and App. 1A.1). Good elementary presentations are to be found in Refs. 7 (p. 634) (very easy to read), 5 (p. 235*), 2 (p. 164*), and 4 (p. 7). (An asterisk next to the page number means that this reference assumes some previous familiarity with quantum theory.)

[3] Justification of Pauli's principle and indications to the limits for its validity, including the definition of *fermions* (particles with odd symmetry in their wavefunction) can be found in Refs. 2 (p. 236*) and 8 (p. 525*). (As indicated by the asterisk, both references require some knowledge of quantum theory.)

interest in the following. In the ground state all levels below the valence level are occupied; all levels above it are empty.[4] Figure 1.5.1 does not depict an actually existing atom; it is just an arbitrary example of atomic structure and is assumed to be in the ground state. The valence level is -4.5 units and all levels below it, if any, are supposed to be filled; all levels above it are empty.

If energy is imparted to this atom, the valence electron can store it only if the energy is sufficient to bring it to another permissible and empty state, so that, for instance, in the figure, the minimum energy capable of inducing a transition is 1.8 units, corresponding to a jump of the electron to the next higher permissible state. In the light of the previous example of a discrete energy system, it might be expected that the electron could absorb any amount of energy equal to—or greater than—this minimum quantity, undergo a transition to the closest lower level, and reject the extra energy. In reality, atomic energy exchanges are governed by a *resonance phenomenon*.

As the student already knows from previous courses, quantum physics postulates that energy is imparted to—and released by—a system in finite amounts, or *quanta*. Each quantum (or *photon* in the case of radiant energy) can be considered as a localized packet of energy. If the energy contained in the packet is equal, or at least very close, to the energy required by a permissible transition, then the packet is said to *resonate* with the transition. It can be shown [9, p. 614*] that the probability that an electron will absorb the energy of a given photon depends on the photon energy and on the energy jump characterizing the transition. For the case of the one-electron atom this probability turns out to be proportional to the sharply resonant function:

$$\frac{\sin^2\left[\pi(f_{k0} - f)t\right]}{\left[\pi(f_{k0} - f)\right]^2} \tag{1.5.2}$$

which is plotted in Fig. 1.5.2. In this expression the photon energy is expressed in terms of the photon frequency $f = E_{ph}/h$ and the transition energy by the frequency $f_{k0} = |E_k - E_0|/h$, the transition resonant frequency, while t represents the duration of the interaction between the atom and the photon. As shown by Fig. 1.5.2, for reasonable interaction durations, the probability that a photon will

---

[4] The above model is very simplified. The student knows that, in reality, a state is determined not only by its energy level but by the whole set of the four *quantum numbers*, that the permissible states are accordingly grouped in *shells* and *subshells*, and that several other conditions prevail. The above simplified model is sufficient for our immediate purposes. The interested student is referred to Sec. 1.4, App. A1.1, and the references, especially Refs. 4 (p. 7), 5 (p. 239*), 6 (p. 135) (easy to read), 2 (p. 246*), and 7 (p. 637) (very easy to read). (As usual, an asterisk after the page number indicates that the reference requires some knowledge of quantum theory.)

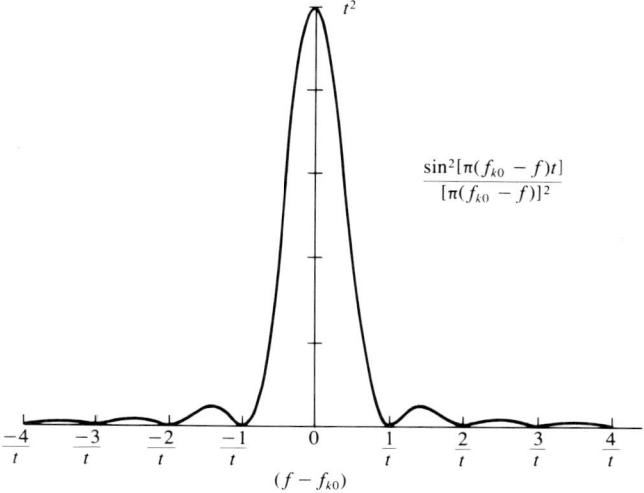

$$\frac{\sin^2[\pi(f_{k0} - f)t]}{[\pi(f_{k0} - f)]^2}$$

$$(f - f_{k0})$$

FIGURE 1.5.2

Probability that an atomic transition from level E_0 to level E_k is stimulated by a photon of frequency f. As usual $f_{k0} = |E_k - E_0|/h$. Notice that the transition probability is essentially zero for all photons of frequency $f \neq f_{k0}$ [9].

stimulate a transition is negligible for all photons that do not resonate with a permissible transition. In other words, only resonant photons are absorbed; all others are rejected.[5]

If such a resonant packet is absorbed by the electron, a transition occurs, bringing the atom from the ground state to an *excited state*. Figure 1.5.3 schematically depicts an atom in which absorption of a photon of 3 energy units has raised the valence electron from the -4.5 to the -1.5 level. This is symbolized in the figure by the vertical bar of length equal to the energy of the input photon. This bar, beginning at the original electron level, ends at the level occupied by the electron in the excited state, indicating the transition with which the absorbed photon resonated.

The atom can exist in the excited state only for a finite amount of time, which is a characteristic of each specific excited state. After this time has elapsed,

---

[5] Proof of the above very important property of quantum systems can be obtained by applying first-order, time-dependent perturbation theory to the system as discussed in Refs. 9 (p. 614*) and 2 (pp. 118* and 231*). (These references assume a moderately advanced familiarity with modern quantum theory.)

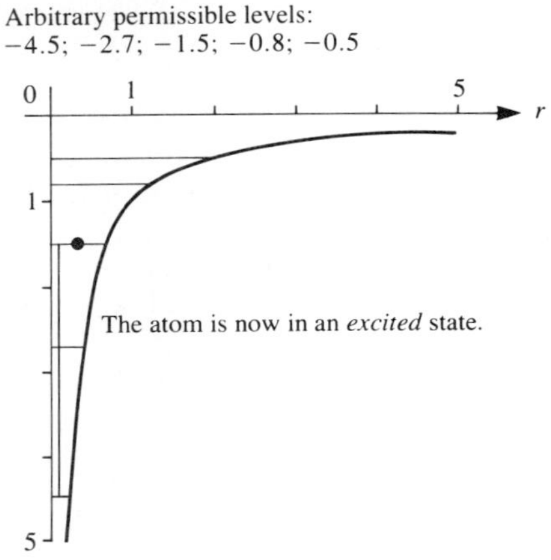

Arbitrary permissible levels:
-4.5; -2.7; -1.5; -0.8; -0.5

The atom is now in an *excited* state.

FIGURE 1.5.3
Hydrogenoid atom energy-position diagram: excited state. The vertical line starting at the level of the initial state indicates the energy imparted to the electron at the beginning of the transition.

the atom returns to the ground state.[6] In doing so, the electron releases the excess stored energy in one of several possible forms. This is known as a *spontaneous decay transition*. If the energy is released in the form of radiation, then the transition is classified as a *radiative transition*.

It is also possible that, while the atom is in the excited state, some external physical event, such as the influence of a resonant packet of energy, may cause the atom to undergo the transition before the characteristic time has elapsed. In this case the event constitutes a *stimulated transition* and the radiation emitted has very special properties relating it to the characteristics of the perturbing energy packet. This phenomenon is at the basis of the behavior of LASERs (Light Amplifiers by Stimulated Emission of Radiation).

Finally, the energy imparted may be equal to—or greater than—the absolute value of the energy of the valence electron in the ground state. This amount of energy (which, as the student should realize, is sufficient to bring the total electron energy to a positive value) is called the atom's *work function*. As previously stated, the positive energy region is densely populated with permissible states so that any energy packet more energetic than the work function in practice always resonates with a permissible transition.

After such a transition, in accordance with the previous considerations, the electron is free to move anywhere in space, irrespective of how far it is from the

---

[6] The amount of time in which the electron remains in the excited state cannot be exactly predicted, but varies randomly from one occurrence of the decay to another. However, the statistics of this random delay are a fixed characteristic of each excited state. The statistical average of this time is known as the *lifetime* of the specific state. We shall often have occasion to take this quantity into consideration. It is important for the reader to realize that this is *not* the time actually spent by the electron in the state during the transition, but a statistical quantity, obtained as the average of the random times spent in the excited state during a number of transitions.

nucleus: the electron is *free from the atom*. This is the phenomenon of *ionization* and results in a free electron and a positively charged atom, generally referred to as an *ion*.

Example 1.5.1. For the atom of Fig. 1.5.1, in which the valence electron in the ground state is indicated by the small disc shown at level -4.5, evaluate the work function and the energy levels resonating with permissible transitions from the ground state. What will be the result of the interaction of this atom with a photon (energy packet) of energy: 1, 2, 3, 4, 5 arbitrary units?

Solution

1. The work function is 4.5 units.
2. Resonating energy levels from the ground state are (in arbitrary units): $4.5 - 2.7 = 1.8$; $4.5 - 1.5 = 3$; $4.5 - 0.8 = 3.7$; $4.5 - 0.5 = 4$.
3. Interaction with energy packets of 1 and 2 energy units results in no transition: the nonresonant energy is rejected.
4. A packet of 3 units is absorbed and generates first an excited state with the electron at level -1.5 and then, after the characteristic time for this transition, a spontaneous decay either directly to the ground state, releasing 3 units of energy, or in two steps to level -2.7 (releasing $2.7 - 1.5 = 1.2$ units) and then to ground level, releasing $4.5 - 2.7 = 1.8$ units.
5. Absorption of a 4-unit resonant packet has similar results, with an excited level at -0.5 energy units.
6. Finally a packet of 5 energy units has enough energy to overcome the work function and results in ionization, yielding a positive ion and a free electron with residual kinetic energy $E_k = 5 - 4.5 = 0.5$ energy units.

1.6 SUMMARY

Physical *systems* consist of *components*, generally interconnected to each other, generating *constraints*, which, in turn, permit the system to exist only in certain permissible *states*.

To each state there corresponds a given amount of *stored energy*, so that a *transition* from one state to another characterized by a different amount of stored energy requires an energy *input* (positive or negative).

System behavior can be represented in *state diagrams*, such as the energy-position diagram, showing *instantaneous system conditions, state lines, permissible and forbidden states*. These diagrams can be continuous or discontinuous. In the latter case a transition may require a *quantum energy jump*.

In dynamic systems, the potential energy line constitutes the boundary of a *forbidden region* and therefore acts as a *potential energy barrier*.

The energy-position diagrams of atomic systems plot total electron energy vs. position. Application of an electric field having an electric potential distribution $\psi(x)$ results in a *distortion of the energy-position diagram*, each point being displaced in the vertical direction by an amount $-q\,\psi(x)$.

Atomic systems, in accordance with the laws of quantum physics, are generally discontinuous. The permissible states are described by the Bohr model in terms of permissible orbits, resonating with the electron's associated wave. In multielectron atoms, the permissible orbits are arranged in shells and subshells, the chemical behavior of the atom being determined mostly by the structure and occupation of the outer subshell. In these systems, certain particles, such as electrons, obey *Pauli's principle*, so that no more than one particle in a system can occupy any one state.

An atom can interact only with *resonant energy packets*, closely approximating the amount of energy necessary for a permissible transition. Nonresonant energy packets are rejected.

Normally an atomic system finds itself in the *ground state*, characterized by a minimum of stored energy. By absorbing resonant energy the system can undergo a transition to an *excited state*, in which it can generally remain only for a finite time, characteristic of the state, after which it returns to the lower state, releasing the corresponding extra energy. This transition can be *radiative* or not, *spontaneous* or *stimulated*.

If the energy absorbed is greater than—or equal to—the *work function* of the atom, the transition can result in *ionization*.

APPENDIX 1A
ATOMIC STRUCTURE AND QUANTUM PHYSICS—
AN INTRODUCTORY SURVEY

1A.1 THE BOHR ATOM

Rather than repeat the train of reasoning that historically led Bohr to the computation of the radii of the stable atomic orbits (cf. Ref. 7, p. 624), we shall instead interpret his hypothesis in the light of a later theory. As predicted by Planck's theory, the energy of a photon is proportional to its frequency:

$$E = hf \tag{1A.1.1}$$

where $h = 6.625 \times 10^{-34}$ J · s is Planck's constant.

The photon, as proved by Einstein's photoelectric Gedank experiment, displays both particle properties (position, energy, mass, etc.) and wave properties (frequency, wavelength, etc.). De Broglie computed the relationship between the particle and wave properties of the photon. From Planck's hypothesis,

$$E = hf = mc^2 \tag{1A.1.2}$$

in accordance with the theory of relativity. From this, the momentum of the photon is

$$p = mc = h\frac{f}{c} = \frac{h}{\lambda} \tag{1A.1.3}$$

Convinced that the symmetry of nature could not permit this dual nature to be the exclusive privilege of the photon, De Broglie then postulated that wave and particle characteristics always appear together, so that all particles animated by a momentum p are characterized by an *associated wave* of wavelength

$$\lambda = \frac{h}{p} \tag{1A.1.4}$$

In the light of this concept, Bohr's privileged orbits suggest a resonance between the orbit and the wave associated with the orbiting electron; the orbit length should be an integer multiple of the associated wavelength:

$$2\pi r = n\lambda = n\frac{h}{p} = n\frac{h}{m_e v} \tag{1A.1.5}$$

where n is an integer number and m_e the electron mass. However, for the orbiting electron the centrifugal force must be balanced by the centripetal electrostatic attraction, so that

$$\frac{m_e v^2}{r} = \frac{q^2}{4\pi\varepsilon_0 r^2} \tag{1A.1.6}$$

Solving for v, substituting in (1A.1.5), and reordering, the radii of the stable orbits are seen to satisfy the equation

$$r_n = \frac{nh}{2\pi m_e} \sqrt{\frac{4\pi\varepsilon_0 m_e r_n}{q^2}} \tag{1A.1.7}$$

where r_n is the radius of the nth orbit in the series, or

$$r_n = n^2 \frac{h^2 \varepsilon_0}{\pi m_e q^2} \approx 0.52 \times 10^{-10} \, n^2 \qquad \text{m} \tag{1A.1.8}$$

Notice that the smallest radius ($n = 1$) of about 0.5 Å corresponds to the known radius of the hydrogen atom.

Computing the total energy of the electron in orbit n as the sum of the potential and kinetic energy [calculated using the value of v implicit in (1A.1.6)],

$$E_n = -\frac{q^2}{4\pi\varepsilon_0 r_n} + \frac{q^2}{8\pi\varepsilon_0 r_n} = -\frac{q^2}{8\pi\varepsilon_0} \frac{\pi m_e q^2}{\varepsilon_0 h^2} \frac{1}{n^2} \tag{1A.1.9}$$

where the value of r_n has been obtained from (1A.1.8). Computing the variation of energy corresponding to a jump from an initial orbit n_i to a final orbit n_f and setting it equal to the energy of the emitted photon,

$$\Delta E = \frac{q^4 m_e}{8\varepsilon_0^2 h^2} \left(\frac{1}{n_f^2} - \frac{1}{n_i^2} \right) = hf = h\frac{c}{\lambda} \tag{1A.1.10}$$

The wave number of the emitted photon can then be computed as

$$\frac{1}{\lambda} = \frac{q^4 m_e}{8\varepsilon_0^2 ch^3} \left(\frac{1}{n_f^2} - \frac{1}{n_i^2} \right) = R_H\left(\frac{1}{n_f^2} - \frac{1}{n_i^2} \right) \qquad \text{cm}^{-1} \tag{1A.1.11}$$

in accord with the known sequence of wave numbers for the Balmer series of H emission lines. Notice that the accord is not merely qualitative: the value of the Rydberg constant $R_H = 109,677 \text{ cm}^{-1}$ is consistent with Eq. (1A.1.11).

1A.2 THE SCHROEDINGER EQUATION

A function representing a sinusoidal wave is known to be of the form [1]:

$$\phi(x, t) = A e^{i(2\pi/\lambda)(x - ct)} + B e^{-i(2\pi/\lambda)(x + ct)} \tag{1A.2.1}$$

where λ is the wavelength of the wave, c its velocity of propagation, and, to simplify matters, the wave is supposed to be plane (one-dimensional symmetry). Such a wave can be represented by the product of two functions of a single variable each:

$$\phi(x, t) = \Psi(x)T(t) \tag{1A.2.2}$$

where

$$T(t) = e^{-i(2\pi/\lambda)ct} \tag{1A.2.3}$$

and

$$\Psi(x) = A e^{i(2\pi/\lambda)x} + B e^{-i(2\pi/\lambda)x} \tag{1A.2.4}$$

Having mathematically imposed wave properties to the physical characteristic represented by $\phi(x, t)$, advantage can be taken of the Planck and De Broglie relationships (cf. App. 1A.1) to impose the corresponding particle properties.

From (1A.2.3), remembering that $c/\lambda = f = E/h$ (Planck's photon energy), $T(t)$ can be expressed in terms of its particle energy as

$$T(t) = e^{-i(2\pi/h)E} \tag{1A.2.5}$$

From (1A.2.4), using the De Broglie relationship $\lambda = h/p$, $\Psi(x)$ can be expressed in terms of the particle momentum p, then, remembering that the kinetic energy equals the total particle energy E minus its potential energy V, or $E_k = p^2/2m = E - V$, finally $\Psi(x)$ can be expressed in terms of its particle energy and mass m as

$$\Psi(x) = A \exp\left[i\frac{2\pi}{h}\sqrt{2m(E - V)}x\right] + B \exp\left[-i\frac{2\pi}{h}\sqrt{2m(E - V)}x\right] \tag{1A.2.6}$$

By differentiating (1A.2.5) with respect to t and (1A.2.6) with respect to x, then, after elementary algebra, the total particle energy can be expressed as

$$E = i\hbar \frac{1}{T}\frac{dT}{dt} \tag{1A.2.7}$$

and

$$E = -\frac{\hbar^2}{2m}\frac{1}{\Psi}\frac{d^2\Psi}{dx^2} + V \tag{1A.2.8}$$

where the usual notation $\hbar = h/2\pi$ has been used. Equating the last two expressions and multiplying by the product $\psi(x)T(t)$:

$$-\frac{\hbar^2}{2m}T\frac{d^2\Psi}{dx^2} + V\Psi T = i\hbar\Psi\frac{dT}{dt} \tag{1A.2.9}$$

Remembering that $\phi(x, t) = \Psi(x)T(t)$ and recognizing its partial derivatives, this can be rewritten as

$$-\frac{\hbar^2}{2m}\frac{\partial^2\phi}{\partial x^2} + V\phi = -\frac{\hbar}{i}\frac{\partial\phi}{\partial t} \qquad (1A.2.10)$$

Assuming three-dimensional spatial distribution of the wave function and repeating for each cartesian component, finally:

$$-\frac{\hbar^2}{2m}\nabla^2\phi + V\phi = -\frac{\hbar}{i}\frac{\partial\phi}{\partial t} \qquad (1A.2.11)$$

Remembering that each side of this equation represents the total energy of the particle and that this quantity, in classical mechanics, is expressed by the particle hamiltonian, Schroedinger postulated that the quantum physical behavior of a particle can be described by substituting the classical hamiltonian expression:

$$\frac{p_x^2 + p_y^2 + p_z^2}{2m} + V = E \qquad (1A.2.12)$$

with equation (1A.2.11). He further noticed that (1A.2.11) could formally be obtained from (1A.2.12) by substituting each component of momentum, the potential, and the total energy, wherever they appear in the formula, with an operator in accordance with the following set of rules:

$$p_x \to \frac{\hbar}{i}\frac{\partial}{\partial x}; \quad p_y \to \frac{\hbar}{i}\frac{\partial}{\partial y}; \quad p_z \to \frac{\hbar}{i}\frac{\partial}{\partial z}; \quad V \to V; \quad E \to -\frac{\hbar}{i}\frac{\partial}{\partial t} \qquad (1A.2.13)$$

The resulting expressions are, of course, operators and, according to Schroedinger's postulation, the equation obtained by operating with them on the wave function describes the quantum behavior of the particle. Applying these rules to Eq. (1A.2.12), the result can be expressed in operational terms by

$$\mathcal{H}\phi = -\frac{\hbar}{i}\frac{\partial}{\partial t}\phi \qquad (1A.2.14)$$

where $\mathcal{H}$ is the Hamiltonian operator of the particle, which, in accordance with rules (1A.2.13) is

$$\mathcal{H} = -\frac{\hbar^2}{2m}\nabla^2 + V \qquad (1A.2.15)$$

so that equation (1A.2.14) is identical with (1A.2.11). Schroedinger then further postulated that this set of formal rules was valid not only for the photon, as indicated, but also for any other particle or system of particles, provided rules (1A.2.13) were applied to the appropriate hamiltonian expression for the system under study.

In summary, the Schroedinger formulation of quantum mechanics is expressed by a set of formal rules that allow us to transform the classical mechanics formulation of a problem into a valid quantum mechanical equation: the Schroedinger equation. The Schroedinger rules are:

1. Write the classical hamiltonian expression for the system in question.
2. Transform this expression according to rules (1A.2.13); this yields the hamiltonian quantum operator of the system.
3. Use this operator $\mathcal{H}$ in the Schroedinger equation, (1A.2.14), and solve the resulting partial differential equation. This yields the system's wave function.

Physical Meaning of the State Function

Sommerfeld has proved [3, p. 63] that the square magnitude of the (usually complex) wave function indicates the probability density distribution of the particle position in space, so that

$$\int_{\mathscr{V}} |\phi|^2 d\mathscr{V} = \int_{\mathscr{V}} \phi\phi^* d\mathscr{V} = P(\mathscr{V}) \qquad (1A.2.16)$$

meaning that the integral measures the probability that the particle finds itself within the volume $\mathscr{V}$.

Actually, by the use of appropriate operators, the wave function allows determination of the probability density of any characteristic of the system in question, so that the integral

$$\int_{\text{all space}} \phi^* Q \phi d\mathscr{V} = \langle Q \rangle \qquad (1A.2.16a)$$

represents the expected value (statistical average, or most probable value) of the physical quantity Q, provided Q is the appropriate Schroedinger operator corresponding to that quantity and obtained by applying the rules (1A.2.13) to the classical expression of the physical quantity.

Tunneling

Consider the energy-position diagram of Fig. 1A.2.1: the distribution of the potential energy V divides the one-dimensional position domain into three ranges: for $x < 0$ the potential energy is zero, for $0 < x < x_0$ the potential energy is V_2, and for $x > x_0$ the potential energy is V_3. A particle of energy $E < V_2$ in the first range, according to classical physics, cannot overcome the higher potential barrier and so will rebound from it, never reaching range 3. From the quantum physical point of view, instead, there is a finite probability that the particle can reach the third range. This phenomenon, which cannot be explained in terms of classical physics, is called *tunneling*, because the particle appears to tunnel under the potential barrier, rather than surmount it. The phenomenon plays an important role in semiconductor behavior.

To discuss the problem in quantum physical terms, we should write and solve the Schroedinger equation (1A.2.10) and then use the result in Eq. (1A.2.16)

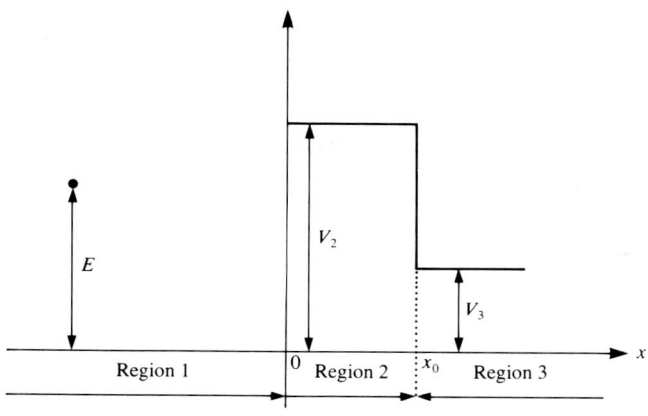

FIGURE 1A.2.1
Energy-position diagram of potential barrier and particle energy for tunneling phenomenon.

to compute the probability of finding the particle in $x > x_0$. However, remembering the form of $T(t)$, the time-dependent part of $\phi(x, t)$, it is immediately apparent that $\phi\phi^* = \Psi\Psi^*$ and, to compute (1A.2.16), it is sufficient to find $\Psi(x)$, instead of $\phi(x, t)$. This reduces the problem to the solution of Eq. (1A.2.8) (the time-independent Schroedinger equation), an ordinary differential equation, rather than a partial differential equation.

We shall divide the field of integration into its three component parts, solve (1A.2.8) in each of them, and use the continuity of the time-independent state function and of its derivative as boundary conditions.

In region 1, $V = 0$ and the solution of (1A.2.8) is

$$\Psi_1 = A \sin (\alpha x) + B \cos (\alpha x) \tag{1A.2.17}$$

where A and B are arbitrary constants to be determined from the boundary conditions and

$$\alpha = \frac{\sqrt{2m_e E}}{\hbar} \tag{1A.2.18}$$

Similarly, in region 2, where $V = V_2 > E$, the solution is

$$\Psi_2 = Ce^{-\beta x} + De^{\beta x} \tag{1A.2.19}$$

with C and D arbitrary constants, $D = 0$, as Ψ_2 must be finite irrespective of the magnitude of x_0, and

$$\beta = \frac{\sqrt{2m_e(V_2 - E)}}{\hbar} \tag{1A.2.20}$$

Finally, in region 3, $V = V_3 < E$ and

$$\Psi_3 = F \sin (\gamma x) + G \sin (\gamma x) \tag{1A.2.21}$$

with F and G constants and

$$\gamma = \frac{\sqrt{2m_e(E - V_3)}}{\hbar} \tag{1A.2.22}$$

Imposing the continuity conditions at the boundaries, for $x = 0$:

$$\Psi_1(0) = \Psi_2(0) \rightarrow B = C$$

$$\left.\frac{\partial \Psi_1}{\partial x}\right|_{x=0} = \left.\frac{\partial \Psi_2}{\partial x}\right|_{x=0} \rightarrow A = -\frac{\beta}{\alpha} C$$

Imposing analogous continuity conditions on Ψ_2 and Ψ_3 at $x = x_0$, after elementary algebra, the three wave functions are seen to be

$$\Psi_1 = C\left[-\frac{\beta}{\alpha} \sin(\alpha x) + \cos(\alpha x) \right]$$

$$\Psi_2 = Ce^{-\beta x}$$

$$\Psi_3 = Ce^{-\beta x_0}\left\{ \left[\sin(\gamma x_0) - \frac{\beta}{\gamma} \cos(\gamma x_0) \right] \sin(\gamma x) \right. \tag{1A.2.23}$$

$$\left. + \left[\cos(\gamma x_0) + \frac{\beta}{\gamma} \sin(\gamma x_0) \right] \cos(\gamma x) \right\}$$

Comparing the amplitudes of Ψ_1 and Ψ_3, it is possible to determine the probability that a particle, originally in region 1, will reach region 3:

$$|\Psi_1|^2 = C^2\left(1 + \frac{\beta^2}{\alpha^2} \right)$$

$$|\Psi_3|^2 = C^2 e^{-2\beta x_0}\left[1 + \left(\frac{\beta}{\gamma}\right)^2 \right] = C^2 e^{-2\beta x_0}\left(1 + \frac{V_2 - E}{E - V_3} \right) \tag{1A.2.24}$$

showing that the probability of tunneling $|\Psi_3|^2/|\Psi_1|^2$ is very small, unless x_0 is small (narrow tunneling gap), V_2 is not too much larger than E (comparatively low barrier), and V_3 is very close to E (tunneling between two states with essentially the same energy).

PROBLEMS

1.1. For the static system of Fig. 1.1.1 assume that the potential energy line in the energy-position diagram has equation $E_p = 625 \, h$ in MKS units.
(a) Find the mass.
(b) Draw the state line for the state corresponding to a total energy $E = 3125$ J.
(c) Does point $h = 3$ m, $E = 3125$ J represent a permissible instantaneous condition?

1.2. In the discontinuous static system of Fig. 1.2.1 the weight has a mass of 7 kg.

(a) Draw the energy vs. position diagram.

(b) If the weight was originally on the first step at a height of 2 m and 200 J of energy are imparted to it, describe the sequence of events and quantitatively determine all energy transfers, the final condition of the system, and the total net energy absorbed by it. Follow the phenomena on the energy vs. position diagram.

(c) Repeat for an energy input of 300 J.

1.3. For the dynamic system of Fig. 1.3.1 assume that the potential energy line has equation $E_p = 625\,h$ in MKS units.

(a) Find the mass.

(b) Draw the state line for the state corresponding to a total energy $E = 3125$ J.

(c) Consider points $h = 3$ m, $E = 3125$ J and $h = 6$ m, $E = 3125$ J. Do they represent permissible conditions? For each point compute the instantaneous potential and kinetic energies and the speeds. Comment.

(d) What is the maximum height reached by the weight in the state corresponding to total energy $E = 3125$ J?

1.4. A point mass of 3 kg is tied to a rotating platform with angular velocity $\omega = 5$ rad/s. The platform is horizontal at a height $h = 4$ m above the floor. Assuming as reference the energy of the weight lying at rest on the floor, draw:

(a) The energy-position diagram of the system as a function of the distance r of the mass from the center of rotation of the platform, for $0 < r < 3$ m

(b) The energy-momentum diagram of the system for the same limits

1.5. Consider an atom in which the electron orbits around the nucleus in a circular orbit of radius r.

(a) Show that, for such an atom, the total energy E is related to the kinetic energy E_k and to the potential energy E_p by

$$E = E_k + E_p = \frac{1}{2}\,E_p = -\frac{q^2}{8\pi\varepsilon_0 r}$$

(b) Comparing this atom with the one-dimensional interpretation of Fig. 1.5.1 in which the electron oscillates about the nucleus in a straight line trajectory, prove that the orbit radius r equals half the maximum distance from the nucleus reached by the electron in the linear atom.

(c) Draw the state line for this electron at the energy level $E = -13.54$ eV.

1.6. Measurements on hydrogen yield an ionization energy (work function) of about 13.54 eV and an atomic diameter of about 1 Å. Show that these results agree with the theory developed in Sec. 1.5 for an electron in a circular orbit around the nucleus.

Hint: Use the result of Prob. 1.5.

1.7. Figure P1.7 represents the electron energy (in electronvolts) vs. position of a one-dimensional system placed inside a parallel plane capacitor with armatures perpendicular to the x direction. The plot of Fig. P1.7 corresponds to zero voltage on the capacitor. Draw the modified energy-position diagram supposing the capacitor is charged to 25 V by switching S to position 2.

1.8. State the values of quantum numbers n and l corresponding to the following subshells:

$$2f;\ 3f;\ 5g;\ 2p;\ 3d;\ 3g$$

and indicate which of these subshells are allowable and which are not allowable.

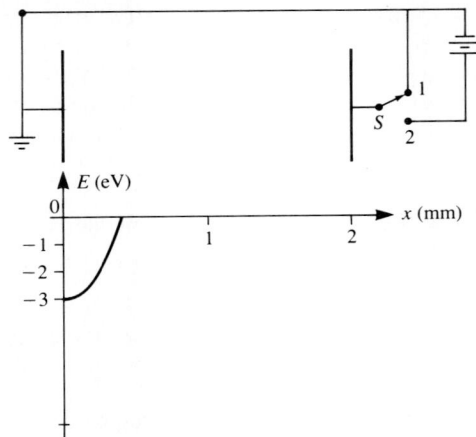

FIGURE P1.7

REFERENCES

1. Hayt, William H.: *Engineering Electromagnetics*, 4th ed., McGraw-Hill Book Company, Inc., New York, 1981.
2. Leighton, Robert A.: *Principles of Modern Physics*, McGraw-Hill Book Company, Inc., New York, 1959.
3. Slater, John C.: *Quantum Theory of Matter*, McGraw-Hill Book Company, Inc., New York, 1951.
4. Langmuir, David B., and W. D. Hershberger: *Foundations of Future Electronics*, McGraw-Hill Book Company, Inc., New York, 1961.
5. Fowles, Grant R.: *Introduction to Modern Optics*, Holt, Rinehart and Winston, Inc., New York, 1968.
6. Masterton, William L., and E. J. Slowinski: *Chemical Principles*, 4th ed., W. B. Saunders Company, Philadelphia, 1977.
7. Holton, Gerald, and H. D. Roller: *Foundations of Modern Physical Science*, Addison-Wesley Publishing Company, Inc., Reading, Mass., 1958.
8. Dekker, Adrianus J.: *Solid State Physics*, 3d ed., Prentice Hall, Inc., Englewood Cliffs, N.J., 1959.
9. Zambuto, Mauro: In *Handbook of Electrical and Computer Engineering* (ed. Sheldon S. L. Chang), vol. 1, pp. 597–623, John Wiley & Sons, Inc., New York, 1984.

ADDITIONAL READING

Fong, Peter: *Elementary Quantum Mechanics*, 2d ed., Addison-Wesley Publishing Company, Inc., Reading, Mass., 1962.
Halliday, David, and R. Resnick: *Physics for Students of Science and Engineering*, John Wiley & Sons, Inc., New York, 1960.
Nanavati, Rajendra P.: *Semiconductor Devices*. Harper & Row Publishers, New York, 1975.
Pantell, Richard H., and H. E. Puthoff: *Fundamentals of Quantum Electronics*, John Wiley & Sons, Inc., New York, 1969.
Sears, Francis W., M. W. Zemansky, and H. D. Young: *College Physics*, 4th ed., Addison-Wesley Publishing Company, Inc., Reading, Mass., 1974.
Yariv, Amnon: *Quantum Electronics*, 2d ed., John Wiley & Sons, Inc., New York, 1975.

CHAPTER
2

ATOMS IN CRYSTALS

2.1 THE LINEAR CRYSTAL—
COMPOSITE POTENTIAL DISTRIBUTION
AND BAND THEORY

In modern electronic devices, atoms very seldom appear as independent separate entities, but rather as constituents of complex multiatomic systems.

When a large number of atoms are assembled they interact, exerting powerful forces of *cohesion* on each other. Under appropriate conditions these forces compel the atoms to become arranged in space in a remarkably regular array, called a *lattice*, the geometric characteristics of which constitute a specific property of the material. Under these conditions the atoms form a *crystal* [1–4].

Without entering into an exhaustive discussion, we shall offer here only some plausibility considerations to indicate qualitatively why and how this happens. We shall limit our discussion to crystals of group IV of the table of elements, because of the special interest of these crystals in semiconductor theory.

As is well known (for specific references cf. Ref. 1 of Chap. 1 and App. 1A.2), atomic configurations are most stable when the outer shell is "filled" by eight electrons and atomic systems tend to approximate this condition as closely as possible.

Group IV elements have four electrons in their outer shell, so that, in order to achieve a "filled" outer shell, each atom should somehow "acquire" four additional electrons in it. This condition can be approximated, albeit vicariously, if each atom shares one of its outer electrons with four of its neighbors. This

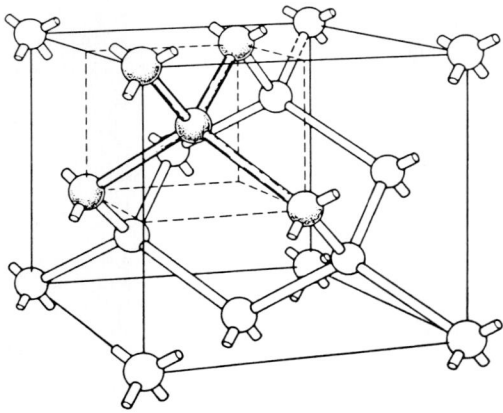

FIGURE 2.1.1
Diamond crystal structure (common to C, Ge, and Si) [1].

sharing can be achieved when the outer shell of each atom is partially superimposed in space on those of these four neighbors. In a crystal of group IV the atoms accordingly tend to arrange themselves so that their outer shells can share electrons. As shown in Fig. 2.1.1, in C, Ge, Si, and a few other elements of group IV, this occurs when the atoms are arranged in a cubic lattice with atomic spacing of about 0.25 nm between atom centers.

It will be useful, for the purposes of visualization, to schematize in two dimensions the complex three-dimensional atomic arrangement of Fig. 2.1.1. To this end, an atom of group IV of Mendeleyev's table will be represented as in Fig. 2.1.2a. Only the features most pertinent to the discussion of the crystalline structure are represented in a symbolic way: the orbits of the four outer shell electrons are symbolized by the dotted ellipses around the small central nucleus, while the large circle centered in the nucleus represents the boundary of the outer atomic shell.

In Fig. 2.1.2b several such atoms are shown in a two-dimensional arrangement approximating some of the salient features of the diamond lattice of Fig. 2.1.1. It will be noticed that the outer shells overlap in such a way that *each atom's outer shell shares some region of space with four of its neighboring atoms.*

The fact that portions of the outer orbits extend to regions of space common to adjacent atoms is meant to indicate that, in its orbiting motion around the nucleus of its atom, each outer shell electron spends some time in a region of space that also belongs to an adjacent atom. While in this region the electron is part of both atomic systems, so that, somehow, during a portion of the orbiting period, each atom shares its four outer electrons with its neighbors, resulting in a maximally stable structure. It should be reiterated here that Fig. 2.1.2 is just a symbolic schematization of a rather complex phenomenon, exclusively for the purposes of visualization. It can, however, be used to achieve an intuitive insight into the crystalline structure and its behavior.

The strong interactions between atoms in the crystalline state result in a *perturbation* of the quantum conditions determining the properties of each atom, so that, finally, the characteristics of the crystal (mechanical, optical, electrical,

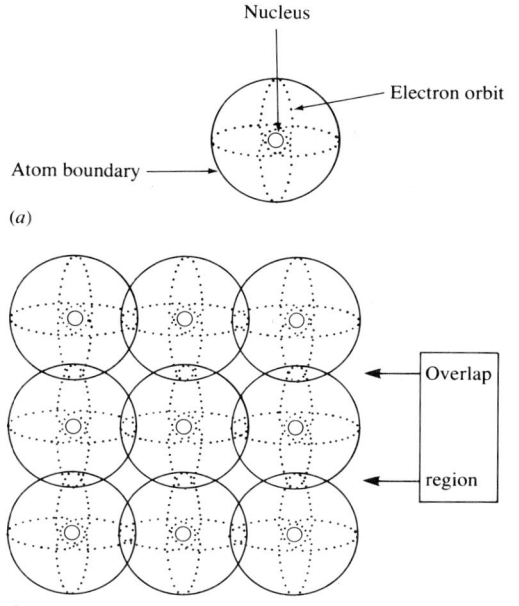

Nucleus

Electron orbit

Atom boundary

(a)

Overlap

region

(b)

FIGURE 2.1.2
Symbolic two-dimensional representation of atomic systems of group IV. (a) Single atom. (b) Atoms in a covalent bond lattice. In some regions of space the outer shells overlap.

etc.), i.e., its physical properties, are closely related to the geometrical arrangement of the atoms in the lattice.

In real crystals, the lattice is, of course, three-dimensional and much more complex than the two-dimensional schematization of Fig. 2.1.2. In such a structure the distances between each atom and its neighbors (and so its interaction with them) is generally not the same in all directions, so that physical properties may vary with crystal orientation. This *anisotropic* behavior of the crystal is of importance in the technology of semiconductor devices. We shall therefore present a brief introduction to the crystallography of semiconductor materials.

Lattices are spatially periodic because they are obtained by repeating a *unit cell* structure over and over in three dimensions, so that the unit cell completely describes the lattice. We shall limit our presentation to cells with cubic symmetry because semiconductor crystals are of this type.

Figure 2.1.3 shows the fcc (face centered cubic) cell with atoms at the eight corners of a cube of side a and at the center of each of the six faces of the cube. As unit cells are adjacent to each other in the crystal, then each cell contains only one-eighth of each corner atom and one-half of each face atom, or $\frac{1}{8} \times 8 + \frac{1}{2} \times 6 = 4$ atoms.

Most of the crystals used in semiconductors have unit cells of the type shown in Fig. 2.1.1. Such unit cells correspond to two interpenetrated fcc cells, displaced from each other by a translation represented by a vector

$$\mathbf{\Delta} = \frac{a}{4} \left(\mathbf{u}_x + \mathbf{u}_y + \mathbf{u}_z \right)$$

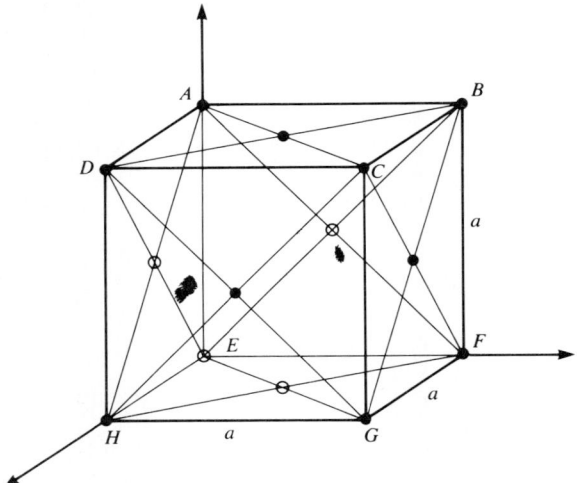

FIGURE 2.1.3
Unit cell of fcc (face centered cubic) structure of side a.

as shown in Fig. 2.1.4. When all the atoms in the cell are the same, as in Si or Ge crystals, then this cell has a diamond structure; conversely, when the atoms of the fcc cells are different from each other, as in GaAs, where one is made of group III and the other of group IV elements, then the cell has a zincblende structure.

Crystalline structures determine crystallographic planes along which the crystal can be easily and very regularly cleaved. It is easy to see (cf., for instance, Fig. 2.1.3) that, as previously stated, the relative positions of the atoms in different planes depend on the plane orientation, resulting in anisotropic crystal behavior. Orientation with respect to the crystal axes is usually identified by the

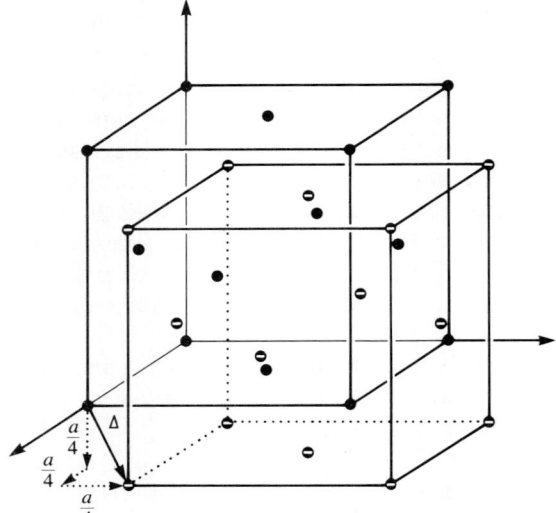

FIGURE 2.1.4
Generation of the diamond and zincblende unit cells from the interpenetration of two fcc cells. The relative displacement corresponds to the vector $\Delta = (a/4)(\mathbf{u}_x + \mathbf{u}_y + \mathbf{u}_z)$.

Miller indices, which can be obtained as follows:

1. Draw the plane with the desired orientation with respect to the crystal coordinate axes.
2. Find the intercepts with the coordinates and compute the reciprocals of these numbers.
3. Find the three smallest integers having the same ratios. These are the Miller indices of the plane orientation.

> **Example 2.1.1.** For the cubic structure of Fig. 2.1.3, find the Miller indices of planes $ABCD$, $EFGH$, $DBFH$, AHF, $ACGE$.
>
> *Solution*
> $ABCD$. Intercepts with the x, y, z crystallographic axes are ∞, ∞, and a respectively, so that the Miller indices are $\langle 0, 0, 1 \rangle$.
> $EFGH$. As this plane contains two axes, the intercepts are indeterminate. Draw a parallel plane (i.e., with the same orientation), such as $ABCD$. From the previous computation, the Miller indices are $\langle 0, 0, 1 \rangle$.
> $DBFH$. Intercepts: a, a, ∞; inverses: $1/a$, $1/a$, 0; indices: $\langle 1, 1, 0 \rangle$.
> AHF. Intercepts: a, a, a; inverses: $1/a$, $1/a$, $1/a$; indices: $\langle 1, 1, 1 \rangle$.
> $ACGE$. Contains the z axis; draw the parallel plane through point H; intercepts: a, $-a$, ∞; inverses: $1/a$, $-1/a$, 0; indices: $\langle 1, \bar{1}, 0 \rangle$. (Notice that a negative index is written with a negative sign above it.)

In the following chapters we shall often reference a semiconductor's properties to the Miller indices of the orientation of its surface.

As already noticed, because of the interactions among neighboring atoms, the behavior of atoms in crystals differs appreciably from the single-atom behavior previously described. Two phenomena are of particular interest here: (1) a *modification of the potential distribution* and (2) the *band structure* of crystals.

Composite Potential Distribution

Atoms in the lattice are generally so close that the potential distribution generated by one nucleus is superimposed in space on that generated by the adjacent atom. It is well known that potential distributions obey the principle of superposition, so that, in the crystalline state, the atomic potential profile in the immediate neighborhood of the atom is distorted. As this constitutes the *potential barrier* holding the valence electron within the confines of the atom, important consequences can be expected from this phenomenon.

In a three-dimensional crystal the changes in the potential distribution can be rather difficult to visualize because of the complexity of the geometrical patterns. To simplify matters, in the following it will be assumed that the crystal structure consists of a single line of atoms strung at a constant interatomic distance from one another. This idealized one-dimensional structure constitutes the *linear crystal*, depicted in Fig. 2.1.5 together with the potential profiles of the single atoms and the composite potential distribution in the crystal.

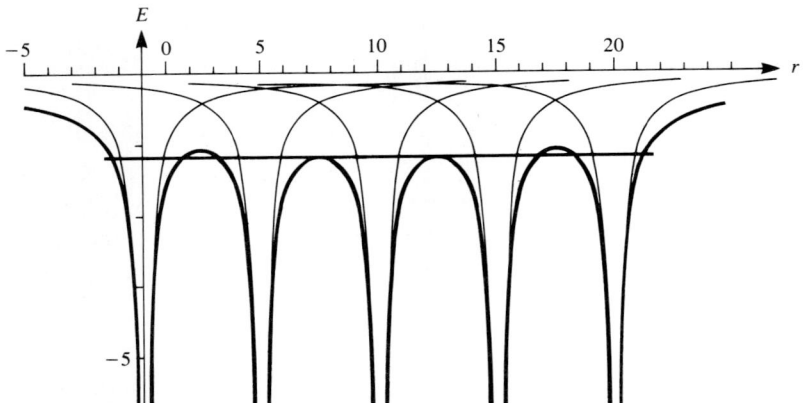

FIGURE 2.1.5
Composite potential distribution (heavy lines) and single atom potential barriers (light lines) of ideal linear crystal. The horizontal heavy line indicates the energy level of the tops of the potential distribution.

It should be noticed that, as the potentials are negative, superposition results in a *lowering of the potential barriers* in the neighborhood of each atom. The resulting composite potential distribution inside the crystal displays a series of maxima well below the potential level for electrons outside the crystal. Only atoms near the surface of the crystal are subject to potential barriers reaching this maximum level.

In Fig. 2.1.5, the level corresponding to the tops of the composite potential energy distribution is indicated by the heavy horizontal line. Any electron with energy above this level is free to move anywhere within the crystal. This should be compared with the energy distribution at the outer edges of the crystal, where there are no more atoms to perturb the individual atom's potential. Here the barrier is seen to rise up to the vacuum level (zero energy).

The interaction of the individual atomic potential configurations and the creation of the composite crystal potential distribution are evidently such that tightly packed structures more significantly lower the potential barriers that keep the electrons bound to their nuclei and oppose ionization.

The student may wish to experiment with different interatomic distances and notice their effect on the composite barrier configuration.

In accordance with the concept of potential barrier, any electron represented in the energy-position diagram by a point *above the top of the potential barrier* is now free to leave the confines of the single atom. Such an electron can freely roam throughout the crystal. Only at the crystal's surface will it encounter a barrier high enough to prevent it from leaving the crystal.

It may be useful to restate the above fundamentally important phenomenon in a slightly different, and more valid, way. In the crystalline state, the *ionization energy* necessary to make a valence electron free of its atom is significantly lowered by the perturbing effects of the neighboring atoms. Electrons that in a single atom were tied to the nucleus may then have enough energy to *ionize the atom*, leaving it. These electrons have enough energy to move throughout the crystal, although a *surface potential barrier* (due to the interruption of the lattice structure at the crystal surface) prevents them from escaping, compelling them to remain within the confines of the crystal.

There is an important difference between these electrons and the ones bound to their atoms. If a voltage were applied to the crystal, an electric field would be established and a force would therefore act on *all* electrons. The bound electrons could not, however, move away from their atom following this force. The free electrons, on the other hand, could. As electrons are electrically charged, their motion constitutes an *electric current*. Because of this, charges that are free to move inside the crystal are called *current carriers*.

Energy Bands

Consider now the permissible states. As previously stated, Pauli's principle does not allow two electrons to occupy the same state in the same system. As long as the atom constitutes an independent system, then each of the permissible states can be occupied by an electron. When atoms form a crystal, however, *the whole crystal constitutes one quantum system*. If a given state were occupied by an electron in two or more atoms, there would be two or more electrons in the same state in the same system!

The perturbing influence of adjacent atomic systems accordingly modifies the quantum conditions characterizing a single atom standing alone, unperturbed, outside the crystal. Each level of the structure of such an atom now *splits into a band of levels*, containing a large number of closely spaced permissible states (as many as there are atoms in the crystalline quantum system) characterized by essentially the same energy. In terms of the energy-position diagram, this *band structure* of the crystal can be represented as in Fig. 2.1.6.

Above the tops of the composite potential distribution there appears a band of closely spaced permissible levels, the *conduction band*. As indicated in Fig.

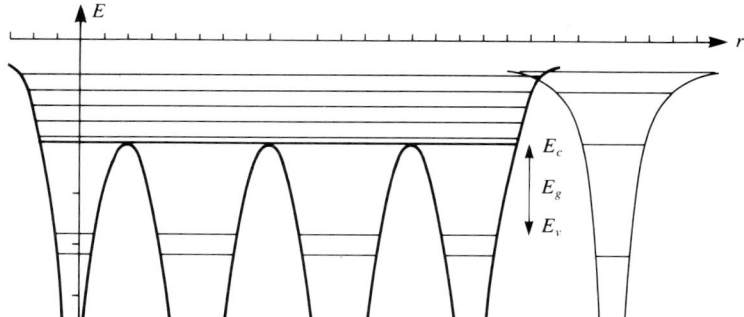

FIGURE 2.1.6
Bands in a linear crystal. E_c indicates the bottom of the conduction band, E_v the top of the valence band, and E_g the gap energy. The single atom potential profile and permissible states are drawn (qualitatively) in lighter lines on the right of the figure.

2.1.6, the *bottom of the conduction band* is characterized by an electron energy level E_c. This is an important parameter of the material.

A more rigorous quantum physical analysis of the behavior of an electron inside a crystal starts from the observation that, in the lattice, the atoms are arranged in an orderly configuration in which the structure of the unit cell is repeated in space at regular intervals. It is then reasonable to assume that the potential of the field generated by such a periodical distribution of charge will also be periodically distributed in space.

The solution of the Schroedinger equation applied to an electron in a periodic potential (under some simplifying assumptions) indicates that the permissible energy levels for the electron are grouped in bands, separated by forbidden gaps, in agreement with the previous intuitive conclusions.

The states contained within these permissible energy bands are characterized by a specific relationship between energy and momentum, so that the state diagram can be drawn in energy-momentum space, as shown in Fig. 2.1.7, which, however, in accordance with common practice, plots electron energy E vs. wave number K (instead of momentum p). The wave number (or propagation constant) is related to the momentum by

$$K = \frac{p}{\hbar} \qquad (2.1.1)$$

where $\hbar = h/2\pi$.

Clearly, a more detailed discussion of this theory, which goes under the name of the *band theory of solids*, is beyond the scope of this text. The interested

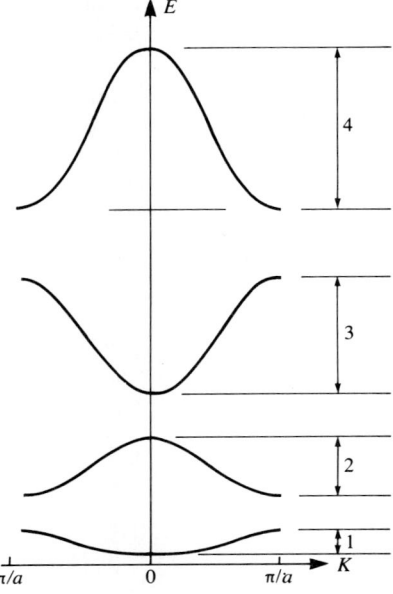

FIGURE 2.1.7
Energy–wave number state diagram showing Brillouin bands.

student is remanded to the abundant literature, which, however, requires some knowledge of elementary quantum theory.[1]

Because of this internal structure, different materials behave in drastically different ways when subjected to an electric field. This behavior depends on the specific combination of parameter values. Of particular importance is the location of the top of the valence band (commonly indicated by the symbol E_v) with respect to the bottom of the conduction band. Several typical possibilities are discussed in the following.

2.2 CONDUCTORS

If the energy level of the atom's outer electron is above the tops of the crystal's composite potential distribution, then as soon as the conduction band is generated, the outer electrons find themselves within it and occupy some of the permissible empty states with which the conduction band is very densely populated. These outer electrons consequently break free from their respective atoms, ionizing them and becoming *current carriers*. They now roam throughout the crystal, constituting an atmosphere of free electrons endowed with mobility and animated by a random distribution of velocities.

The resulting configuration of the crystal's energy-position diagram is depicted in Fig. 2.2.1, showing (at right) the single atom potential profile with occupied levels in relation to the composite potential distribution. The location of E_c (the bottom of the conduction band) is also indicated. Notice that, in this case, as E_c is below the level of the outer electron in the single atom, this electron finds itself within the crystal's conduction band and, in the crystal structure, it cannot any longer be considered a valence electron. Instead, the valence band of

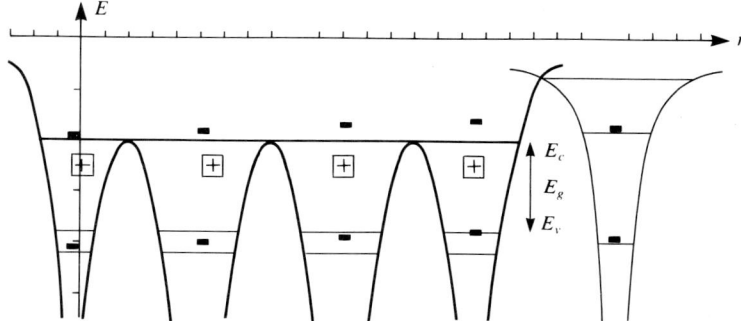

FIGURE 2.2.1
Band structure and electron occupancy in a conductor. The + signs enclosed in squares symbolize the "frozen" ionic charges.

[1] A very readable presentation of Brillouin's band theory is offered in Refs. 3 (p. 238*) and 4 (p. 270*). (The asterisk indicates that some knowledge of quantum theory is required.)

the crystal (as shown by E_v in Fig. 2.2.1) now corresponds to the next inner electron of the single atom configuration.

It should be remembered that the single atom outer electrons (now current carriers) do not remain at the position of their respective generating atoms. Instead, as previously stated, they move around at random throughout the crystal. As these electrons move away, each of them leaves behind an *ionized atom*, i.e., a positive charge. This charge is bound to the atom, and so to the lattice; hence it has no mobility and is frozen in space and cannot contribute to conduction. The position of the bound charges is also indicated schematically in the figure by a + sign within a square.[2] Notice that, because of the negative charge of the conduction electrons, the total charge of the crystal, as a whole, is zero: the crystal as a whole is *electrically neutral*.

Being free of their atoms, the current carriers are able to follow the force acting on them by the presence of any external voltage applied to the crystal, giving rise to a *current*. The crystal, therefore, behaves as a *conductor*. Further analysis of the conditions making conduction possible is in order, from a quantum physical point of view.

If a charge accelerates under the pull of an electric field, thereby increasing its momentum, it must acquire the corresponding additional energy from the electric field. In accordance with the laws of quantum physics, however, this energy transfer can take place only if the energy acquired resonates with a *permissible transition*. For electrons in the conduction band this is always the case because of the large number of very closely spaced states in it, so that it is always possible to reach some permissible unoccupied state without appreciable expenditure of energy.

This condition should be compared with the case of an electron in a *full band*: the only unoccupied permissible states available to it would be in an adjacent higher band, so any permissible transition would require an energy equal to the quantum jump to the next band, across the corresponding energy gap. In most practical cases, the probability that such a large amount of energy can be absorbed by the electron from the electric field is quite remote, so this electron would not be available for conduction.

If, however, the band is not completely filled (so that some permissible states are empty, creating vacancies in the band), there is a high probability that some valence electron could acquire the small amount of energy necessary to fill one of these vacancies. Notice that the vacancy is localized in a given atom; therefore, in order to fill it, the electron should migrate from the atom in which it finds itself originally to the one where the vacancy is.

In this motion, the electron leaves a vacancy in the atom it has abandoned; therefore the final result of the transition is not to make the vacancy disappear,

---

[2] From now on, whenever possible, we shall follow the convention that charges without mobility will be enclosed within square shapes, mobile charges within round shapes.

but only to *move it from one atom to another*. The vacancy in the new position can now be filled by another valence electron from a neighboring atom, so there is a high probability that the position of the vacancy will keep changing from atom to atom. In each transition a negatively charged electron moves from one atom to another: this motion of a charged particle, of course, constitutes a current. This mechanism of current generation will be analyzed in greater detail in Secs. 2.4 and 3.1.

In conclusion, if an energy band does not contain charges (i.e., if all its permissible states are empty), it evidently cannot contribute to conduction. If, on the contrary, it contains so many electrons that all of its permissible states are filled, then these electrons cannot move and again the band cannot contribute to conduction. It must, therefore, be concluded that *an energy band can contribute to conduction only if it is neither empty nor full, but partially occupied.*

As seen, electrons in the conduction band satisfy the above requirement, so it can safely be concluded that materials for which the valence band overlaps the conduction band are *conductors*. Notice that, in these materials, all bands below the conduction band are full, so all other electrons have a negligible probability of contributing to conduction and conduction is mediated exclusively by the electrons in the conduction band.

2.3   INSULATORS—ENERGY IN CRYSTALS

If the *valence band lies entirely below the conduction band*, i.e., if

$$E_v < E_c \qquad (2.3.1)$$

then, after the crystal is formed, the maximum energy level of the electrons in the crystal is lower than the minimum energy required to be in the conduction band. What happens in this case is shown by Fig. 2.3.1.

At the right of the figure is shown the energy-position diagram of the isolated atom, indicating the various permissible state lines and also the electrons occupying each state. On the left, the energy-position diagram for a linear crystal

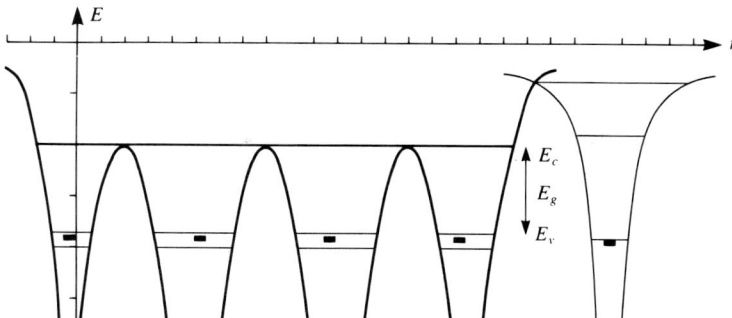

FIGURE 2.3.1
Band structure and electron occupancy in an insulator.

displays the composite potential energy barriers and the energy bands with the electrons occupying them. Also shown are the energy levels for the bottom of the conduction band E_c and the top of the valence band E_v.

In order for valence electrons to become current carriers, they would have to acquire enough energy to bring them at least to the level E_c at the bottom of the conduction band, i.e., an energy

$$E_g = E_c - E_v \tag{2.3.2}$$

This is known as the *energy gap* (or *bandgap*) between the conduction and the valence band. How can a valence electron acquire such energy?

Interaction between Energy and Crystals

In most practical cases, energy is imparted to the entire bulk of a crystal in macroscopic quantities. This comparatively large amount of energy then distributes itself among the microscopic particles in a random way. Under these conditions it is impossible to predict the amount of energy absorbed by any one particle.

To visualize the phenomenon, consider a classroom occupied by, say, 20 students and suppose that an unknown benefactor throws into it $1000 in pennies. After the initial transient, probably of a rather violent nature, each student will possess a certain amount of pennies, but not all students will necessarily have acquired the same sum. Indeed, it would be very difficult, if not impossible, to predict exactly how much money each student has grabbed.

Yet we do know something about the distribution of the sum among the students. For instance: (1) the amount held by any one student must be between $0 and $1000, (2) it must be an integer multiple of 1 cent, (3) if all the money has been picked up, then the average amount is $50 per student, and so on. Notice that none of these predictions can determine precisely the amount acquired by any one student. *All of these statements are statistical in nature.* They refer exclusively to properties of the distribution in general.

Such statistical predictions depend on the conditions under which the phenomenon has occurred and are the more accurate the more is known about the phenomenon and the larger the population among which the distribution has taken place.

In the case of random energy distribution among microscopic particles, the total particle population is very large and, provided the mechanism of interaction between particles and energy is well understood, it is possible to predict very reliably the *distribution density function* $n_d(E)$.

The meaning of this fundamental statistical function should be clearly understood. *For each energy level E, the number of particles per unit volume possessing an energy between E and E + dE is given by $n_d(E) \, dE$.* This implies that the distribution density has the dimension of particles per unit volume and per unit energy range. It also indicates that the number of particles per unit volume having energy levels between E_1 and E_2 is

$$\int_{E_1}^{E_2} n_d(E) \, dE \tag{2.3.3}$$

Thus, in a plot of $n_d(E)$ vs. E, such as Fig. 2.3.2a, the above number of particles per unit volume is indicated by the shaded area.

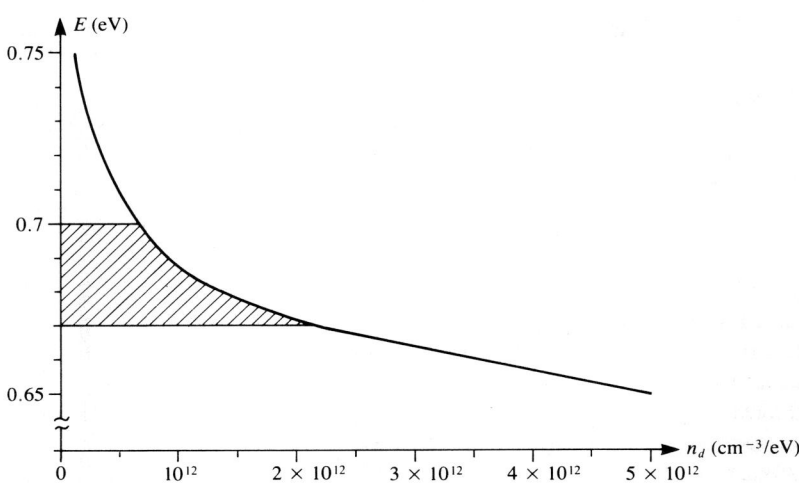

(a)

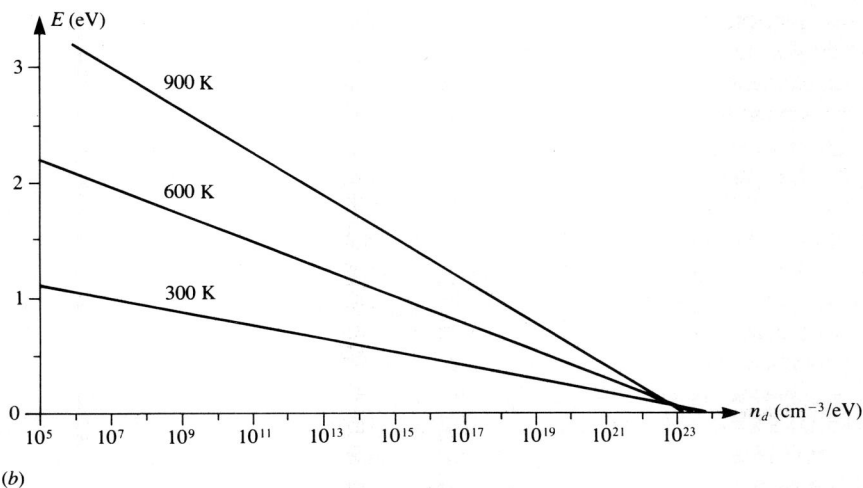

(b)

FIGURE 2.3.2
Boltzmann distribution density function for $N = 10^{22}$ cm^{-3}. (a) Linear scales and $T = 300$ K. The graph must be limited to a small energy range. The shaded area measures the concentration of all particles having energy between 0.67 and 0.7 eV. (b) Semilogarithmic scales and $T = 300$, 600, and 900 K.

In many cases of practical importance, the energy available to the particles is in the form of heat. Then the energy distribution is determined by the temperature of the crystal. Under these conditions application of classical statistical particle mechanics yields, for the distribution, the expression

$$n_d(E) = \frac{N}{kT} e^{-E/kT} \tag{2.3.4}$$

known as *Boltzmann's distribution*, where N is the total number of particles per unit volume of the crystal, k is Boltzmann's constant:

$$k = 1.38 \times 10^{-23} \text{ J/K} = 8.625 \times 10^{-5} \text{ eV/K} \tag{2.3.5}$$

and T is the temperature in kelvin.[3]

Expression (2.3.4) has been used in computing the graph of Fig. 2.3.2a. Because of the exponential nature of the distribution, at room temperature, a minuscule variation of E results in an enormous change in n_d. Indeed, to accommodate a range of energies of 3 eV, semilogarithmic plotting is used, as in Fig. 2.3.2b, where the curves for 300, 600, and 900 K are indicated.

Example 2.3.1. A crystal with an atomic concentration of 10^{22} cm^{-3} is at a temperature of 300 K. Assuming the Boltzmann distribution of Eq. (2.3.4), compute the percentage of atoms that have absorbed a thermal energy equal to or greater than 1 eV. What is the concentration of such atoms?

Repeat for a temperature of 600 K.

Solution. From (2.3.3) and (2.3.4), the concentration of atoms having energy ≥ 1 eV is

$$\int_1^\infty \frac{10^{22}}{8.625 \times 10^{-5} \times 300} e^{-E/0.0259} \, dE$$

or a concentration of 1.7×10^5 cm^{-3}, corresponding to a percentage of 1.7×10^{-15} percent.

At 600 K, analogously, the concentration becomes 4.1×10^{13} cm^{-3} with a percentage of 4.1×10^{-7} percent.

Notice that, at moderate temperatures, only very few particles per unit volume of the crystal possess energies in excess of a few electronvolts. For the number of such particles to be at all significant, the absolute temperature would have to be very high indeed, generally well above the melting point of the specimen. This proves that, if the energy gap is greater than about 2 eV, there is a negligible probability for a valence electron to receive enough thermal energy to *jump the energy gap* and enter the conduction band.

---

[3] The student interested in this topic is urged to start by reviewing the kinetic theory of matter, an elementary discussion for which is offered in Ref. 5 (p. 429); more advanced discussions are to be found in Refs. 6 (pp. 489 and 589), 7 (p. 322), and 8 (p. 15), and after that in Refs. 8 (pp. 19 and 24), 9 (p. 104), and 3 (p. 525) for classical mechanics and Boltzmann statistics.

Consequently, for the energy band configuration of Fig. 2.3.1, if the gap energy is greater than about 2 eV, then for temperatures below the specimen's melting point, the conduction band is essentially empty and the valence band essentially full. In accordance with previous considerations (see Sec. 2.2) it must be concluded that the material behaves as an *insulator*.

2.4 SEMICONDUCTORS—ELECTRONS AND HOLES

The crystals of some elements falling in group IV of Mendeleyev's table and of some compounds of group III and V elements are characterized by gap energies in the neighborhood of 1 eV. For instance, the gap energy is 0.68 eV in Ge, 1.12 eV in Si, and 1.42 eV in GaAs. For these crystals, then, even at room temperature, there is a nonnegligible probability that a valence electron can acquire enough thermal energy to jump the energy gap and enter the conduction band. A sizable number of atoms then ionize, each resulting in a free electron (in the conduction band) and a positively charged ion.

Figure 2.4.1*a* shows the band structure and electron occupancy correspond-

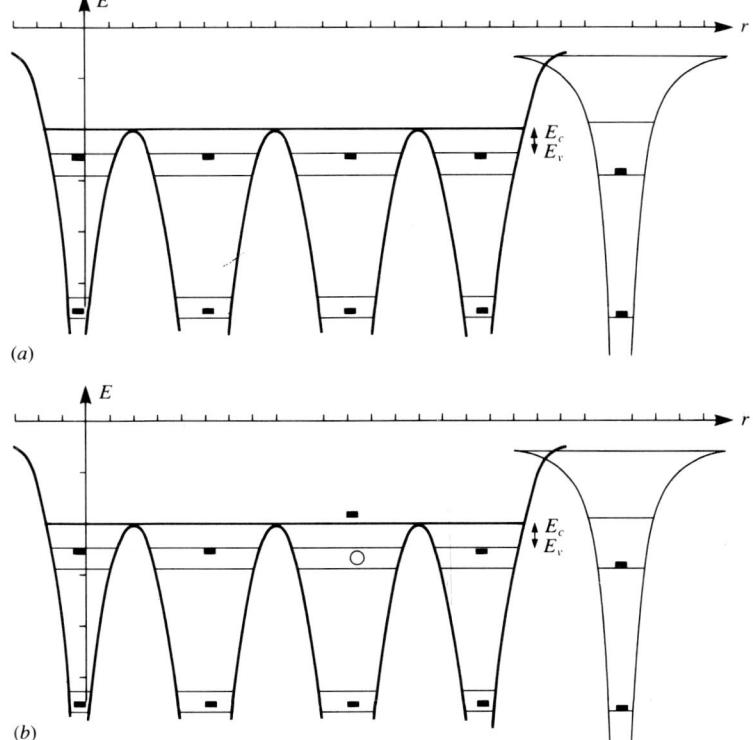

FIGURE 2.4.1
Band structure and electron occupancy of an intrinsic semiconductor (*a*) before any ionization has occurred and (*b*) after ionization of one atom (at center of picture), with the generation of an electron-hole pair.

ing to this case. The state diagram of the unperturbed single atom is shown on the right. Notice the small difference $E_c - E_v$ and the large number of valence electrons close to the conduction band. In Fig. 2.4.1b, one of these electrons is supposed to have absorbed sufficient energy to jump the gap into the conduction band, so that an electron-ion pair has been generated.

Notice that the positive ion corresponds to a *localized vacancy in the valence band* where the electron has escaped the potential barrier holding it tied to its nucleus and migrated into the conduction band. The final result in this sequence of thermally induced transitions is a crystal with a partially filled conduction band and a partially empty valence band. As discussed in Sec. 2.2, both bands contribute to conduction. At absolute zero temperature there is no thermal energy available to ionize the atoms, so that the crystal, which is a conductor at moderate temperatures, behaves as an insulator at absolute zero. These materials are called *semiconductors*.

Electrons and Holes

It has just been recognized that the valence band of a semiconductor at moderate temperatures contains some localized vacancies, corresponding to the thermally ionized atoms, and that, not being full, the valence band can now contribute to conduction. In Sec. 2.2 the mechanism permitting conduction in an almost full band was introduced, but not analyzed in sufficient detail. Consider an analogy.

Figure 2.4.2 represents schematically a familiar puzzle game: each disc in the figure represents a plastic chip. The chips can slide on a support divided into squares, as shown. Each square can be either empty or occupied by no more than one chip (this rule is evidently equivalent to Pauli's exclusion principle). The aim of the game is to move the chips around until they are arranged in a desired order.

Suppose now that all the squares are occupied. Evidently no chip can be moved because the chips are in each other's way. If, however, even a single chip is removed (leaving behind a *vacancy*), then the chips acquire some mobility. For the same reason, as stated in Sec. 2.2, the electrons in a full band have no mobility, but acquire it as soon as vacancies appear in the band.

Notice that, because all chips are identical and it is impossible to tell them

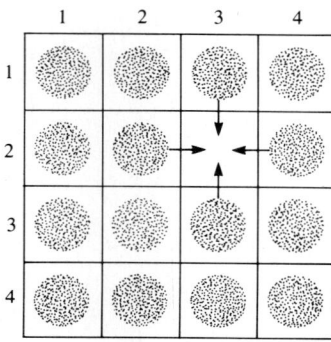

FIGURE 2.4.2
Model of electron motion in partially filled band. This system simulates Pauli's principle.

apart, an observer has a choice of two equally correct ways of describing the motion: (1) the chips move around or (2) it is the vacancy that moves. In solid state physics, this second point of view turns out to be the most useful.

In a semiconductor, each vacancy is localized at the position of a positive ion, yet, in the motion, it is not the ion that moves from place to place: *it is the phenomenon of ionization that migrates from one atom to another.* Also, a net positive charge is always present at the location of the ionization, moving from atom to atom as the ionization phenomenon migrates.

To sum up, in a semiconductor, thermal energy generates pairs of charges of opposite sign, both endowed with mobility. The negative charges are *electrons;* the positive charges, corresponding to a vacancy, are called *holes.* At temperatures higher than absolute zero, both electrons and holes, enjoying mobility, are animated by random motion.

Generation and Recombination

Later on, to determine quantitatively the electrical properties of semiconductors, it will become important to compute the concentration of the current carriers in the crystal (this concentration, as shall be seen, determines the conductivity of the material). To this end, consider another aspect of the phenomenon.

The availability of thermal energy (heat) in the crystal, as evidenced by a temperature above absolute zero, generates a significant probability of ionization of the crystal's atoms. This means that, every so often, an atom will ionize, the *rate of ionization* being an increasing function of temperature. If this were the only phenomenon present, the atoms would continue to ionize, producing each time an electron-hole pair, and so the current carrier concentration (their number per unit volume) would keep increasing until all the crystal's atoms were ionized. This evidently does not occur, indicating the presence of some other mechanism limiting the carrier population. This mechanism is *recombination.*

As soon as electron-hole pairs appear in the crystal there arises the possibility that, in their motion, an electron may meet a hole. When this occurs, because of their opposite charges, the two neutralize each other, the electron filling the vacancy in the ion's valence band, and both disappear as free charges. The probability of such an encounter and recombination evidently increases with the concentration of electron-hole pairs.

As the carrier population increases, therefore, the rate of recombination increases, while the rate of generation remains essentially constant (as long as the temperature does not change). The *net rate of increment* of the carrier population is evidently the difference between the rates of generation and recombination; therefore, when the increasing rate of recombination finally becomes equal to the constant rate of generation, the electron-hole pair population remains stationary, because the number of electron-hole pairs generated each second now equals the number recombining in the same time. A condition of *dynamic equilibrium* has been reached.

The corresponding electron-hole pair concentration is the *equilibrium concentration* for the specific material at the given temperature. As indicated by the mechanism just described, the equilibrium concentration is an increasing function of temperature. Consequently, a semiconductor crystal behaves as a conductor at moderate temperatures, although it is an insulator at absolute zero.

2.5 EXTRINSIC SEMICONDUCTORS

Up to now the crystals we have considered consisted of identical atoms, all of the same element. Suppose now that, at the moment of crystal formation, some other type of atom is present as a very small percentage *impurity* among the *host* material. The characteristics of the crystal lattice are determined exclusively by the nature of the majority atoms. As the impurity atoms are too few appreciably to perturb the lattice constants they fit in the lattice of the host material, occupying the same positions that would normally have been filled by majority atoms. These impurity atoms, however, have totally different quantum properties, so their behavior in the lattice also differs from that of the host atoms.

At this point, attention will be concentrated on host materials of group IV elements of Mendeleyev's table, with impurities belonging to either group V or group III. These combinations result in two types of material called *extrinsic semiconductors*. Let us analyze the two cases separately.

Group V Impurities

As indicated schematically in Fig. 2.5.1, the configuration of permissible states of an element of group V (shown at the center of the figure) is such that, when this impurity becomes a constituent of a group IV crystal, *one of its occupied states occurs at an energy level within the gap* of the host crystal's composite energy-position diagram. Of course, such a state would be forbidden in a crystal composed exclusively of group IV atoms, but the presence of each impurity atom creates an occupied permissible state within the crystal gap and very close to the bottom of the conduction band.[4] These special states are localized in space, existing only at the position of the group V impurity atom, as shown in Fig. 2.5.1*a*.

Consider now the valence electrons of the impurity. In order to move to the next available permissible state, they must overcome only a very small energy gap (generally, as we have seen, of the order of 0.01 eV), so this transition requires the absorption of only an extremely low amount of energy, easily available, even at moderate temperatures; therefore in semiconductor crystals *at room temperature essentially all group V impurity atoms are ionized.* Figure 2.5.1*b* shows the final

---

[4] For instance, the energy gap of the valence electron of As in Si is 0.013 eV; of P in Si it is 0.012 eV; etc.

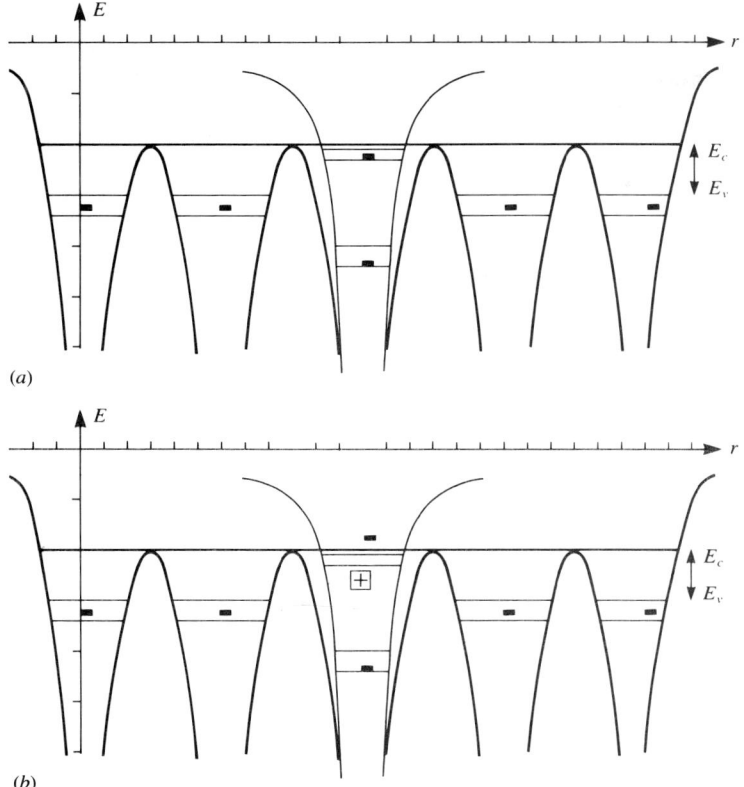

(a)

(b)

FIGURE 2.5.1
Band structure and electron occupancy for N-type semiconductor. (a) Before impurity ionization, a localized, full, permissible state is created within the gap near E_c at the location of the group V impurity. (b) After impurity ionization, a conduction electron has been generated together with a localized, positive ion "frozen" at the donor location. The permissible state near the bottom of the conduction band is empty.

result. Because each group V impurity atom contributes one electron to the conduction band, these impurities are known as *donors*.

As shown in Fig. 2.5.1b, the ionization process, of course, also produces a positive ion, due to the vacancy left in the valence level of the ionized atom. This, however, is the valence level of the impurity atom and is therefore not within the valence band of the group IV crystal, so the vacancy will not behave as a hole. Indeed, it can be computed that the probability of migration of this vacancy is negligible. As usual, in the figure, this is conventionally symbolized by the square box enclosing the charge sign (cf. Sec. 2.2). The localized positive charge generated by ionization *remains frozen in the impurity atom* and is not endowed with mobility, so *it does not contribute to conduction*.

In conclusion, in a semiconductor crystal doped with donor impurities, conduction electrons are generated by both impurity and intrinsic atoms, while holes are generated only by intrinsic atoms (when the impurity atoms ionize, they generate "frozen" positive ions bound to the lattice). Therefore, in these extrinsic semiconductors, electrons outnumber holes. Conduction is mediated prevalently by electrons, which are said to be *majority carriers*, while holes are *minority carriers*. The result is an *N-type semiconductor*.

Group III Impurities

As shown in Fig. 2.5.2*a*, the configuration of the permissible levels of an element of group III is such that, when this impurity becomes a constituent of a group IV crystal, one of its *unoccupied states occurs at an energy level within the gap* of the

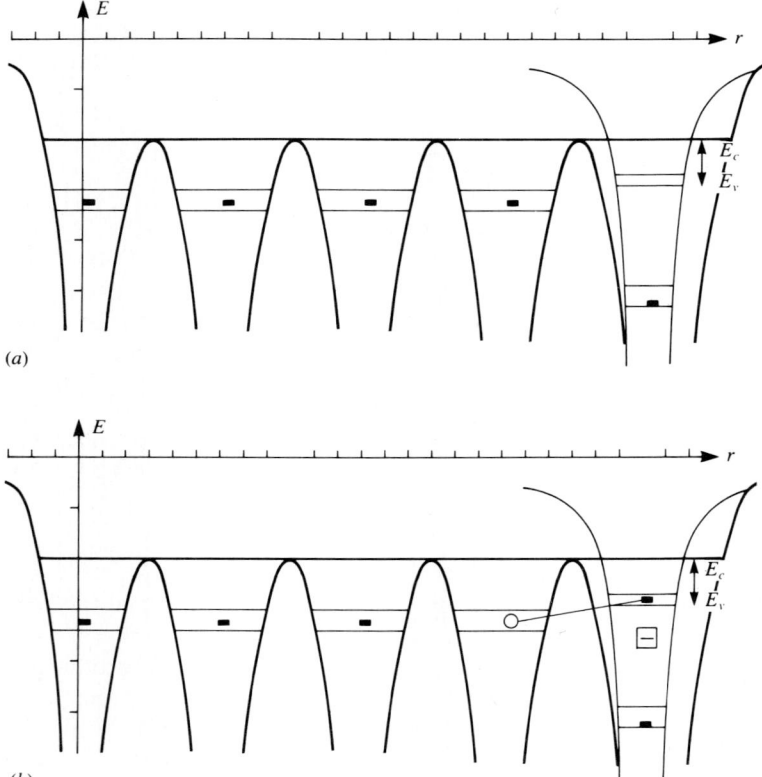

(a)

(b)

FIGURE 2.5.2
Band structure and electron occupancy of P-type semiconductor. (*a*) Before impurity ionization, a localized, permissible state is created by each impurity atom. This state is at an energy level within the gap, very close to E_v, it is originally empty, and exists only at the locations of the group III impurity atoms. (*b*) After impurity ionization, the impurity has accepted an electron from a neighboring valence band. The transition occurs by tunneling, as symbolized by the electron "trajectory" in the figure. The result is a hole and a negative ion "frozen" at the location of the donor impurity.

host crystal's energy-position diagram. As a consequence, an empty permissible state is created within the crystal's gap. This state is also localized in space, existing only at the positions of the impurity atoms, and is very close to the top of the valence band.[5]

Electrons in the valence band of the crystal must overcome only a very small energy gap (about 0.01 eV) in order to move into the impurity's empty state within the gap, so this transition, requiring the absorption of only an extremely low amount of energy, is very probable, even at room temperature.

Actually, inspection of Fig. 2.5.2b shows that, although the difference between the electron energies in the initial and final states of this transition is extremely small, in order to move from one atom to the other the electron must overcome a comparatively large ridge of potential. Climbing over this potential barrier would apparently require considerable energy. This, however, is not the case, thanks to a peculiar quantum physics phenomenon: *tunneling*.

This phenomenon was introduced in Chap. 1 and App. 1A.2, where it was shown that displacement over even a classically impenetrable potential barrier may not require the absorption of large quantities of energy and have a large probability of occurrence provided: (1) the difference in energy between the initial and final positions is negligible and (2) the distance between the two positions is small enough.

Under these conditions the electron can undergo the displacement without actually "climbing" over the potential barrier, but simply by "tunneling" under it. The conditions for tunneling and its probability of occurrence can be quantitatively determined by quantum physical considerations (the interested student with some quantum background will find an elementary quantitative discussion in App. 1A.2). For the purposes of this presentation the above qualitative statement of conditions 1 and 2 will suffice.

In Fig. 2.5.2b the tunneling nature of this transition is evidenced by the line indicating, in a purely symbolic way, that in the energy-position diagram the displacement does not occur by climbing over the potential barrier, but rather by burrowing under it. Notice that, in accordance with the observations of Sec. 1.3, a portion of this path entering a region below the potential barrier would correspond to a motion with imaginary velocity. This by itself is evidence of the purely symbolic nature of the path indicated and of the expression (tunneling) used to designate the phenomenon. What is really meant is that, under favorable conditions, the laws of quantum physics assign a finite probability to a transition between two states separated by a potential barrier, without requiring (as do the laws of classical physics) the acquisition of sufficient energy to reach the top of the barrier.

This transition removes an electron from the crystal's valence band, so that the vacancy left behind constitutes a *hole*. At the same time it adds an electron to

---

[5] For B in Si this gap turns out to be 0.01 eV; for In in Si it is 0.011 eV; etc.

the electron atmosphere of the trivalent impurity, negatively ionizing the impurity atom. The resulting configuration is shown in Fig. 2.5.2*b*.

The energy required for ionization is very low so that in semiconductor crystals *at room temperature essentially all group III impurities are negatively ionized*. Because each of these atoms receives one electron from the valence band of the crystal, group III impurities are known as *acceptors*.

As seen, the ionization process also produces a negative ion, due to the addition of an electron to the originally vacant state of the impurity atom. This state, again, does not fall within the conduction band of the crystal, so the ionization electron is not a current carrier. The probability of migration of this negative charge is negligible. To all intents and purposes *this charge remains "frozen" in the impurity atom* and is not endowed with mobility (notice the square shape of its representation in the figure, in accordance with our convention), so *it does not contribute to conduction*.

In conclusion, in a semiconductor crystal doped with acceptor impurities, holes are generated by both impurity and intrinsic atoms and conduction electrons only by intrinsic atoms (when the impurity atoms ionize, they generate "frozen" negative ions bound to the lattice). Therefore, in these extrinsic semiconductors, holes outnumber electrons. Conduction is mediated prevalently by holes, which are said to be *majority carriers*, while electrons are *minority carriers*. The result is a *P-type semiconductor*.

In Sec. 2.4 it was noticed that the probability of a collision (and consequent recombination) of an electron and a hole increases with increasing carrier concentration. This is self-evident in the case of an intrinsic semiconductor, in which electron and hole concentrations are the same. However, in the case of extrinsic semiconductors, it is possible that the concentration of minority carriers (thermally generated from the intrinsic material atoms) may be very small, even in the presence of a large concentration of majority carriers from the impurity atoms. In this case, the carrier concentration might be considerable, but the probability of a collision might still be small, for lack of carriers of the other type (holes in N-type and electrons in P-type semiconductors).

Without entering here into more rigorous considerations, it is easy to understand intuitively that the probability of a collision (and so the rate of recombination) increases with *the product np* of the concentrations of the majority and minority carriers. As dynamic equilibrium is reached when the rate of recombination equals the rate of generation, it can finally be concluded that *the product of the equilibrium concentrations of the two types of carriers depends exclusively on the rate of carrier generation*. The alert student will recognize this result as a special case of the general *mass law*.

2.6 SUMMARY

Group IV elements form cubic crystal lattices by sharing outer shell electrons with four neighboring atoms. When these crystals are formed, a new potential distribution results from the superposition of the potential barriers of the component atoms. This distribution results in a *lowering of the barriers inside the crystal.*

The permissible energy levels of the isolated atoms forming the crystal are transformed into *bands* of closely spaced permissible states. The *conduction band* is located above the top of the composite potential distribution. The *valence band* is the next lower band of the crystal.

The electron energy level of the *bottom of the conduction band* is E_c. *The top of the valence band* is at energy level E_v. These are characteristic parameters of the material.

If a band is either empty or full it cannot contribute to conduction. *Only partially filled bands contribute to conduction.*

When energy is imparted to the bulk of a crystal, it is distributed among the atoms in a random way. The statistics of this distribution determine the *distribution density function* $n_d(E)$.

Conductors and Insulators

Materials in which the energy level of the outermost electron is above E_c have a plentiful supply of current carriers in the form of electrons in the conduction band. They are, therefore, *conductors*. In them, the positive charges associated with ionized atoms are *bound to the lattice* and cannot contribute to conduction, which is mediated by electrons.

If the top of the valence band is much below the bottom of the conduction band (more than about 2 eV), the *gap energy E_g is large*, the probability of ionization is negligible, there are essentially no current carriers, and the material is an *insulator*.

Intrinsic Semiconductors

In intrinsic semiconductors, because of the comparatively small energy gap E_g, thermal energy causes *ionization* of some of the crystal atoms. This results in a continuing process of *generation* of negative *electrons* in the conduction band and positive *holes* in the valence band. These charges can move freely throughout the crystal and are therefore *current carriers*. At nonzero temperatures these current carriers are animated by random motion.

In their motion, electrons and holes can meet and *recombine*, disappearing as free charges. The rate of generation is an increasing function of temperature, while the rate of recombination is an increasing function of the product np of the current carrier concentrations. Consequently, the current carrier concentrations continue to vary until a condition of dynamic equilibrium is reached. This condition is determined by the *electron-hole concentration product np*. For a given semiconductor (intrinsic or extrinsic) the equilibrium value of the concentration product is an increasing function of temperature.

At absolute zero no electron-hole pairs are produced and the semiconductor behaves as an insulator. Above absolute zero, electron and hole concentrations are equal and current is mediated by both types of current carriers.

Extrinsic Semiconductors

When a semiconductor crystal is *doped* with pentavalent or trivalent impurities it becomes an *extrinsic semiconductor*.

In *N-type semiconductors* doped with pentavalent *donor impurities*, each impurity atom creates a localized permissible state near the top of the gap. This state is originally occupied, but at room temperature all impurity atoms ionize, each contributing one electron to the conduction band and generating a localized *positive ion* bound to the crystal lattice and therefore not endowed with mobility. Electrons outnumber holes and conduction is prevalently mediated by electrons (*majority carriers*) rather than by holes (*minority carriers*).

In *P-type semiconductors* doped with trivalent *acceptor impurities*, each atom creates an empty permissible state near the bottom of the gap. All impurity atoms soon ionize, receiving, by a *tunneling* transition, one electron each from the crystal's valence band, resulting in a *hole* and an ion. The negative charge of this ion, being localized in the impurity atom, is bound to the crystal lattice and has no mobility. Holes outnumber electrons and conduction is prevalently mediated by holes (*majority carriers*) rather than by electrons (*minority carriers*).

PROBLEMS

2.1. Compare Fig. 2.4.1*a* with Fig. 2.3.1.
  (*a*) What is the most significant similarity between the two?
  (*b*) In what do they differ?
  (*c*) What consequence can be expected in the electrical behavior of the two crystals?

2.2. In which way can adjacent atoms in a crystal be said to share their outer electrons?

2.3. By what mechanism are conduction electrons generated in a conductor?

2.4. Compare the ionization potential of an isolated atom with that of the same atom in a crystal lattice. How do these ionization potentials compare with the crystal's work function?

2.5. Why are the electron potential energies within a crystal negative?

2.6. What is the surface potential barrier, how is it generated, and what is its effect?

2.7. What are E_c and E_v? How are they related to the gap energy? What phenomenon does E_g control?

2.8. A certain material at 0 K has a full valence band and an empty conduction band. The gap energy is 1 eV. At any given moment, what percentage of the valence electrons of this material has enough energy to generate ionization at 300 K?

2.9. Discuss qualitatively what happens as larger and larger voltages are applied to an insulator.
  Hint: Consider the energy that an electron can acquire from the electric field.

2.10. What conditions determine carrier generation and recombination rate in a semiconductor and why?

2.11. When is equilibrium reached between generation and recombination? After equilibrium is reached, what happens if the temperature T is increased?

2.12. In an extrinsic semiconductor with As impurities in Si compute the percentage of As atoms that have enough energy to ionize at 300 K. What determines the concentration of ionized As atoms?

2.13. Why does an ionized As impurity atom not behave as a hole?

2.14. (a) What is the potential barrier to be overcome by a Si valence electron in order to ionize a B impurity atom?
   (b) Why is the probability of B impurity ionization much higher than that of Si?
   (c) Why is this transition limited essentially to Si atoms adjacent to the impurity?

2.15. In a cartesian coordinate system, sketch the plane orientations corresponding to the following Miller indices: (a) [0, 1, 0]; (b) [0, 1, 2]; (c) [3, 6, 2]; (d) [1, 0, 4]; (e) [2, 1, 0].

REFERENCES

1. Shockley, W.: *Electrons and Holes in Semiconductors*, Van Nostrand, Inc., New York, 1950.
2. Yang, Edward S.: *Fundamentals of Semiconductor Devices*, McGraw-Hill Book Company, Inc., New York, 1978.
3. Dekker, Adrianus J.: *Solid State Physics*, 3d ed., Prentice Hall, Inc., Englewood Cliffs, N. J., 1959.
4. Kittel, Charles: *Introduction to Solid State Physics*, 2d ed., John Wiley & Sons, Inc., New York, 1959.
5. Holton, Gerald, and H. D. Roller: *Foundations of Modern Physical Science*, Addison-Wesley Publishing Company, Inc., Reading, Mass., 1958.
6. Halliday, David, and R. Resnick: *Physics for Students of Science and Engineering*, John Wiley & Sons, Inc., New York, 1960.
7. Sears, Francis W., M. W. Zemansky, and H. D. Young: *College Physics*, 4th ed., Addison-Wesley Publishing Company, Inc., Reading, Mass., 1974.
8. Ramey, Robert E. L.: *Physical Electronics*, Wadsworth Publishing Company, Belmont, Calif., 1961.
9. Adler, R. B., A. C. Smith, and R. L. Longini: *Introduction to Semiconductor Physics*, John Wiley & Sons, Inc., New York, 1964.

ADDITIONAL READING

Fortino, A. G.: *Fundamentals of Integrated Circuit Technology*, Reston Publishing Company, Inc., Reston, Va., 1984.
Sze, S. M.: *Physics of Semiconductor Devices*, 2d ed., John Wiley & Sons, Inc., New York, 1981.
Sze, S. M.: *Semiconductor Devices, Physics and Technology*, John Wiley & Sons, Inc., New York, 1985.

CHAPTER

3

THERMAL EQUILIBRIUM

We have obtained a qualitative insight into the phenomena occurring in semiconductor crystals at the atomic level. For engineering purposes, however, mere qualitative understanding is not sufficient: the problems of design demand computation of quantitative data. For instance, it is not enough to know that, at moderate temperatures, because of the presence of current carriers, a semiconductor crystal behaves as a conductor: the relationship between voltage and current must be determined quantitatively for each value of temperature and doping.

As current density is, by definition, the product of particle velocity times particle concentration and particle charge, $J = \rho_v V = qnv$ [1], then it will be necessary to compute (1) the concentration of holes and of conduction electrons required to reach equilibrium between generation and recombination at any given temperature and (2) the laws of motion of the current carriers in the presence of an applied electric field.

To this end we shall first establish some fundamental relationships describing the general behavior of electrons in crystals.

3.1 ELECTRONS AND HOLES— EFFECTIVE MASS

In the preceding chapter it was shown that, in a semiconductor, carrier motion can occur within two bands:

1. The conduction band, *because it is not quite empty*. Here charge motion is mediated by *conduction electrons*, which are not bound to the lattice atoms.

56

2. The valence band, *because it is not quite full.* Here *valence electrons,* bound to their atoms, move to fill vacancies in neighboring atoms.

In each of these two cases, it is evidently important to find out how the electrons absorb the energy required for their motion and what kind of motion is imparted to them. On the other hand, one cannot expect that the laws governing electron motion will be the same under the very different conditions prevailing in the two bands, so electron motion should be analyzed separately for each case.

Conduction Electrons

The motion of a charged particle in empty space can be predicted by the laws of classical mechanics, but the "free" conduction electrons are not in empty space: they find themselves inside a crystal and are therefore subjected to a number of forces exerted on them by the other crystal charges, such as the nuclei, other electrons, etc. In other words, they move within a nonuniform potential field. They cannot be expected to move as if they were in empty space any more than a billiard ball moving on a corrugated surface can be expected to behave as if it were on a flat pool table.

The situation is further complicated by the fact that the interaction of the electrons with the lattice follows the laws of quantum physics, so that the only way rigorously to figure out how conduction electrons absorb different types of energy and what are the characteristics of the resultant motion is by the use of quantum theory.

Fortunately, if such an analysis is carried out, it shows that, within reasonable approximations, *everything happens as if the electron were a classical charged particle having an effective mass m_e to be determined* in each case by some quantum considerations. The value of this mass is different from that of the electron in empty space and depends on type of crystal, type and amount of impurities present, etc. It has been computed for most common conditions and the results are available in tabulated form (cf. Table 3.1.1).

The quantum determination of m_e proceeds from the energy-momentum relationship resulting from the band theory and plotted in Fig. 2.1.5.

In classical mechanics, as is well known, the kinetic energy of a particle of mass m and speed v is

$$E_k = \tfrac{1}{2}mv^2 \tag{3.1.1}$$

and its momentum is

$$p = mv \tag{3.1.2}$$

By elementary algebra, kinetic energy and momentum are seen to be related by

$$E_k = \frac{1}{2}\frac{p^2}{m} \tag{3.1.3}$$

TABLE 3.1.1

Characteristics of Ge, Si, and GaAs at 300 K

Characteristic	Symbol	Units	Ge	Si	GaAs
Effective density of states	N_c	cm^{-3}	1.04×10^{19}	2.8×10^{19}	4.7×10^{17}
	N_v	cm^{-3}	6.1×10^{18}	1.02×10^{19}	7×10^{18}
Energy gap	E_g	eV	0.68	1.12	1.42
Electron affinity	χ	V	4.13	4.01	4.07
Intrinsic carrier concentration	n_i	cm^{-3}	2.4×10^{13}	1.5×10^{10}	10^7
Effective mass	m_e	um†	0.22	0.33	0.068
	m_h	um†	0.31	0.56	0.56
Mobility	μ_e	cm^2/(V·s)	3900	1350	8600
	μ_h	cm^2/(V·s)	1900	480	250
Dielectric constant	ε_R		16.3	11.8	10.9
Atomic concentration		cm^{-3}	4.42×10^{22}	5×10^{22}	4.42×10^{22}
Breakdown field	ε_{bk}	V/cm	10^5	3×10^5	3.5×10^5

† One um (unit of mass) equals the mass m_{e0} of the free electron at rest *in vacuo*, $m_{e0} = 9.11 \times 10^{-31}$ kg.

yielding the familiar formula $m = p^2/2E_k$. However, m could also be obtained by differentiating (3.1.3) twice:

$$m = \frac{1}{d^2 E_k/(dp^2)} \tag{3.1.4}$$

A similar reasoning, in quantum terms, shows that (3.1.4) remains valid for electrons in bands, taking the form:

$$m_e = \frac{\hbar^2}{d^2 E/(dK^2)} \tag{3.1.5}$$

which, remembering Eq. (2.1.1), is seen to be equivalent to (3.1.4).

Considering the shape of the energy–wave number plots of Fig. 2.1.5, the effective mass is seen to vary from state to state within a band, as shown in Fig. 3.1.1, which, in Fig. 3.1.1a, reproduces one of the E vs. K curves of Fig. 2.1.5 and, in Fig. 3.1.1b, plots m_e, following Eq. (3.1.5). It should be noticed that the effective mass increases from a minimum for electrons at the bottom of the band to a maximum for electrons at the center of the band. The effective mass of electrons in the top half of the band is negative.

Any further discussion is beyond the scope of this text; as usual, the student with some understanding of quantum theory is referred to the literature (cf. Refs. 2, p. 248*, and 3, p. 288*).

It is usually easier and more reliable to obtain the effective mass from measurements of related crystal characteristics (such as, for instance, the mobility, Hall effect, etc.—cf. Chap. 5). Values of m_e are generally expressed in units of mass (1 um $= m_{e0}$, the rest mass of an electron *in vacuo;* see Table 3.1.1) and can range from several to $\frac{1}{100}$ um.

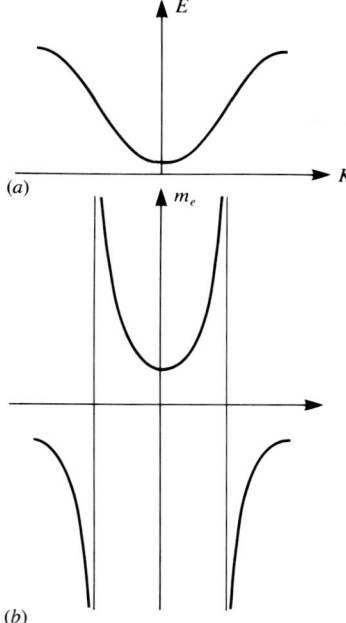

(a)

(b)

FIGURE 3.1.1
Energy and effective mass vs. wave number K within a Brillouin band [cf. Eq. (3.1.5)]. The effective masses are positive for electrons in the lower half of the band, negative in the upper half.

Valence Electrons

The motion of the bound electrons is even more complex. It will be remembered that it has been described in Secs. 2.2 and 2.4 as a "jump" of an electron from the valence orbit of one atom to the permissible orbit vacated in an adjacent ion at the moment of ionization. The analogy of Fig. 2.4.2 was used to show how the presence of a vacancy in a system subject to Pauli's exclusion principle could make such a motion possible. Still, mostly for mnemonic purposes, it may be desirable to visualize a plausible mechanism for such a transition. To achieve this end, the phenomenon will be simulated by a mechanical model, even though this deterministic model is rather questionable and not scientifically rigorous.

Figure 3.1.2 repeats the two-dimensional representation of a semiconductor crystal bond arrangement previously introduced in Fig. 2.1.2. As before, the dotted ellipses represent "electron orbits" and the fact that orbits of adjacent atoms overlap indicates that their respective electrons are "shared" by the adjacent atoms.

To avoid clutter, the electrons moving around these orbits are not shown, with the exception of one, labeled e, in atom 1 at center left in Fig. 3.1.2a. Atom 2 at the center of the picture is supposed to be ionized, as symbolized by the missing orbit and by the $+$ sign (enclosed in a circle to indicate mobility) representing its ionic charge.

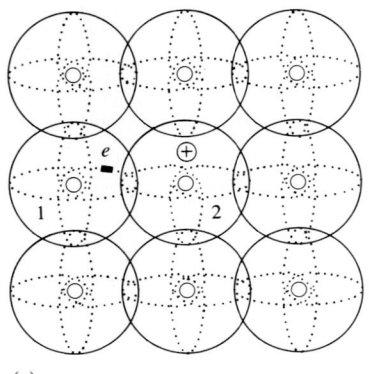

(a)

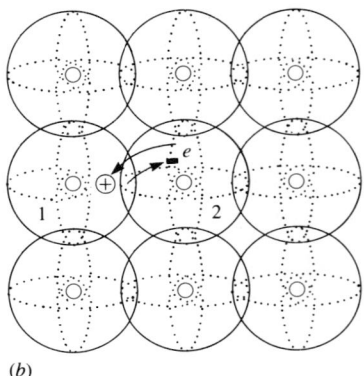

(b)

FIGURE 3.1.2
Two-dimensional model of hole displacement. The hole is designated by the encircled + sign. (a) The hole is located in atom 2 where one outer shell electron is missing (missing orbit). Part of the orbit of electron e shown in atom 2 lies within the region of space common to atoms 1 and 2. (b) The outer shell electron e switched orbit from atom 1 to atom 2 while it found itself in the region common to the two atoms; consequently the hole has now moved to atom 1.

In this *purely symbolic, deterministic system*, an electron "shared" by two atoms, while orbiting, periodically reaches the region of space where the outer shells of the two adjacent atoms overlap. While in this common region of space, the electron finds itself within the spheres of influence of *both* atoms and can be considered a part of each of them. To this extent, the electron can be said to be "shared" by the outer shells of both atoms.

Once in this region (where the electron is under the influence of both nuclei), it appears plausible that the electron may continue its course following orbits belonging to either of the two atoms. Indeed, there should be a good probability that the attraction of the vacancy in the adjacent atom could cause the electron to leave its original orbit and move to an orbit encircling the neighboring nucleus, filling the vacancy. Such a transition would result in a spatial displacement of the localized charge (hole) associated with the vacancy.

In Fig. 3.1.2b this transition of the valence electron orbit from one atom to the adjacent one is symbolized by the arrows, indicating the displacement of electron e and its orbit from atom 1 to 2, and of the ionic charge (encircled + sign) from 2 to 1. Compare Fig. 3.1.2a and b.

To avoid serious misinterpretations, it is imperative to realize that analogies of this type have a purely pedagogical value, possibly helping the student to visualize *some phenomenon that would behave similarly to the one under discussion*, but that this classical, and therefore more easily visualized, model is fundamentally different from what actually takes place, to the point where, for instance, even such basic concepts as electron position, orbit, and momentum may be too deterministic and have little or no meaning in the reality of quantum phenomena.

Indeed, for certain phenomena (such as the Hall effect), if the above model were to be used to predict the behavior of a crystal, one might come to conclusions that are directly opposite to the actual results. An accurate prediction of behavior, especially if quantitative, requires the use of quantum theory which, in this case, would result in the concept of negative mass.

Fortunately, however, it can be shown that the motion of valence band electrons can be successfully discussed using classical physics by assuming that the moving object is a *positively charged hole* of mass m_h appropriately computed. Evidently such an element is a fiction of our imagination and it might prove difficult to attach a conventional physical meaning to the "mass" of a vacancy. The importance of the concept of hole results from the fact that, in many instances, "it works" in predicting with reasonable accuracy what actually occurs, using the more familiar techniques of classical physics instead of quantum theory.

From now on we shall analyze the motion of current carriers on the basis of classical physics applied to two types of particles: negatively charged conduction electrons and positively charged holes. Both are endowed with mass, although the numerical value of this physical property depends on the conditions under which the phenomena occur. In practice such masses are calculated from measurements made in each condition and are tabulated for future use (cf. Table 3.1.1). In spite of its limitations and theoretical shortcomings, this method yields valid results in most cases of interest and can be very useful in intuitively visualizing phenomena. In any case, numerical answers obtained in this way are to be considered "ballpark values" and no greater precision than order of magnitude should be expected.

For the last time we shall alert the student that, like all models, these concepts have, of necessity, limited validity. As the technology develops, it is to be expected that more and more often the scientist and the engineer will be confronted with phenomena for which the model fails. Then it will be necessary to look at the facts more rigorously and, at the present writing, the only techniques at our disposal for this task are those of quantum physics. In simple words, the student who is interested in doing profitable work in this field in the future is urged to obtain the conceptual and theoretical tools of quantum physics and review all of this material in their light.

3.2 FERMI-DIRAC STATISTICS

Electrons and holes, as previously defined, can absorb and store energy in several different forms. For our purposes, the most important are thermal and electrical energy, which are present when the material is at nonzero temperature and/or under the influence of an electric field.

In Sec. 2.3 some preliminary considerations were offered regarding the problem of estimating how energy imparted to bulk material is distributed among the component atoms. The concept of *distribution density function* $n_d(E)$ was defined and the Boltzmann distribution was introduced in (2.3.4). As stated, this distribution is obtained by using classical statistical mechanics, remembering the principles of thermodynamics.

However, electrons do not behave as classical particles. They are subject to Pauli's exclusion principle; consequently they can exist only in a set of *discrete permissible quantum states and no more than one electron can occupy any one state in the system.* They cannot be expected to follow the Boltzmann distribution any more than one could expect a group of people in a theatre to distribute themselves as if there were no seats, with each person free to occupy any position at all in the theatre.

If statistical considerations are applied under the constraint of Pauli's exclusion principle, it is found that not all of the permissible states are occupied, but, instead, within any range of energy considered, the percentage of occupied states depends on the energy level, just as, in a stadium, the percentage of seats occupied might vary depending, for instance, on the cost of the seat.

This investigation can be carried out quantitatively and the attendant mathematics do not require the use of quantum techniques, provided Pauli's principle is accepted *a priori.* Valid results can be obtained by the straightforward application of elementary probability theory. An easy to read account is given in Ref. 4 (p. 114).[1]

The percentage of permissible states occupied by an electron at energy level E turns out to be expressed by the *Fermi-Dirac function:*

$$f(E) = \frac{1}{1 + e^{(E - E_f)/(kT)}} \qquad (3.2.1)$$

where k is Boltzmann's constant and T the temperature in kelvins. The mathematical discussion introduces and defines the energy E_f, designated as the *Fermi level,* an important characteristic of the material, depending on temperature and the number of electrons in the system but not on the system state. An interpretation of its physical meaning can be obtained from a discussion of (3.2.1).

At absolute zero temperature, the exponential in the denominator becomes infinity [and $f(E) = 0$] if $E > E_f$, but if instead $E < E_f$, then the infinitely large exponent becomes negative, the exponential equals zero, and $f(E) = 1$. The Fermi-Dirac function therefore predicts that *at absolute zero temperature all states characterized by energies greater than the Fermi level are empty and all states below the Fermi level are fully occupied by electrons.* This is shown by the curve for $T = 0$ in Fig. 3.2.1. Notice that, conversely, classical physics expects all particles to have zero energy at absolute zero and the Boltzmann distribution accordingly states that, under classical conditions, at $T = 0$, all electrons would be at zero energy level.

---

[1] More sophisticated discussions are offered in Refs. 5 (p. 327*) and 3 (pp. 251 and 590*), but, as indicated by the asterisks, they involve quantum theory.

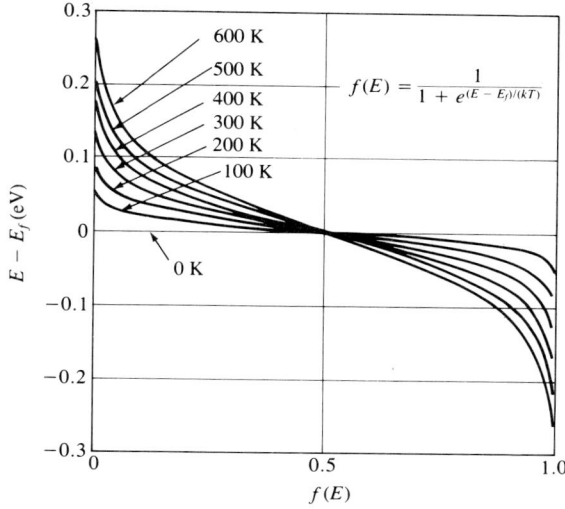

FIGURE 3.2.1
Fermi–Dirac function $f(E)$ vs. $E - E_f$ at various temperatures [6].

When the temperature is above absolute zero, states below the Fermi level are no longer fully occupied; some of the electrons acquire some energy and move to states above the Fermi level. As the temperature increases, occupancy of the states above the Fermi level increases at the expense of the lower levels, as shown in Fig. 3.2.1. Notice that, at all temperatures, if $E = E_f$, the exponent becomes zero, the exponential equals 1, and $f(E) = \frac{1}{2}$, indicating that states at the Fermi level are always 50 percent occupied.

From the above discussion, the Fermi level can be defined on the basis of two of its physical properties: (1) it is the level *above which no state is occupied at absolute zero temperature*; (2) it is the level which is *50 percent occupied at all temperatures*.

A deeper interpretation of the physical meaning of the Fermi level as the electrochemical potential of the electron is pointed out in the original paper [7].

Example 3.2.1

(a) What percentage of the permissible states at level $E = E_f + 0.5$ eV is occupied if: (1) $T = 0$ K, (2) $T = 300$ K?

(b) Under the same conditions (1 and 2), what percentage of the permissible states at level $E = E_f - 0.5$ eV is vacant?

Solution

(a) (1) 0%.
   (2) At $T = 300$ K: $kT = 1.38 \times 10^{-23} \times 300 = 4.14 \times 10^{-21}$ J $\Rightarrow 0.0259$ eV. Therefore,

$$\frac{100}{1 + e^{0.5/0.0259}} = 4.13 \times 10^{-7}\%$$

(b) (1) 0%.

(2) The probability of a vacancy is $1 - f(E)$, so

$$\left(1 - \frac{1}{1 + e^{-0.5/0.0259}}\right)100 = 4.13 \times 10^{-7}\%$$

Such extremely low percentages may appear negligible, but, considering the extremely high concentration of atoms in crystals and the low average conductivity of semiconductors, they are significant and cannot be discounted.

It now becomes evident that, if the Fermi level falls within the conduction band, then, even at absolute zero temperature, the band is not empty. If it falls within the valence band, then, at absolute zero, the valence band is not full. In both cases conduction starts at zero temperature and the crystal behaves as a conductor. *In insulators and semiconductors, therefore, the Fermi level must fall within the gap.*

To understand how the Fermi-Dirac function can be used to compute the distribution density function, consider the audience distribution in a theatre, where, of course, the rule of no more than one person to a seat (Pauli's exclusion principle) holds true. One way to compute the number of spectators within any range of admission prices (the density distribution function) could be to count the number of seats in the theatre within that price range (the seat density distribution function) and multiply this number by the percentage occupation of the range.

The same logical reasoning leads to the conclusion that the *electron density distribution function* $n_d(E)$ can be computed as the product of the number of states per unit volume and per unit energy range (the *state density distribution function*) times the Fermi-Dirac function. The state density distribution[2] within the conduction band can be shown to be

$$n_{sc}(E) = 4\pi\left(\frac{2m_e}{h^2}\right)^{3/2}\sqrt{E - E_c} \qquad \text{cm}^{-3}/\text{J} \tag{3.2.2}$$

so that the number of conduction electrons per unit volume and per unit energy range around the energy level E is

$$n_d(E) = n_{sc}(E)f(E) \qquad \text{cm}^{-3}/\text{J} \tag{3.2.3}$$

Similarly, to compute the *hole density distribution function*, the state density distribution in the valence band is computed as

$$n_{sv}(E) = 4\pi\left(\frac{2m_h}{h^2}\right)^{3/2}\sqrt{E_v - E} \qquad \text{cm}^{-3}/\text{J} \tag{3.2.4}$$

---

[2] For specific references to the computation of the state density distributions n_{sc} and n_{sv}, cf. Refs. 8 (p. 328), 9 (p. 80*), 10 (p. 126*), and 6 (p. 15).

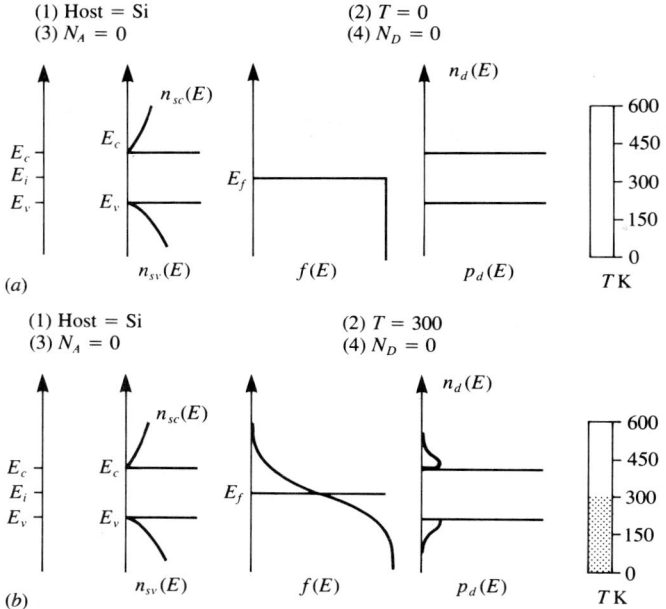

(a)

(b)

FIGURE 3.2.2
State density distributions $n_{sc}(E)$ and $n_{sv}(E)$, Fermi function $f(E)$, and carrier density distributions $n_d(E)$ and $p_d(E)$ in intrinsic Si (a) at 0 K and (b) at 300 K.

Next, remembering that a hole is a vacancy in a permissible state, the percentage of *nonoccupied states* is calculated from the Fermi-Dirac function as $1 - f(E)$, so that, finally, the desired hole density distribution function is shown to be

$$p_d(E) = n_{sv}(E)[1 - f(E)] \qquad \text{cm}^{-3}/\text{J} \qquad (3.2.5)$$

The above concepts and computations are represented in graphical form by Fig. 3.2.2, which is divided into two parts: Fig. 3.2.2a for a temperature $T = 0$ K and Fig. 3.2.2b for a temperature $T > 0$ K.

In each part are shown, from left to right:

1. The energy band diagram of the intrinsic semiconductor in question, indicating the edges of the conduction and valence bands E_c and E_v, the band gap E_g, and the location of the Fermi level. It should be noticed that the Fermi level is designated here as E_i (intrinsic Fermi level) and is located at the center of the gap midway between the bottom of the conduction band E_c and the top of the valence band E_v. This choice will be justified later for an intrinsic semiconductor (cf. Sec. 3.4). Notice that it falls within the bandgap, as previously discussed for a semiconductor.
2. The state density distribution diagram for both the conduction [$n_{sc}(E)$] and valence [$n_{sv}(E)$] bands, displaying the half-power dependence of Eq. (3.2.2).
3. The Fermi-Dirac function [$f(E)$] for the given temperature.

4. The carrier density distributions for electrons $[n_d(E)]$ and holes $[p_d(E)]$. In accordance with Eqs. (3.2.3) and (3.2.5), the electron density distribution $n_d(E)$ is the product of the state density distribution in the conduction band $n_{sc}(E)$ times the Fermi function. The hole density distribution $p_d(E)$ is the product of the state density distribution in the valence band $n_{sv}(E)$ times one minus the Fermi function.

5. A schematic representation of a thermometer as a reminder of the choice of temperature.

The graphic presentation is limited to qualitative indications only, because the actual quantitative variations cannot be accurately represented within the limited size of the graph. It can, however, be used to clarify the concepts embodied in Eqs. (3.2.1) through (3.2.5). Notice, for instance, that at a temperature of 0 K (Fig. 3.2.2a) both the conduction and the valence band display a carrier density of zero. Both bands are empty of carriers, confirming the previous statement that semiconductors behave as insulators at absolute zero. At a higher temperature (Fig. 3.2.2b), carriers are present in both bands, indicating conduction mediated by both electrons and holes, as predicted. A comparison of carrier density vs. state density functions at $T > 0$ K clearly shows that both bands are neither empty nor completely occupied, which can be confirmed by observation of the Fermi-Dirac function.

Similar qualitative diagrams will be presented later (cf. Figs. 3.4.2 and 3.4.3).

3.3 EQUILIBRIUM CARRIER CONCENTRATION

In a crystal, all electrons having energies above the bottom of the conduction band E_c are conduction electrons. Consequently, using (3.2.3) in (2.3.3) and extending the integration to all levels above E_c, it can be concluded that, at equilibrium at temperature T, the *conduction electron concentration* n (total number of conduction electrons per unit volume of the crystal) is

$$n = \int_{E_c}^{\infty} n_{sc}(E) f(E) \, dE = N_c e^{-(E_c - E_f)/(kT)} \qquad \text{cm}^{-3} \qquad (3.3.1)$$

where N_c, the *conduction band effective density of states*, is

$$N_c = 2\left(\frac{2\pi m_e kT}{h^2}\right)^{3/2} \qquad \text{cm}^{-3} \qquad (3.3.2)$$

In evaluating the integral of (3.3.1), the Fermi-Dirac function has been approximated by $\exp[-(E - E_f)/(kT)]$, under the assumption that $E_c \gg E_f$, so that, over the whole interval of integration, the exponential in the denominator of (3.2.1) is much larger than 1. As will soon become apparent, this assumption is unwarranted for conductors, but it is well justified in the case of intrinsic semiconductors.

Similarly, all vacancies below the top of the valence band are holes, so that the *hole concentration p* is

$$p = \int_{-\infty}^{E_v} n_{sv}(E)[1 - f(E)] \, dE = N_v e^{-(E_f - E_v)/(kT)} \qquad \text{cm}^{-3} \qquad (3.3.3)$$

in which N_v, the *valence band effective density of states*, is

$$N_v = 2\left(\frac{2\pi m_h kT}{h^2}\right)^{3/2} \qquad \text{cm}^{-3} \qquad (3.3.4)$$

Again, the assumption that $E_f \gg E_v$, under which the integral of (3.3.3) was evaluated, proves to be valid for semiconductors. N_v and N_c are tabulated, together with other characteristic constants of the most common types of semiconductor materials, in Table 3.1.1.

From (3.3.1) and (3.3.3), remembering that the difference $E_c - E_v$ equals the gap energy E_g,

$$np = N_c N_v e^{-E_g/(kT)} \qquad \text{cm}^{-6} \qquad (3.3.5)$$

It should be realized that the ratio $E_g/(kT)$ determines the rate of thermal generation of electron-hole pairs. (Indeed, this is the ratio of the energy required to ionize an intrinsic atom to the average thermal energy available at the equilibrium temperature.) This shows that (3.3.5) is in agreement with the statement of Sec. 2.5 and that the rate of electron-hole recombination depends on the product of the equilibrium concentrations of electrons and holes.

The remarkable fact about (3.3.5) is that the Fermi level E_f of the material does not enter into the expression for the product np, so that, as long as equilibrium conditions prevail, this product *is independent of the doping of the sample and of the individual electron and hole concentrations.*

Intrinsic Semiconductors

As electrons and holes are generated in pairs, their concentrations are equal, so that

$$n = p = n_i = \sqrt{np} = \sqrt{N_c N_v} \, e^{-E_g/(2kT)} \qquad \text{cm}^{-3} \qquad (3.3.6)$$

where n_i is the *equilibrium intrinsic carrier concentration*. This is also tabulated in Table 3.1.1 for Ge, Si, and GaAs.

It should be kept in mind that experimental values listed in tabulations, such as Table 3.1.1, or plotted in graphs, usually have limited validity and may require modification before they can be used in specific practical cases. For instance, Table 3.1.1 lists values at 300 K (room temperature), so it would be incorrect to use these values if the temperature is different from 300 K.

Appropriate modifications can be computed on the basis of the mathematical expressions of the characteristic quantities involved as a function of temperature. As an example, consider the equilibrium intrinsic carrier concentration

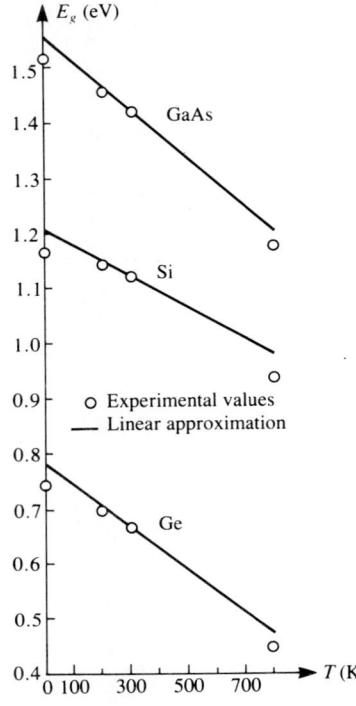

FIGURE 3.3.1
Gap energy E_g vs. temperature for Si, GaAs, and Ge.

n_i listed in the table as 1.5×10^{10} cm^{-3} for Si, 2.4×10^{13} cm^{-3} for Ge, and 10^7 cm^{-3} for GaAs at 300 K. How can its temperature dependence be computed?

From (3.3.6), n_i is seen to depend on N_c and N_v, the temperature dependence of which is, in turn, expressed in Eqs. (3.3.2) and (3.3.4). The other parameter of interest is E_g, listed as 1.12 eV for Si, 0.68 eV for Ge, and 1.42 eV for GaAs at 300 K. Although E_g undergoes only moderate variations with T, so that it can often be considered a constant, measurements show its temperature dependence to be approximated by the empiric formulas:

$$E_g(T) \approx 1.205 - 2.8 \times 10^{-4} \, T \qquad \text{eV for Si}$$

$$E_g(T) \approx 0.782 - 3.9 \times 10^{-4} \, T \qquad \text{eV for Ge} \qquad (3.3.7)$$

$$E_g(T) \approx 1.55 - 4.3 \times 10^{-4} \, T \qquad \text{eV for GaAs}$$

Using the above formulas and after some algebra, it can be shown (cf. Prob. 3.7) that the temperature dependence of n_i is expressed by

$$n_i = 3.93 \times 10^{16} \, T^{3/2} e^{-7000/T} \qquad \text{cm}^{-3} \text{ for Si}$$

$$n_i = 1.78 \times 10^{16} \, T^{3/2} e^{-4550/T} \qquad \text{cm}^{-3} \text{ for Ge} \qquad (3.3.8)$$

and $\qquad n_i = 9 \times 10^{15} \, T^{3/2} e^{8976/T} \qquad \text{cm}^{-3} \text{ for GaAs}$

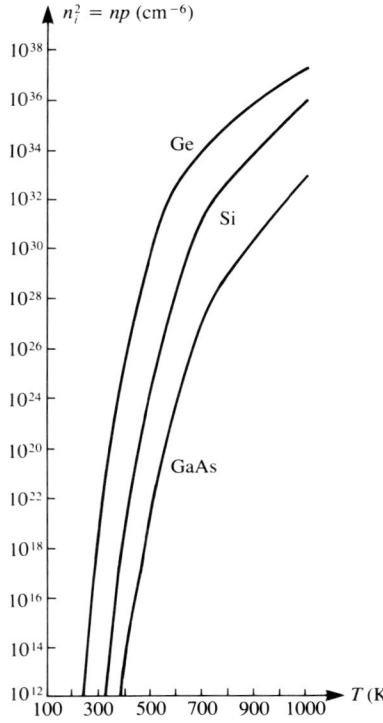

FIGURE 3.3.2
Carrier concentration product $np = n_i^2$ vs. temperature for Si, GaAs, and Ge.

These formulas are in reasonably good agreement with experimental results, as shown by Fig. 3.3.1, plotting E_g vs. T, and Fig. 3.3.2, plotting $n_i^2 = np$ vs. T for Si, Ge, and GaAs.

It should be emphasized here that computations of the above type can be relied upon to yield exclusively order of magnitude results. They are of high conceptual value, indicating general trends of the dependence of important parameters on conditions of operation, but, for accuracy of computations, it is best to use experimentally obtained values. So, for practical purposes, theoretically obtained formulas are often empirically modified to conform to experimental facts.

Indeed, in working out problems of this type (e.g., Probs. 3.5 and 3.6), it becomes immediately apparent that the theoretical formulas and the published experimental values agree only in an approximate, order of magnitude, sense and would be considered incompatible, easily by as much as 200 or 300 percent, if precise results were expected.

N-type Semiconductors

In this case, holes are produced exclusively by thermal ionization of intrinsic atoms, which generates equal concentrations of electrons and holes. The hole concentration in an N-type semiconductor is therefore

$$p_N = p_{int} = n_{int} \tag{3.3.9}$$

where the subscript N is used to indicate N-type semiconductor characteristics and the subscript int to designate the characteristics of ionized intrinsic atoms.

Conduction electrons are generated by both impurity and intrinsic atoms. If N_D is the *net* concentration of donor impurities, then the impurities, being essentially 100 percent ionized, contribute N_D electrons per unit volume to the conduction band. The total conduction electron concentration is therefore seen to be

$$n_N = N_D + n_{int} = N_D + p_N \tag{3.3.10}$$

Remembering that, at equilibrium, the product of the concentrations of holes and electrons equals the square of the intrinsic concentration,

$$p_N = \frac{n_i^2}{n_N} \tag{3.3.11}$$

Substituting this expression into (3.3.10), it is easy to obtain the equation $n_N^2 = N_D n_N + n_i^2$, from which, solving for n_N,

$$n_N = \frac{N_D}{2}\left(\sqrt{1 + \frac{4n_i^2}{N_D^2}} + 1\right) \tag{3.3.12}$$

so that, if $n_i \ll N_D$, then, to a good approximation,

$$n_N = N_D$$
$$p_N = \frac{n_i^2}{N_D} \tag{3.3.13}$$

P-type Semiconductors

Analogously, if the net acceptor concentration is much greater than n_i,

$$p_P = N_A$$
$$n_P = \frac{n_i^2}{N_A} \tag{3.3.14}$$

It should be emphasized here that N_A and N_D in the above formulas refer to the *net* impurity concentration. If impurities of both types are present in a semiconductor crystal, the semiconductor behaves as if only the impurity of highest doping were present, with a *net concentration equal to the difference of the partial concentrations* of impurities of the opposite type. Certain semiconductor properties, however, depend on the *total* impurity concentration, equal to the sum of the partial concentrations of any and all impurity atoms in the crystal, irrespective of whether they are donors or acceptors (for an example cf. Sec. 6.7).

Equations (3.3.13) and (3.3.14) are simple, useful expressions which can be used in the majority of practical cases. However, it must be remembered that they are predicated on the condition that the net impurity concentration be much smaller than the intrinsic carrier concentration *at the temperature of operation*. When this condition is not valid, then the more accurate expressions (3.3.11) and (3.3.12) must be used.

Example 3.3.1. A Si semiconductor is doped with 10^{17} atoms of B/cm³ and with 9×10^{16} atoms of As/cm³. Compute the concentrations of majority and minority carriers at temperatures of 300 and 600 K.

Solution. From Table 3.1.1, at 300 K the intrinsic carrier concentration is $n_i = 1.5 \times 10^{10}$ cm⁻³. The net concentration of impurities is $N_A = 10^{17} - 9 \times 10^{16} = 10^{16}$ cm⁻³; therefore, from (3.3.13),

$$p_P = N_A = 10^{16} \text{ cm}^{-3}$$

$$n_P = 2.25 \times 10^{20}/10^{16} = 2.25 \times 10^4 \text{ cm}^{-3}$$

At a temperature of 600 K, using (3.3.8) or Fig. 3.3.2,

$$n_i = 4.89 \times 10^{15} \text{ cm}^{-3}$$

so that (3.3.13) would yield

$$p_P = 10^{16} \text{ cm}^{-3}$$

$$n_P = 23.9 \times 10^{30}/10^{16} = 2.39 \times 10^{15} \text{ cm}^{-3}$$

However, the approximation assumed for expressions (3.3.13) is no longer valid, so the computation of n_P must be done on the basis of (3.3.10):

$$n_P = 1.99 \times 10^{15} \text{ cm}^{-3}$$

from which

$$p_P = 23.9 \times 10^{30}/1.99 \times 10^{15} = 1.2 \times 10^{16} \text{ cm}^{-3}$$

Notice the large increment of the minority carrier population at the higher temperature and also that the more precise computation yields a majority carrier population appreciably higher than the impurity concentration.

3.4 LOCATION OF THE FERMI LEVELS

For an intrinsic material, electron and hole concentrations are equal, so the location of the intrinsic Fermi level can be computed by equating (3.3.1) to (3.3.3). After some elementary algebraic manipulations, this results in

$$E_i = \frac{E_c + E_v}{2} + \frac{1}{2} kT \ln \frac{N_v}{N_c} \tag{3.4.1}$$

but from (3.3.2) and (3.3.4) the ratio N_v/N_c equals $(m_h/m_e)^{3/2}$ and this is very close to 1 (indeed, the temperature dependence of the intrinsic Fermi level in Si results in a slope smaller than 3.5×10^{-5} eV/K—cf. Prob. 3.13), so that, finally:

$$E_i \approx \frac{E_c + E_v}{2} \tag{3.4.2}$$

The intrinsic Fermi level is located at about the center of the gap between the conduction and valence bands.

The electron and hole concentrations, (3.3.1) and (3.3.3), can also be expressed in terms of intrinsic concentration and intrinsic Fermi level to yield the useful forms:

$$n = n_i \, e^{(E_f - E_i)/(kT)} \tag{3.4.3}$$

$$p = n_i \, e^{(E_i - E_f)/(kT)} \tag{3.4.4}$$

To determine the location of the Fermi level in extrinsic semiconductors, using (3.3.13) and (3.3.14) in (3.3.1) and (3.3.3), one obtains, for *N-type semiconductors*,

$$n_N = N_D = N_c \, e^{-(E_c - E_f)/(kT)} \tag{3.4.5}$$

from which

$$E_{fN} = E_c + kT \ln \frac{N_D}{N_c} \tag{3.4.6}$$

Analogously, using (3.4.3):

$$E_{fN} = E_i + kT \ln \frac{N_D}{n_i} = E_i - V_{if} \, q \tag{3.4.7}$$

For *P-type semiconductors*, similarly,

$$E_{fP} = E_v - kT \ln \frac{N_A}{N_v} \tag{3.4.8}$$

or, using (3.4.4),

$$E_{fP} = E_i - kT \ln \frac{N_A}{n_i} = E_i - V_{if} \, q \tag{3.4.9}$$

Equations (3.4.7) and (3.4.9) implicitly define a voltage

$$V_{if} = \frac{E_i - E_f}{q} \tag{3.4.10}$$

measuring, in electronvolts, the difference in energy between the intrinsic Fermi level and the Fermi level of the extrinsic semiconductor being considered. This quantity depends both on doping and on temperature and is intimately connected with many important properties of extrinsic semiconductors. In the future V_{if} will often be encountered and its computation on the basis of (3.4.7) and (3.4.9) will be used in determining some of the characteristics of several semiconductor devices.

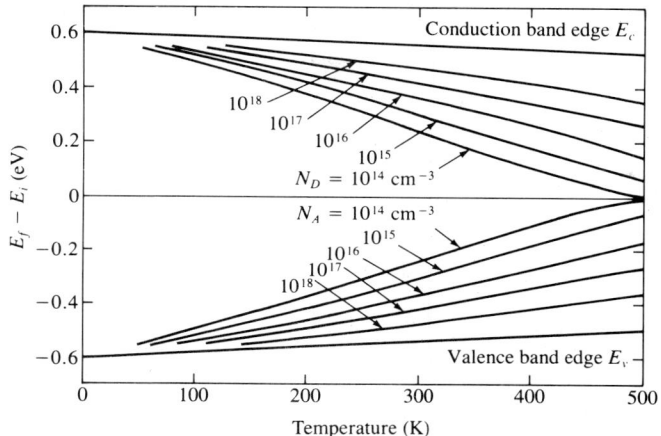

FIGURE 3.4.1
Fermi level location in the gap for Si as a function of temperature for different doping concentrations
[12].

From the above it appears evident that V_{if} is positive for P-type semiconductors and negative for N-type.

The dependence of E_f and V_{if} on temperature and doping is plotted in Fig. 3.4.1.

It should be realized that expressions (3.4.6) through (3.4.9), having been obtained through the use of Eqs. (3.3.13) and (3.3.14), are subject to the hypothesis that the net impurity concentration should be much smaller than the intrinsic carrier concentration. If the hypothesis is not valid, then N_D and N_A in the above equations must be substituted by n_N and p_P respectively. These quantities must, of course, be computed on the basis of (3.3.12) (cf. Prob. 3.18).

Figure 3.4.1 shows how the position of the Fermi level within the gap changes with variations in temperature for various net doping concentrations; consequently it also implicitly yields V_{if} as a function of the same variables. The figure also indicates the variation of the bandgap E_g with temperature, in accordance with Eq. (3.3.7). As previously stated, the slope of the intrinsic Fermi level curve is extremely small and is neglected in the figure.

For extrinsic semiconductors with *impurity concentrations smaller than the effective density of states*, as the temperature increases, the location of the Fermi level is seen to approach the intrinsic level (it should be remembered that the logarithm of a number less than 1 is negative). This, in turn, indicates that the higher the temperature, the more the specimen behaves like an intrinsic semiconductor and the less like an extrinsic one.

According to (3.4.6) and (3.4.7), it appears that, when the temperature is high enough, the locus of the Fermi level should cross the intrinsic level line. This would indicate that the specimen is now behaving as an extrinsic semiconductor *of the opposite type*, which is manifestly impossible. The paradox is easily resolved by reference to the assumptions under which (3.4.6) and (3.4.7) were obtained.

(1) Host = Si   (2) $T = 450$
(3) $N_A = 0$ (4) $N_D = 1E + 16$

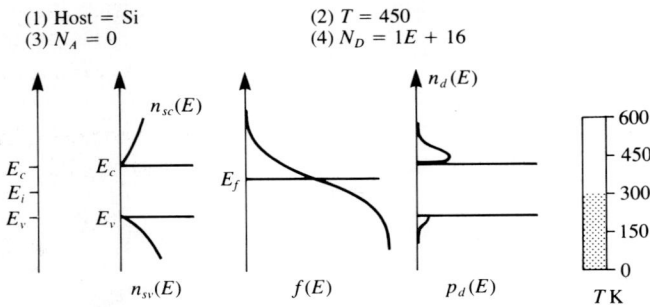

FIGURE 3.4.2
Qualitative state density distributions $n_{sc}(E)$ and $n_{sv}(E)$, Fermi function $f(E)$, and carrier density distributions $n_d(E)$ and $p_d(E)$ at 300 K in Si with net donor impurity concentration $N_D = 10^{14}$ cm^{-3}. Carrier concentrations are numerically computed in Example 3.4.1.

Physically it is easy to visualize what happens. As the temperature increases, the number of current carriers produced by the impurity atoms does not vary, but the concentration of carriers produced thermally by ionization of intrinsic atoms increases exponentially, as shown by (3.3.6) together with (3.3.11) and (3.3.12). When the temperature is high enough, the intrinsic carriers finally outnumber the impurity-produced carriers, so that the electron and hole populations become practically the same. The condition that the impurity concentration should be much larger than the intrinsic carrier concentration does not hold any more and the conclusions drawn under this assumption cease to be valid. At these high temperatures the Fermi level does not follow Eqs. (3.4.6) through (3.4.9), but instead approaches the intrinsic level asymptotically. The majority of the carriers is now generated from intrinsic atoms and the specimen behaves as an intrinsic semiconductor.

Contrary to what might appear from a superficial inspection of (3.4.6) and (3.4.8), the Fermi level does not vary strictly linearly with temperature. Indeed, from (3.3.2) and (3.3.4), N_c and N_v are seen to vary with the 3/2 power of T. Expanding the expressions obtained by substituting these equations into (3.4.6) and (3.4.8) an extra, nonlinear term appears: $1.5\,kT\ln(T/300)$ (cf. Prob. 3.19). This extra term, however, is usually comparatively small, so the Fermi level can be assumed to vary linearly with T, except in the neighborhood of the extrinsic Fermi level.

An interesting situation arises when the impurity concentration is so high that it exceeds the effective density of states. In accordance with (3.4.6) and (3.4.7), the Fermi level should now be *within the conduction band* for N-type semiconductors and *within the valence band* for P-type semiconductors, so that, in both cases, the specimen should behave as a conductor rather than a semiconductor. Actually this condition holds not only when the effective density of states is exceeded but also as soon as it is approached, so that the Fermi level is within a few kT of E_c or E_v. *When the impurity concentration reaches levels close to—or higher than*—the effective density of states, the semiconductor is said to be degenerately doped. This turns out to be desirable in several cases of practical importance; thus, in practice, degenerate doping is encountered comparatively often. For one very usual case, cf. Sec. 7.8.

The above theory is graphically summarized in Figs. 3.4.2 and 3.4.3. The figures are similar to Fig. 3.2.2, qualitatively showing the energy diagram, the state density distributions, the Fermi function, the density distribution functions for electrons and holes, and finally a thermometer.

Figure 3.4.2 refers to an N-type Si semiconductor at $T = 300$ K, as indicated. Notice that, although the graphs are quantitatively incorrect (n_d and p_d could not be significantly represented in the same scale and the Fermi function could not show its conceptually important characteristics), yet the location of the Fermi level, closer to E_c than to E_v, and the general shape of the Fermi function, together with the much larger size of the n_d graph relative to the p_d graph, are all qualitatively in agreement with the above theory.

An elementary application of the quantitative theory to this case is shown in the following example.

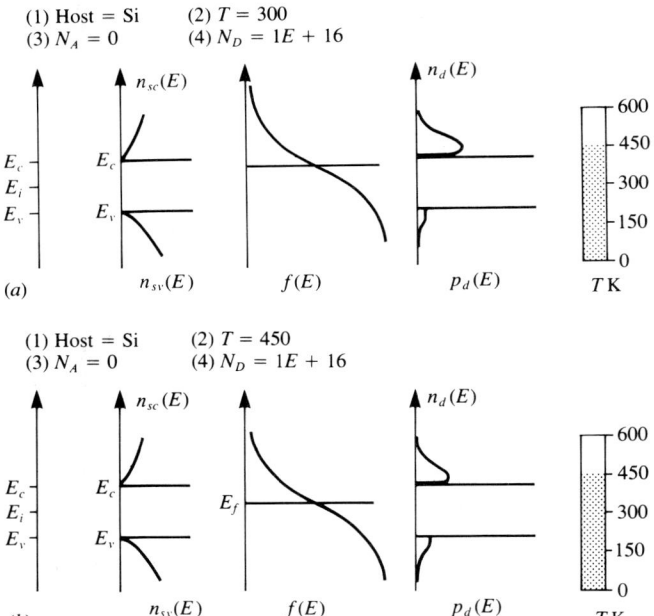

FIGURE 3.4.3
Same characteristics as in Fig. 3.4.2 for $T = 450$ K. (*a*) For pedagogical purposes (see text), only the variation of the shape of the Fermi function due to a change in temperature is taken into account. The Fermi level is artificially maintained at the 300 K level. The figure shows that, if this were the case, the majority carrier density function would be much increased above its 300 K value. (*b*) The previous increase in majority carrier concentration is compensated for by the Fermi level displacement (now taken into account). Essentially only the minority carrier density varies from Fig. 3.4.2. The pertinent parameters are computed in Example 3.4.2.

Example 3.4.1. A Si crystal at 300 K is doped with 10^{16} atoms of As/cm^3. For these conditions repeat the qualitative display of Fig. 3.2.2 and compute the Fermi level and the majority and minority carrier concentrations.

Solution. Figure 3.4.2 represents the new conditions. In comparison with intrinsic conditions, notice how, in accordance with (3.4.6) and (3.4.7), the doping forces an upward displacement of the Fermi level (and consequently of the whole Fermi-Dirac function), resulting in a large increment of the electron density and decrement of the hole density distributions.

Quantitatively, from (3.4.6), the Fermi level is

$$E_f = E_c + 0.0259 \ln \left(\frac{10^{16}}{2.8 \times 10^{19}} \right) = E_c - 0.21 \qquad \text{eV}$$

which should be compared with the location of the intrinsic Fermi level, which can be computed from (3.4.2) as

$$E_i = E_c - \frac{E_g}{2} = E_c - 0.56$$

From (3.3.13), the carrier concentrations are

$$n_N = N_D = 10^{16} \text{ cm}^{-3}$$

$$p_N = \frac{2.25 \times 10^{20}}{10^{16}} = 2.25 \times 10^4 \text{ cm}^{-3}$$

The conditions depicted in Fig. 3.4.3 are discussed in the following example.

Example 3.4.2. The semiconductor of Example 3.4.1 is brought to a temperature of $T = 450$ K. Display the plots for the new condition and compute Fermi level and electron and hole densities.

Solution. The variation in temperature affects the shape of the Fermi function and the location of the Fermi level. The results of these two phenomena will be displayed separately, one after the other, in the qualitative graphs, to illustrate the purport of the Fermi level displacement and to comment on its consequences.

In accordance with the discussion of Eq. (3.2.1), as also depicted in Fig. 3.2.1, the increment in temperature results in a further deviation of the Fermi function plot from its shape at 0 K. This greatly increases the percentage of occupied states above the Fermi level, correspondingly decreasing the percentage occupation of states below it. Figure 3.4.3a shows what would result if this shape variation were not accompanied by a lowering of the Fermi level. As shown, both the electron and the hole density distribution functions (in the conduction, respectively valence band) would be greatly increased in comparison with Fig. 3.4.2. This would, of course, result in a large increment of *both* electron and hole concentrations. However, in accordance with (3.4.6), the Fermi level is displaced to

$$E_f = E_c + 0.0358 \left(\ln \frac{10^6}{2.8 \times 10^{19}} - 1.5 \ln \frac{450}{300} \right) = E_c - 0.306 \qquad \text{eV}$$

in which, in the computation of ln (N_D/N_c), the temperature dependence of N_c [cf. Eq. (3.3.2)] has been taken into account (also cf. Prob. 3.17). This value of E_f should be compared with the results of Example 3.4.1.

Displacing the Fermi function diagram accordingly, the carrier density function graphs are redimensioned, as shown in Fig. 3.4.3b, with the result that the increment of the electron concentration over the value at the previous temperature becomes negligible. In accordance with (3.3.13) this concentration remains:

$$n_N \approx N_D = 10^{16} \text{ cm}^{-3}$$

while the hole concentration increases still more, to the value predicted by (3.3.13), in which, however, n_i must be computed for the new temperature. From (3.3.7):

$$n_i = 3.93 \times 10^{16} \times 450^{1.5} \, e^{-(7000/450)} = 6.58 \times 10^{13} \quad \text{cm}^{-3}$$

Using this value in (3.3.7),

$$p_N = \frac{4.33 \times 10^{27}}{10^{16}} = 4.33 \times 10^{11} \quad \text{cm}^{-3}$$

Notice than $p_N \ll n_N$, so that the approximate equation (3.3.7) is valid. On the other hand, the increment in temperature, while leaving the majority carrier concentration essentially *unchanged*, as noted, has resulted in an increment of *seven orders of magnitude* in the minority carrier population!

By working out a few numerical examples, the student should observe how increments in doping displace the Fermi level and with it the Fermi-Dirac function, increasing the area of the majority carrier density distribution and correspondingly decreasing the minority carrier concentration, while keeping the product of the concentrations constant. Conversely, variations of temperature change the shape of the Fermi function *and, at the same time, the position of the Fermi level*, increasing the concentration of the minority carriers while keeping essentially constant the concentration of the majority carriers.

Some experimentation with different parameter values, making appropriate quantitative *predictions*, should reinforce the visualization of the phenomena involved and confirm intuitive expectations of semiconductor behavior.

3.5   SUMMARY

Although carriers are electrons following the laws of quantum mechanics, carrier motion can be satisfactorily discussed by assuming that:

1. Electrons in the conduction band behave as negatively charged classical particles endowed with an *effective mass* m_e, appropriately computed for each condition.

2. The motion of electrons in the valence band can be simulated by the motion of *holes:* positively charged classical particles with effective mass m_h also appropriately computed in each case.

The distribution of electrons among the permissible states in a system at equilibrium is described by the *Fermi-Dirac function* [Eq. (3.2.1)] defining a *Fermi level* E_f, 50 percent occupied at all temperatures.

Fermi-Dirac statistics predict that *all states below the Fermi level are 100 percent occupied at absolute zero*. Classical theory, instead, predicts that, at $T = 0$, all electrons have zero energy.

In semiconductors at equilibrium, the *conduction electron and hole concentrations* predicted by Fermi-Dirac statistics are given by Eqs. (3.3.1) and (3.3.3) in which there appear the *effective concentrations of states* N_c and N_v [Eqs. (3.3.2) and (3.3.4)]. The *product np* of the electron and hole concentrations is *independent of the doping* of the sample, being determined only by the temperature and the gap energy, not by the Fermi level.

The *intrinsic carrier concentration* n_i is given by Eq. (3.3.6) and the *extrinsic carrier concentrations* can be computed from Eqs. (3.3.13) for N-type materials and Eqs. (3.3.14) for P-type materials. Electron and hole concentrations can also be expressed in terms of intrinsic concentration and intrinsic Fermi level by Eqs. (3.4.3) and (3.4.4). These concentrations respectively correspond to *Fermi level positions in the gap* given by Eq. (3.4.2) for intrinsic material, Eq. (3.4.6) or (3.4.7) for N-type materials and Eq. (3.4.8) or (3.4.9) for P-type materials.

These equations are displayed graphically in Fig. 3.4.1, indicating that, as temperature increases, extrinsic semiconductors behave more and more like intrinsic ones.

Degenerately doped semiconductors, with doping levels so high that they exceed or nearly approximate the effective concentration of states, behave similarly to conductors.

Carrier concentration, Fermi level position, Fermi-Dirac function shape, etc., all vary with material, temperature, and doping.

PROBLEMS

3.1. If it is known that, in a certain material, 1 percent of the states at level $E = E_f - 0.25$ eV are vacant, what is the temperature?

3.2. Compute the number of states per cubic centimeter in the conduction band of Si that exist within the range of energies from E_c to $E_c + 0.12$ eV.

3.3. Prove Eq. (3.3.1).

3.4. Compute the concentration of electrons in the conduction band within a range of energies from E_c to $E_c + 0.1$ eV for Si at 300 K.

Hint: From the definition of error function,

$$\int_0^x e^{-t^2/\alpha} \, dt = \frac{\sqrt{\alpha\pi}}{2} \, \text{erf}\left(\frac{x}{\sqrt{\alpha}}\right)$$

The error function is tabulated in several reference tables, e.g., cf. Ref. 13.

3.5. Compute n_i for Si and Ge at 300 K using the values of N_c, N_v, and E_g from Table 3.1.1 and compare with the values of n_i appearing in the same table. Comment.

3.6. Compute the carrier concentrations at 600 K for Si doped with 10^{17} As atoms/cm^3 and 8.5×10^{16} B atoms/cm^3.

3.7. Verify Eq. (3.3.8) for Si assuming that the value of n_i at 300 K shown in Table 3.1.1 is correct and using the temperature dependence implied in Eqs. (3.3.6), (3.3.7) and (3.3.2), (3.3.4).

3.8. Assuming the values of the effective densities of states shown in Table 3.1.1 for Si to be correct at $T = 300$ K, compute their values at $T = 600$ K.

3.9. Equations (3.3.13) are obtained from (3.3.12) as a zero-order approximation (discounting $4n_i^2/N_D^2$ with respect to 1). From (3.3.11) and (3.3.12) compute the first-order approximations to n_N and p_N and comment on the physical meaning of the result.

3.10. A sample of Si at 500 K is doped with 9×10^{14} B atoms/cm^3 and 1.4×10^{15} As atoms/cm^3.

    (a) Is the sample N, P, or intrinsic?

    (b) Compute net and total doping.

    (c) Compute majority and minority carrier populations.

3.11. In a Si sample at 300 K, the hole concentration is $p = 2.25 \times 10^3$ cm^{-3}. Both In and As impurities are present and the In doping is known to be 5×10^{16} cm^{-3}.

    (a) Is the sample P, N, or intrinsic?

    (b) Compute the As impurity concentration.

    (c) Compute the conduction electron concentration.

    (d) Compute the majority and minority carrier concentrations at 450 K.

3.12. Prove Eq. (3.4.1).

3.13. Prove that the slope of the E_i vs. T curve in Si is less than 3.5×10^{-5} eV/K.

3.14. Prove Eq. (3.4.3).

3.15. For a Si sample doped with 10^{16} donor impurity atoms/cm^3 at 400 K compute the voltage V_{if} defined in Eq. (3.4.10). Verify the computed value using Fig. 3.4.1.

3.16. Draw a curve of V_{if} vs. T for Ge doped with 10^{17} As atoms/cm^3, using appropriate data from the text. Show a temperature interval from 100 to 600 K. Repeat for a doping of 10^{16} B atoms/cm^3. Notice that the linear approximation yields an inversion of the semiconductor type at high temperatures. Find a more accurate formula and plot the curve.

3.17. Equations (3.4.6) and (3.4.8) seem to indicate a linear relationship between E_f and T. Show that actually a small nonlinear term $1.5\, kT \ln (T/300)$ must be added to the linear term to take into account the temperature dependence of the effective density of states.

3.18. A P-type Si sample has $N_A = 10^{14}$ cm^{-3}. At 300 and 600 K find out whether the semiconductor behaves as intrinsic, N-, or P-type material and compute the hole and electron concentrations and V_{if} (the location of the Fermi level expressed in electronvolts from the intrinsic level).

REFERENCES

1. Hayt, William H., Jr.: *Engineering Electromagnetics*, 4th ed., McGraw-Hill Book Company, Inc., New York, 1981.
2. Dekker, Adrianus J.: *Solid State Physics*, 3d ed., Prentice Hall, Inc., Englewood Cliffs, N. J., 1959.
3. Kittel, Charles: *Introduction to Solid State Physics*, 2d ed., John Wiley & Sons, Inc., New York, 1959.
4. Adler, R. B., A. C. Smith, and R. I. Longini: *Introduction to Semiconductor Physics*, John Wiley & Sons, Inc., New York, 1964.
5. Leighton, Robert B., *Principles of Modern Physics*, McGraw-Hill Book Company, Inc., New York, 1959.
6. Sze, S. M.: *Semiconductor Devices, Physics and Technology*, John Wiley & Sons, Inc., New York, 1985.
7. Fermi, Enrico: "Zur Quantelung des idealen einatomiges Gases," *Physik*, vol. 36, pp. 902–912, May 1926.
8. Yang, Edward S.: *Fundamentals of Semiconductor Devices*, McGraw-Hill Book Company, Inc., New York, 1978.
9. Ferry, David K., and R. D. Fanin: *Physical Electronics*, Addison-Wesley Publishing Company, Reading, Mass., 1971.
10. Levine, Sumner N.: *Quantum Physics of Electronics*, The Macmillan Company, New York, 1965.
11. Yariv, Amnon: *Quantum Electronics*, John Wiley & Sons, Inc., New York, 1975.
12. Grove, A. S.: *Physics and Technology of Semiconductor Devices*, John Wiley & Sons, Inc., New York, 1967.
13. Abramowitz and Stegman: *Handbook of Mathematical Functions*, Dover Publications.

4.1 RANDOM THERMAL MOTION

As already mentioned, the availability of thermal energy results in random motion of the carriers. As predicted by the kinetic theory of gases (for references see Sec. 2.3, footnote 3), on the average, at temperature T kelvin, each particle is animated by a temperature-dependent energy

$$E_{th} = \tfrac{1}{2}kT \tag{4.1.1}$$

per degree of freedom. Modeling the current carriers as free particles with an effective mass m_e (or m_h), this energy is stored as kinetic energy of translational motion in each of three dimensions (3 degrees of freedom), yielding a total energy of

$$\tfrac{1}{2}mv_{th}^2 = \tfrac{1}{2}m(v_{thx}^2 + v_{thy}^2 + v_{thz}^2) = \tfrac{3}{2}kT \tag{4.1.2}$$

or a root mean square *thermal speed*

$$v_{th} = \sqrt{v_{thx}^2 + v_{thy}^2 + v_{thz}^2} = \sqrt{3\,\frac{kT}{m}} \tag{4.1.3}$$

At room temperature (300 K), for a free electron, this is of the order of magnitude of $v_{th} \approx 10^7$ cm/s.

To avoid misunderstandings it should be emphasized that this is the *rms*

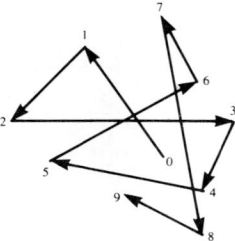

FIGURE 4.1.1
Random thermal motion of a "free" carrier in a crystal.

thermal speed, a totally different quantity from the *average thermal velocity*.[1] As the velocity distribution is random, for each carrier moving in one direction there is almost certainly another moving the opposite way, so that, vectorially, the average velocity is zero, with the result that *thermal motion results in zero average current*.

Thermal energy is also stored by the atoms forming the crystal lattice, but because of the lattice constraints their motion is a vibration around their mean positions in the lattice.

The random dynamic condition described results in frequent *collisions* among the different types of particles, with energy exchanges between them. This helps in maintaining the random nature of the thermal energy distribution. The laws governing such energy exchanges are rather complicated and no practical purpose would be served by analyzing them here. In the following, all transport phenomena will be discussed on the assumption that carriers and lattice atoms behave as classical particles, colliding at random.

Figure 4.1.1 qualitatively represents the motion of one such classical free particle moving in the lattice, colliding with—and rebounding from—the lattice atoms and other free particles. In most collisions the particles involved generally rebound elastically, but other phenomena (such as recombination or ionization) may also occasionally happen when conditions warrant it.

4.2 DRIFT

If an electric field is applied to the crystal, each current carrier is subject to an electrostatic force ($q\mathscr{E}$ for holes, $-q\mathscr{E}$ for electrons) so that another component is superimposed on the random thermal motion of the particles. The direction of the motion induced by the field is not random: classically each particle acquires an acceleration *in the direction of the electrostatic force*, so that the average veloc-

---

[1] The student is reminded that speed is a scalar quantity whereas velocity is a vectorial quantity. Also the difference between rms and average quantities should be kept in mind.

ity is no longer zero and, as the particles are charged, a net electric current flows.[2]

At first sight it might appear that the current carriers would continuously and indefinitely accelerate in this direction, gathering energy from the field, so that their velocity (and, with it, the electric current intensity) would keep increasing without bounds. This, of course, is not the case because the carriers do not behave as if they were in a vacuum: they are instead moving inside a crystal, where they collide with one another and with the lattice atoms. Whenever such collisions occur, as previously stated, complex energy exchanges take place and the particle velocity is again randomized.

When it is desired to investigate macroscopic events, such as the flow of current through the crystal, all microscopic phenomena should be described on the basis of a statistical average taken over a macroscopic volume of the crystal and over a reasonably long period of time. Considering the above carrier motion from this statistical point of view (and concentrating attention only on the electrons for easier visualization), the following description is adequate:

1. Before the electric field is applied, the free electrons are in random motion with rms speed v_{th} and zero average velocity, as schematically indicated in Fig. 4.1.1.
2. When the electric field is applied, the electrostatic force adds a nonrandomly directed component of motion with acceleration $-q\mathscr{E}/m_e$. Without losing their random thermal motion, all elections now also keep accelerating in the direction opposite to the field.
3. For each electron, however, this pickup of speed in one direction lasts only until it collides with some other particle or, on the average, for a time τ_c (the *mean free time between collisions*), dependent on material and temperature. Because of the addition of this directed component of motion, just before collision the average electron velocity is no longer zero but has become

$$\mathbf{v}_d = -\frac{q\mathscr{E}}{m_e}\tau_c \qquad (4.2.1)$$

4. Upon collision, the distribution of velocities is again randomized, the average velocity goes back to zero, and the acceleration cycle resumes. The resulting electron motion is depicted in Fig. 4.2.1.

Over a period of time much larger than τ_c, it can *statistically* be said that, when submitted to an electric field, the randomly moving electrons *drift in the direction opposite to the field with a uniform drift velocity* $\mathbf{v}_d$ given by (4.2.1).

---

[2] Of course, holes move in the direction of the field, electrons *in the opposite direction*, as evidenced by the sign of the force. As the charges carried by electrons and holes are opposite in sign, this means that the electric currents generated by these opposite motions are *in the same direction*.

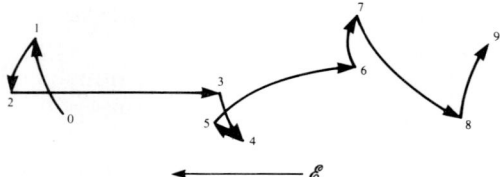

FIGURE 4.2.1
Qualitative representation of "free" electron motion in a crystal under the action of thermal energy and of an applied electric field $\mathscr{E}$.

The same considerations applied to the holes yield a drift velocity

$$\mathbf{v}_d = \frac{q\mathscr{E}}{m_h}\,\tau_c \tag{4.2.2}$$

in the direction of the field.[3] For both electrons and holes, the drift velocity is proportional to the applied field:

$$\mathbf{v}_d = -\mu_e\mathscr{E}$$

$$\tag{4.2.3}$$

$$\mu_e = \frac{q\tau_c}{m_e}$$

for electrons and

$$\mathbf{v}_d = \mu_h\mathscr{E}$$

$$\tag{4.2.4}$$

$$\mu_h = \frac{q\tau_c}{m_h}$$

for holes. The constants μ_e and μ_h are the electron (respectively the hole) *mobility* in square centimeters per volt-second.[4]

From the definition of current density [5], $\mathbf{J} = \rho_v\mathbf{v}$, where ρ_v is the charge density and $\mathbf{v}$ the charge velocity. In our case, the electron charge density is $\rho_v = -qn$ and the hole charge density is $\rho_v = qp$ so that, from (4.2.3) and (4.2.4),

---

[3] A more rigorous discussion can be found in Refs. 1 (p. 238), 2 (p. 235*), 3 (p. 32), and 4.

[4] From an intuitive point of view, the student may find it difficult to accept that a constant force results in a constant velocity. Yet when the space in which an object falls is not empty, but filled with a distribution of objects (such as air molecules), it has to push them away to move through them. In doing so, it must impart to them a velocity equal to its own and so accordingly alter their momentum. By the laws of mechanics, under these conditions, the moving object is subjected to a reaction force equal to the rate of change of momentum. It is then easy to see that an object "falling" under a constant force in a space occupied by a distribution of particles is subject to a "resistance" increasing with its velocity, so that, as amply verified by everyday experience, it accelerates only up to the speed that produces a resistance equal and opposite to the driving force and, from then on, proceeds at a constant speed.

with evident notation,

$$\mathbf{J}_e = qn\mu_e \mathscr{E} \tag{4.2.5}$$

$$\mathbf{J}_h = qp\mu_h \mathscr{E} \tag{4.2.6}$$

and the total conduction current density is

$$\mathbf{J} = \mathbf{J}_e + \mathbf{J}_h = q(n\mu_e + p\mu_h)\mathscr{E} \tag{4.2.7}$$

However, by Ohm's law [5], $\mathbf{J} = \sigma\mathscr{E}$, and, from (4.2.7), the conductance is seen to be

$$\sigma = q(n\mu_e + p\mu_h) \tag{4.2.8}$$

The mobilities in the above expressions depend on material and temperature. For Si, Ge, and GaAs semiconductors, values of the mobility at 300 K as a function of *total* impurity concentration N_t are displayed in Figs. 4.2.2, 4.2.3, and 4.2.4 respectively. The following equations yield the same data for Si with sufficient accuracy for most computer calculations:

$$\mu_h = \frac{495 - 47.7}{1 + [N/(6.3 \times 10^{16})]^{0.72}} + 47.7 \tag{4.2.9}$$

$$\mu_e = \frac{1330 - 65}{1 + [N/(8.5 \times 10^{16})]^{0.76}} + 65.5$$

The values of mobility at 300 K plotted in Figs. 4.2.2 through 4.2.4 and/or computed by Eqs. (4.2.9) are also presented in tabular form in Tables 4.2.1, 4.2.2, and

TABLE 4.2.1
Silicon

Concentration, cm^{-3}	Mobility, cm^2/(V·s)		Diffusivity, cm^2/s	
	Holes	Electrons	Holes	Electrons
10^{14}	490	1320	13	34
2×10^{14}	485	1313	12.7	34
5×10^{14}	480	1304	12.4	33.5
10^{15}	470	1290	12	33
2×10^{15}	460	1280	11.6	32.5
5×10^{15}	439	1216	11.3	31
10^{16}	400	1120	10.4	29
2×10^{16}	359	1000	9.3	26
5×10^{16}	292	826	7.6	21
10^{17}	235	660	6.2	17
2×10^{17}	186	517	4.8	13
5×10^{17}	161	324	4.2	8
10^{18}	100	230	2.6	6
2×10^{18}	82	170	2	4.4
5×10^{18}	68	124	1.7	3
10^{19}	60	100	1.5	2.6

TABLE 4.2.2
Germanium

Concentration, cm^{-3}	Mobility, cm^2/(V·s)		Diffusivity, cm^2/s	
	Holes	Electrons	Holes	Electrons
10^{14}	1900	3900	50	101
2×10^{14}	1870	3850	48	99
5×10^{14}	1830	3790	47	98
10^{15}	1700	3730	45	97
2×10^{15}	1650	3690	43	95
5×10^{15}	1550	3620	40	94
10^{16}	1400	3500	36	90
2×10^{16}	1250	3365	32	87
5×10^{16}	1000	3070	26	80
10^{17}	820	2800	21	72
2×10^{17}	750	2555	20	66
5×10^{17}	450	2248	12	58
10^{18}	350	1900	9	50
2×10^{18}	255	1570	6.6	41
5×10^{18}	170	1180	4.5	31
10^{19}	110	900	3	23

4.2.3 for Si, Ge, and GaAs respectively. The figures and tables also display diffusivity values obtained through the Einstein relationships (cf. Sec. 4.3).

Approximate expressions of the type of (4.2.9) can be obtained for other semiconductors with the aid of Figs. 4.2.3 and 4.2.4. For computer calculations,

TABLE 4.2.3
Gallium arsenide

Concentration, cm^{-3}	Mobility, cm^2/(V·s)		Diffusivity, cm^2/s	
	Holes	Electrons	Holes	Electrons
10^{14}	380	8000	9.8	208
2×10^{14}	360	7930	9.4	205
5×10^{14}	345	7800	9	203
10^{15}	330	7700	8.6	200
2×10^{15}	320	7500	8.4	195
5×10^{15}	310	7400	8	191
10^{16}	300	7250	7.6	185
2×10^{16}	280	6700	7.2	173
5×10^{16}	260	6000	6.8	155
10^{17}	230	5000	5.9	130
2×10^{17}	220	4300	5.5	112
5×10^{17}	190	3000	5	78
10^{18}	160	2100	4	51
2×10^{18}	130	1350	3.3	35
5×10^{18}	78	1000	2	26
10^{19}	50	900	1.3	22

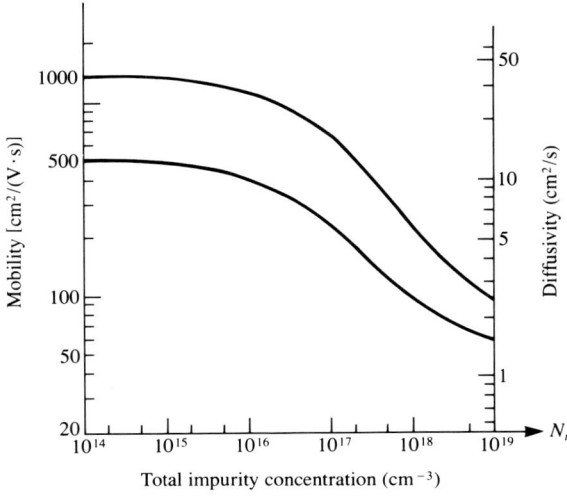

FIGURE 4.2.2
Mobility and diffusivity vs. total impurity concentration in Si at 300 K.

Tables 4.2.1 through 4.2.3 can be used as truth tables and appropriate interpolation can be introduced in programming.

It should be emphasized that, in the above, N stands for the *total*, rather than the net, impurity concentration (cf. Sec. 3.3).

Example 4.2.1. A Si specimen of length 2 cm and cross section 2 mm^2 is at 300 K and is doped with 10^{17} atoms of As and 9×10^{16} atoms of B. Compute the specimen's resistance and compare the contribution to conductivity of the electrons to that of the holes.

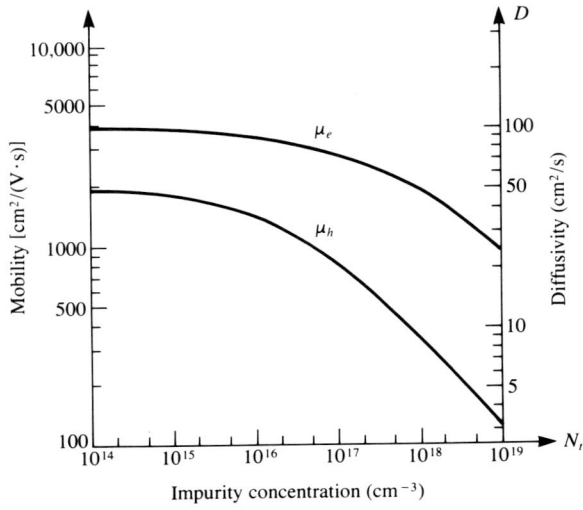

FIGURE 4.2.3
Mobility and diffusivity vs. total impurity concentration in Ge at 300 K.

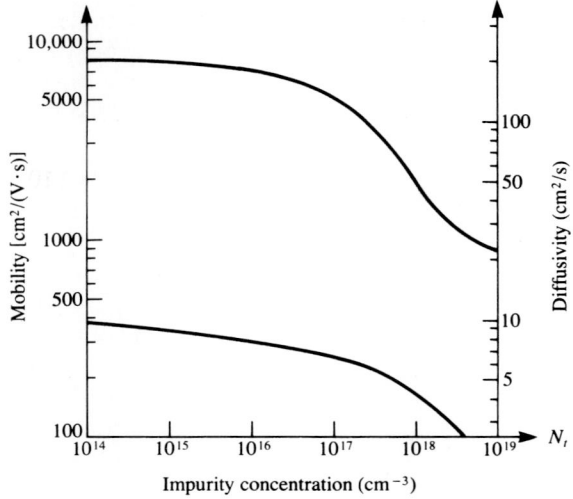

FIGURE 4.2.4
Mobility and diffusivity vs. total impurity concentration in GaAs at 300 K.

Solution. The net doping is $N_D = 10^{17} - 9 \times 10^{16} = 10^{16}$ cm^{-3}. The total doping is $N = 10^{17} + 9 \times 10^{16} = 1.9 \times 10^{17}$ cm^{-3}. Therefore the carrier concentrations are (see Example 3.4.1)

$$n_N = 10^{16} \text{ cm}^{-3} \quad \text{and} \quad p_N = 2.25 \times 10^4 \text{ cm}^{-3}$$

From (4.2.9) (or Table 4.2.1), using $N_t = 1.9 \times 10^{17}$, the mobilities are

$$\mu_e = 510 \text{ cm}^2/(V \cdot s) \quad \text{and} \quad \mu_h = 187 \text{ cm}^2/(V \cdot s)$$

so that the conductivity is

$$\sigma = 1.6 \times 10^{-19}(510 \times 10^{16} + 187 \times 2.25 \times 10^4) = 0.816 \ 1/(\Omega \cdot \text{cm})$$

and finally the resistance:

$$R = \frac{l}{\sigma A} = \frac{2}{0.816 \times 10^{-2} \times 2} = 122.5 \ \Omega$$

The minority carrier concentration is much smaller than the electron concentration. Consequently, the second addend in the expression for the conductivity turns out to be 12 orders of magnitude smaller than the first. It can be concluded that in this case the contribution of the holes to conductivity is negligible and the conductivity could be correctly computed on the basis of the conduction electrons exclusively.

Mobility was defined on the basis of the carrier motion mechanism just described. This same mechanism suggests that the mobility must be temperature dependent. Indeed, intuitively, it can be expected that both the frequency of the collisions and the resulting change of momentum should increase at higher temperatures, when the carriers are faster and the lattice atom vibrations are wider, resulting in lower freedom of motion for the carriers; therefore the mobility should decrease with increasing temperature. This expectation is confirmed by experiment: for intrinsic semiconductors the temperature dependence of mobility can be expressed with reasonable approximation by the empirical formulas [6]:

$$\mu_e \approx 4.9 \times 10^7 \ T^{-1.66}$$
$$\mu_h \approx 1.05 \times 10^9 \ T^{-2.35} \tag{4.2.10}$$

The above formulas are valid for intrinsic Ge between the temperatures of 100 and 300 K. For intrinsic Si between 150 and 400 K, the mobilities can be computed approximately by

$$\mu_e \approx 2.1 \times 10^9 \ T^{-2.5}$$
$$\mu_h \approx 2.3 \times 10^9 \ T^{-2.7} \tag{4.2.11}$$

The influence of temperature on the mobility, however, also depends on the impurity concentration, because carriers interact differently with ions (charged impurity atoms) than with neutral (host material) atoms.

The previous argument, suggesting a decrement of the mobility with rising temperature, is valid for collisions between current carriers and neutral atoms.

Collisions with charged ions, however, are more in the nature of an interaction with their surrounding electric fields. Such interactions last for a shorter time when the carrier velocity is high, so that, at high temperatures, the "resistance" offered by the ions to carrier motion can be expected to be lower and the mobility increases with temperature. A more rigorous analysis [7, 8] predicts that collisions with impurity ions cause the mobility to vary with $T^{3/2}$, while neutral atom collisions yield a mobility proportional to $T^{-3/2}$ [compare this theoretical prediction with the empirical formulas (4.2.10) and (4.2.11)].

The temperature dependence of the mobility therefore varies with the relative concentration of impurity vs. intrinsic atoms, the effect of impurity collisions prevailing at high impurity concentrations and low temperatures, and vice versa.

The theoretical temperature dependence of the mobility in an extrinsic semiconductor can then be expected to be of the type shown in Fig. 4.2.5. At low

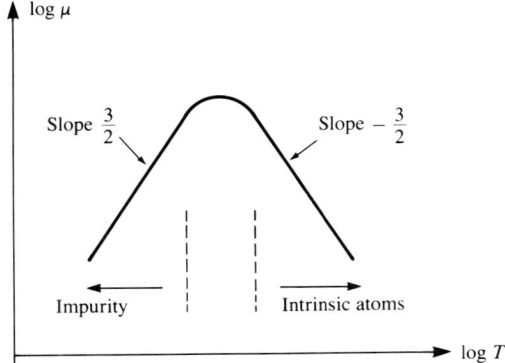

FIGURE 4.2.5
Theoretical, qualitative plot of mobility vs. temperature for an extrinsic semiconductor, showing temperature ranges in which mobility is determined mainly by impurity, respectively intrinsic, atom collisions.

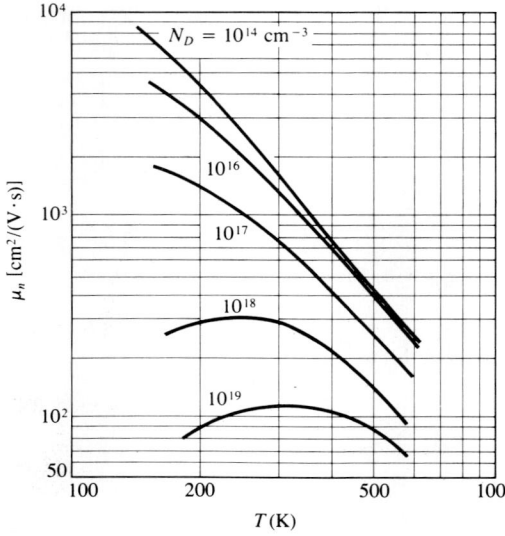

FIGURE 4.2.6
Electron mobility in N-type Si vs. temperature for several doping concentrations [9].

temperatures, collisions with impurity ions are the prevalent mechanism and the graph rises with $T^{3/2}$. At high temperature, intrinsic, neutral atom collisions become the determining mechanism and the mobility drops with increasing temperatures, following a $T^{-3/2}$ curve. In between there is a transition range of temperatures, where the graph reaches a maximum before beginning to decay.

A more comprehensive picture of the phenomenon is given by the empirical graph of Fig. 4.2.6 [9], in which each curve of constant doping is seen to follow roughly the shape of Fig. 4.2.5. Notice that, as expected, with increasing dopant concentrations, the position of the maximum is displaced toward higher and higher temperature ranges.

Example 4.2.2. The same specimen as in Example 4.2.1 is brought to a temperature of 500 K. Compute the resistance and comment on the relative contribution of the electrons and holes to conduction.

Solution. From Fig. 4.2.6, interpolating for $N = 1.9 \times 10^{17}$ cm^{-3} and $T = 500$ K:

$$\mu_e \approx 220 \text{ cm}^2/(\text{V} \cdot \text{s})$$

The intrinsic concentration is computed from (3.3.7):

$$n_i = 3.65 \times 10^{14}$$

This is two orders of magnitude less than N_D, so the approximation of (3.3.13) holds and the minority carrier concentration is

$$p_N = \frac{(3.61 \times 10^{14})^2}{10^{16}} = 1.3 \times 10^{13} \text{ cm}^{-3}$$

as this quantity is still 3 orders of magnitude smaller than the majority carrier concentration, the contribution of the holes to conduction is negligible, and the conductivity can be computed on the basis of the electrons only:

$$\sigma \approx 1.6 \times 10^{-19}(220 \times 10^{16}) = 0.35 \ 1/\Omega \ \text{cm}$$

and finally:

$$R = \frac{2}{0.35 \times 10^{-2} \times 2} = 284 \ \Omega$$

Sheet Resistance

In integrated circuits, resistors are often fabricated as thin sheets of extrinsic semiconductor material. Supposing the sheet has uniform thickness δ and width w, as shown in Fig. 4.2.7a; then its resistance is

$$R = \frac{L}{\sigma w \delta} \ \Omega \qquad (4.2.12)$$

If the resistor is square in shape, as in Fig. 4.2.7b, then $L = w$ and (4.2.12) becomes

$$R_\square = \frac{1}{\sigma \delta} \ \Omega_\square \qquad (4.2.13)$$

Notice that, for a given material and thickness, all square sheets have the same resistance, independently of the size of the square, provided only they have a square shape. The resistance of such square wafers, indicated as $R_\square$ in (4.2.13), is known as the sheet resistance and is measured in ohms per square ($\Omega_\square$); it depends only on the thickness of the sheet and its average conductivity.

For the integrated circuit of Fig. 4.2.7c, the resistance can then be computed by counting the number of squares in the strip and multiplying by the sheet resistance. Conversely, this method can be used to design the resistor.

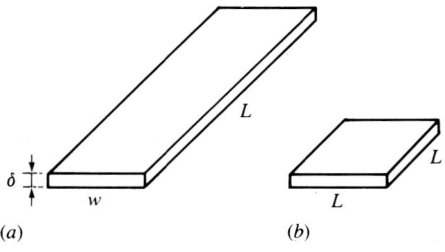

(a)   (b)

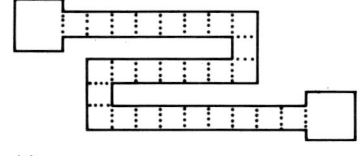

(c)

FIGURE 4.2.7
Sheet resistance. (a) Sheet resistor of uniform doping, width w, and thickness δ. (b) Square sheet resistor of side L. (c) Typical integrated circuit sheet resistor. A division into squares to ease computation of the total resistance is indicated by the dotted lines.

If the semiconductor material is extrinsic with sufficiently large net doping and the doping is uniform throughout the thickness of the sheet, then σ in (4.2.13) can be computed as:

$$\sigma \approx qN_D\mu_e \tag{4.2.14}$$

for N-type semiconductors (and analogous for P-type).

In most integrated circuit sheet resistors, however, the doping is not uniform, because they are obtained by a diffusion process (cf. Sec. 5.5). In that case, the average conductivity $\bar{\sigma}$ should be used in (4.2.13):

$$\bar{\sigma} = \int_0^\delta qN_D(x)\mu_e(x)\,dx \tag{4.2.15}$$

with analogous form for P-type material (cf. Example 5.5.3).

4.3   DIFFUSION

It has already been noticed (see Sec. 4.1) that a system of free particles at a temperature different from absolute zero is animated by a random thermal velocity distribution. If, in such a system, the particle concentration n (particles/cm^3) is not uniformly distributed, but varies from point to point, then *diffusion* occurs: *particles flow away from densely populated regions toward regions of low particle concentration.*

Diffusion phenomena are quite commonly observed: a gas introduced into a container diffuses to fill it with uniform concentration; an ink drop, dropped in water, diffuses, so that, given enough time, the water becomes uniformly colored. All this happens simply because there is a greater probability that particles in their random thermal motion will move away from regions of high concentration than into them. Notice that *no force is applied to the particles:* diffusion can be said to be due exclusively to the *push of probability.*[5]

In diffusion phenomena, the average particle displacement occurs in a well-defined, predictable direction (from low to high concentration). How is this possible when the very randomness of the phenomenon implies that each particle has an equal probability of moving in any direction?

Consider a surface S at the interface between two regions (1 and 2) with different particle concentrations ($n_1 > n_2$), as in Fig. 4.3.1. In the random thermal motion, a particle in region 1 has the same probability of flowing through S into

---

[5] One generally thinks of probability as a mathematical concept, rather than a physical force, and so it might be felt that a phenomenon mediated by it should, at most, represent some rather mild event (such as the spread of an ink drop in a glass of water).

Actually, in diffusion, probability *merely directs the stored thermal energy* toward leveling out variations of concentration. When coupled with large amounts of stored energy, diffusion can be responsible for the most spectacularly disruptive physical events, such as explosions, atomic or otherwise, including that big bang that supposedly started our whole universe on its way.

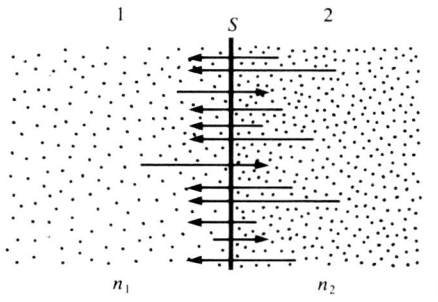

1   S   2

n_1 n_2 $n_1 < n_2$

FIGURE 4.3.1
Interface between two regions with different particle concentrations. If, each
second, the same percentage of the available particles crosses the interface from
the two sides, the flux rate from the large
concentration side is the larger.

region 2 as a particle in region 2 has of flowing into region 1; consequently, if in
any given interval of time a certain percentage of n_1 flows through S into region
2, then in the same period of time the *same percentage* of n_2 flows through S in
the opposite direction. If region 1 is more densely populated, the rate of flow
must be greater in the direction from 1 to 2 than in the opposite one, resulting in
a *net flow* from high to low concentration regions.

Diffusion plays a vital role in determining the behavior of solid state
devices, so that a quantitative study of such behavior requires a mathematical
expression of the diffusion law. To this end the most important diffusion relationships are summarized here, without proof, in a form suitable for application to
elementary solid state device analysis. They are limited to one-dimensional diffusion systems.[6] An elementary mathematical analysis of diffusion phenomena in
three dimensions, leading to this formulation as a one-dimensional limit case, is
offered, by way of proof, in App. 4A.

For a one-dimensional system, the *diffusion flow-rate density* (defined as the
net number of particles per second flowing by diffusion through a unit surface
perpendicular to the direction of diffusion) depends on the spatial rate of variation of the particle concentration n, and is given by

$$F = -D \frac{dn}{dx} \qquad \text{particles/(cm}^2 \cdot \text{s)} \qquad (4.3.1)$$

where the *diffusivity* D in square centimeters per second depends on the material
and on temperature. D can be computed from the mobility μ by means of Einstein's relation:

$$D = \frac{kT}{q} \mu \qquad (4.3.2)$$

---

[6] Notice that the sufficient condition for an otherwise three-dimensional system to behave one dimen-
sionally with regard to diffusion is that its particle concentration varies only in one direction (planar
symmetry in cartesian coordinates).

Thanks to Einstein's relation, Figs. 4.2.2, 4.2.3, and 4.2.4 also plot D vs. the impurity concentration for common semiconductor materials at 300 K. For computer calculations D can be computed from Eq. (4.2.9) (and similar for other materials) by application of Einstein's relation (4.3.2). For temperatures different from 300 K, Fig. 4.2.6 (or analogous) can be used.

The rate of flow through a surface S perpendicular to the x direction is then

$$\int_S F \, dS \qquad \text{particles/s} \tag{4.3.3}$$

If the diffusing particles are charged, their motion results in an electric current density equal, by definition, to the net flow-rate density times the single particle charge. The current density, therefore, may have not only drift but also diffusion components, so that, for a one-dimensional system in the x direction,

$$J_x = q\left(n\mu_e \mathcal{E}_x + D_e \frac{dn}{dx} + p\mu_h \mathcal{E}_x - D_h \frac{dp}{dx} \right) \tag{4.3.4}$$

in which it is easy to identify electron and hole drift and diffusion components.

Example 4.3.1. In a Si semiconductor doped with 5.5×10^{18} B atoms/cm^3 and 4.5×10^{18} As atoms/cm^3 there has occurred an accumulation of minority carriers, resulting, in the region from $x = 0$ to $x = 1$ cm, in a minority carrier concentration distribution:

$$n_p(x) = 10^7(1 - x) + 1.33 \times 10^{11} \text{ cm}^{-3}$$

where x is in centimeters. The temperature is 500 K. Compute the flow-rate density of electron diffusion and the diffusion current density at $x = 0.5$ cm.

Solution. The total doping is $N = 5.5 \times 10^{18} + 4.5 \times 10^{18} = 10^{19}$ cm^{-3}, so that, from Fig. 4.2.6, at 500 K, the mobility is

$$\mu_e = 90 \text{ cm}^2/(\text{V} \cdot \text{s})$$

Using Einstein's relation, $D_e = \mu_e kT/q = 3.885$ cm^2/s. At $x = 0.5$ cm the rate of variation of minority carrier concentration is $dn_p/dx = -10^7$ cm^{-4}, so that, from (4.3.1),

$$F = -3.885(-10^7) = 3.885 \times 10^7 \text{ electrons/(cm}^2 \cdot \text{s)}$$

The diffusion current density is then

$$J = -qF = -6.2 \text{ pA/cm}^2$$

4.4   GENERATION AND RECOMBINATION

Nonequilibrium Conditions

In Sec. 2.4 it was argued that the thermal equilibrium concentration of carriers in a semiconductor depends on the balance between generation and recombination

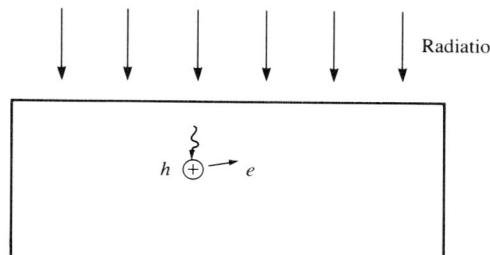

FIGURE 4.4.1
Carrier generation by illumination of a semiconductor surface. A photon is absorbed by an atom of the host semiconductor material. Ionization results in the emission of a conduction electron (e) and production of an ionized intrinsic atom, i.e., a hole (h).

phenomena, the net rate of increment of the population being the difference between the rate of generation and the rate of recombination.

It was further argued, in Sec. 2.5 and confirmed in Sec. 3.3, that the rate of recombination increases with the product np of the electron and hole concentrations. In discussing thermal equilibrium conditions, it was assumed that the energy responsible for carrier generation is thermal and consequently that the rate of generation depends on temperature. The final conclusion reached was that, for each host material (i.e., for each E_g) at equilibrium, the product np depends exclusively on the temperature of the sample [Eq. (3.3.5)].

However, thermal energy is not the only possible source of carrier generation: other forms of energy can contribute to the ionization rate. Transport phenomena of several types (e.g., a current) can also inject extra carriers into a region of the sample.

For instance, ionization of intrinsic atoms can be induced by the absorption of photons of suitable energy. This case is illustrated in Fig. 4.4.1, where the surface of a semiconductor is flooded with light. The radiant energy impinging on the semiconductor surface is composed of photons of energy hf. If the photon frequency is high enough, then photons, upon absorption by semiconductor atoms, may induce ionization transitions in them. Remembering the considerations of Secs. 1.4 and 2.1, it is easily seen that this can happen when the photon energy is greater than the gap energy: $hf > E_g$.

Under proper illumination, the rate of ionization (and so the production of electron-hole pairs) may be *increased above the thermal rate*, because now photon energy can be utilized in addition to the available thermal energy to mediate transitions from the valence band to the conduction band.

This tends to increase the carrier concentration at the surface, resulting in carrier diffusion from the surface into the body of the semiconductor: *the illuminated surface injects extra carriers into the crystal.*

If the semiconductor is extrinsic, practically all the impurity atoms are already thermally ionized, so the incoming photons will act almost exclusively on the intrinsic host atoms, generating electron-hole pairs.

A similar condition may occur if current carriers are injected into the semiconductor by causing an electric current to flow through its surface. As will be discussed in detail in Chap. 6, one way of obtaining this result is to connect an N-type semiconductor to a P-type one and apply a voltage to the system, making

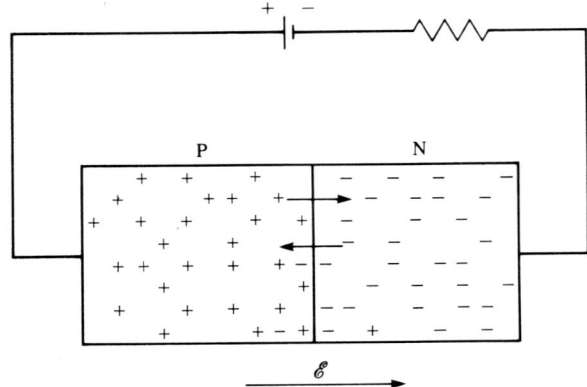

FIGURE 4.4.2
A voltage applied to a PN junction, making the P side positive, creates an electric field $\mathscr{E}$, as shown. This injects holes into the N-type semiconductor and electrons into the P-type semiconductor. The rate at which minority carriers appear in each semiconductor exceeds the thermal generation rate.

the N-type semiconductor negative with respect to the P-type one, as schematically indicated in Fig. 4.4.2. As can easily be inferred intuitively (and will be more rigorously discussed in Chap. 6), electrons are pushed by the applied voltage from the N into the P semiconductor, so that the rate at which free electrons appear in the region near the surface of the P semiconductor is increased above the thermal generation rate, just as in the case of surface illumination. Therefore, in this case too, carrier diffusion will result and *at the PN contact surface extra carriers are injected into the crystal.*

It is customary to use the generic term *injection* to refer to any and all increments in the generation rate due to a nonthermal source, irrespective of the nature of this source, and to differentiate between *high- and low-level injection:*

1. Low-level injection occurs when the number of injected carriers per unit volume is negligible in comparison with the majority carrier concentration. Notice that, in extrinsic semiconductors, the majority carrier concentration is often many orders of magnitude larger than the minority carrier concentration, so that the latter can be very significantly affected, even by low-level injection.
2. High-level injection is so high as to be significant, even with respect to the majority carrier concentration.

Irrespective of the mechanism causing the injection of extra carriers, whenever in a region of a semiconductor the generation rate increases above the thermal level then, in order to reach steady state, the recombination rate must correspondingly increase. The final result is that the steady state product np in the affected region differs from the thermal equilibrium value, as predicted by the considerations of Sec. 2.5.

In the following, whenever necessary, thermal equilibrium values of the electron and hole concentrations will be indicated by adding a 0 subscript to any other subscripts. For instance, the thermal equilibrium concentrations of majority electrons, majority holes, and minority electrons will be indicated as n_{oN}, p_{oP}, and n_{oP} respectively, while the analogous concentrations under conditions other than thermal equilibrium will be characterized by the notation n_N, p_P, and n_P. Generation rates will be indicated by G [particles/(cm$^3 \cdot$ s)] and recombination rates by R [particles/(cm$^3 \cdot$ s)] with appropriate subscripts, when required.

Low-Level Injection

The above concepts can be expressed mathematically: the recombination rate R is proportional to the product of the concentrations:

$$R = \beta p n \qquad (4.4.1)$$

The condition of thermal equilibrium is

$$G_{th} - R_{th} = 0 \qquad (4.4.2)$$

Suppose that some nonthermal cause increases the rate of generation by an additional rate G_A. Some increment in carrier concentrations will result. Assuming, as an example, low-level injection of minority carriers in a P-type semiconductor,

$$n_P = n_{oP} + \Delta n \qquad (4.4.3)$$

$$p_P \approx p_{oP} \qquad (4.4.4)$$

in accordance with the definition of low-level injection. Remembering (4.4.1), this results in a recombination rate higher than the thermal value R_{th}:

$$R = \beta(n_{oP} + \Delta n)p_{oP} = R_{th} + R_A \qquad (4.4.5)$$

where R_A is the additional (or net) recombination rate. By definition, the rate of increment of the minority carriers is now

$$\frac{dn_P}{dt} = G_{th} + G_A - R_{th} - R_A \qquad (4.4.6)$$

where, remembering (4.4.5),

$$R_A = \beta p_{oP}(n_P - n_{oP}) = \frac{n_P - n_{oP}}{\tau_e} \qquad (4.4.7)$$

with

$$\tau_e = \frac{1}{\beta p_{oP}} \qquad (4.4.8)$$

From (4.4.6), using (4.4.2) and (4.4.7),

$$\frac{dn_P}{dt} = G_A - \frac{n_P - n_{oP}}{\tau_e} \qquad (4.4.9)$$

At *steady state* $dn_P/dt = 0$ and, from (4.4.9),

$$n_P = n_{oP} + G_A \tau_e \qquad (4.4.10)$$

To understand the physical meaning of τ_e, assume that the injection rate G_A has been active for a long time, so that steady state has been reached when, at $t = 0$, the injection is suddenly stopped (i.e., the injection rate is expressed by $G_A(t) = G_A[1 - u(t)]$). Then, for $t > 0$, $G_A = 0$ and (4.4.9) becomes

$$\frac{dn_P}{dt} = -\frac{n_P - n_{oP}}{\tau_e} \tag{4.4.11}$$

with boundary conditions $t = 0 \rightarrow n_P = n_{oP} + G_A \tau_e$ and $t = \infty \rightarrow n_P = n_{oP}$. Solving under these conditions gives

$$n_P(t) = n_{oP} + \tau_e G_A e^{-t/\tau_e} \tag{4.4.12}$$

indicating that τ_e is *the time constant of the relaxation of the system to thermal equilibrium*. Consequently, τ_e is often referred to as the relaxation time or *lifetime* of the excess minority carriers. This quantity, in most cases ranging between 0.1 and 100 μs, plays an important part in many semiconductor-related phenomena, so that technologically it is very important to measure it and to control it. Some of these techniques will be discussed in following chapters (cf. Chap. 5).

Example 4.4.1. A Si specimen at 300 K is doped with 5×10^{14} B atoms/cm^3. An outside source of energy has been injecting minority carriers into the specimen at the rate of 10^{15} carriers/(cm$^3 \cdot$ s) over a period of time long enough to reach steady state. At $t = 0$ the external source is deactivated. If the minority carrier lifetime is 10 μs:
(a) Is this low- or high-level injection?
(b) At what time is the minority carrier concentration 10 percent above the equilibrium concentration of the sample at the given temperature?

Solution. The thermal equilibrium minority carrier concentration is

$$n_{oP} = \frac{2.25 \times 10^{20}}{5 \times 10^{14}} = 4.5 \times 10^5 \text{ cm}^{-3}$$

(a) The steady state minority carrier concentration is, from (4.4.10),

$$n_P(0) = 4.5 \times 10^5 + 10^{15} \times 10^{-5} \approx 10^{10} \text{ cm}^{-3}$$

This is 4 to 5 orders of magnitude lower that the majority carrier concentration, so the condition for *low-level injection* is met.
(b) From (4.4.12), after elementary algebra, the time requested is computed as

$$t = -10^{-5} \ln\left(\frac{4.5 \times 10^4}{10^{10}}\right) = 0.123 \text{ ms}$$

Direct and Indirect Transition Mechanisms

In the above it was tacitly assumed that, when an electron and a hole "collide," they automatically neutralize each other, the electron filling the hole; as a result both conduction electron and hole disappear, simply resulting in a neutral atom. Such a transition, however, just as in any other physical phenomenon, must satisfy the laws of physics, which include the conservation of energy and of momentum.

In a recombination transition, the electron jumps down from the conduction band to the valence band to fill the hole. The principle of *conservation of energy* requires that the difference in energy, E_g, between the two states be released in some form (just as the generation transition could not take place without absorption of the energy E_g from some outside source). The energy released during recombination can be dissipated in the form of radiation (as a photon of energy $hf = E_g$), in which case the recombination is said to be *radiative*, or it may be released to the lattice in the form of vibration (as a *phonon*), usually thermal.

To understand the implications of the law of *conservation of momentum* it is necessary to take into account the distribution of momentum among electrons and holes, a quantity that has not yet been considered in this presentation. As in classical mechanics, the momentum is related to kinetic energy by

$$E_{ek} = \frac{(p_e^*)^2}{2m_e} \tag{4.4.13}$$

As indicated by the subscripts e, the above equation refers to electrons; an analogous one holds for holes. As the electrons find themselves within the potential configuration of the lattice, the mass m_e in (4.4.13) must be interpreted as the effective electron mass under the given conditions (cf. Sec. 3.1) and the momentum as the *crystal momentum* (as indicated by the asterisk), a quantity proportional to the quantum propagation constant ($\mathbf{p}_e^* = \hbar\mathbf{K}$) and best described in quantum terms [1, 2, 10, 11].

The relationship between particle energy and momentum is usually expressed by energy–momentum band diagrams, two examples of which are shown in Fig. 4.4.3a and b for GaAs and Si respectively.[7] It is important to notice that in GaAs the electrons at the bottom of the conduction band and the holes at the top of the valence band both have the same crystal momentum $p^* = 0$. In Si, however, this is not the case: the holes at level E_v still have zero crystal momentum, but the electrons at level E_c have a momentum $p_e^* \neq 0$.

It follows that, to undergo a recombination transition, a Si electron, originally at level E_c, would have to *change its momentum*, so that the total system momenta before and after the transition would not be the same. To reconcile the transition with the conservation law *some other element of the system should be*

---

[7] There is a further complication in that the crystal momentum is a vector quantity and the shape of the energy–momentum band diagram depends on the direction along the crystallographic axes. In Fig. 4.4.3 the plots are drawn for K along a $\langle 100 \rangle$ face (to the right of the energy axis) and along a $\langle 111 \rangle$ face (to the left of the energy axis). Fortunately, in this presentation, it is not necessary to discuss these matters in detail. The reader conversant with quantum physics is referred to the literature [1, 2, 10–13].

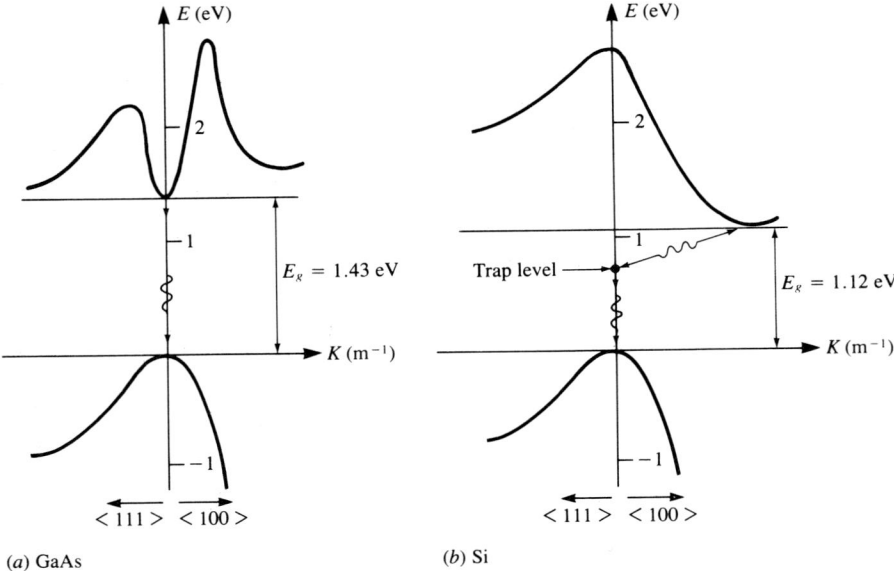

(a) GaAs (b) Si

FIGURE 4.4.3
Energy–momentum band (expressed as energy–quantum propagation constant) diagrams. Each plot is divided in two parts: to the right of the energy axis, **K** varies along a ⟨100⟩ plane; to the left of the energy axis, along a ⟨111⟩ plane. (a) Direct gap semiconductor, showing a direct, band to band transition. (b) Indirect gap semiconductor, showing the two stages of an indirect transition involving a deep gap trap level [12].

forced to take over the lost momentum. This requires interaction with some other particle, i.e., *some intermediate step in the transition.*

For a transition to occur, an intermediate permissible state (where the interaction can occur) must then be available somewhere within the gap. A carrier can be "trapped" by this state, undergoing a nonradiative momentum-balancing transition to the intermediate energy level from which it can then move to the next band, as shown in Fig. 4.4.3b. Because of this action, these intermediate states are often called *traps.* As they must occur somewhere within the forbidden gap of the crystal, they require the introduction of some irregularity in the quantum characteristics of the lattice and, as we saw, this is usually obtained by the inclusion of appropriate crystal impurities.[8]

It can then be concluded that:

1. In crystals like GaAs (electrons at E_c have the same momentum as holes at E_v), recombination transitions can occur directly from band to band, and so

---

[8] Other conditions for momentum-balancing interactions (e.g., with phonons) are possible. Here attention will be concentrated on impurity-induced trap levels.

such crystals are designated as direct gap, direct transition, or simply *direct* semiconductors.

2. In crystals behaving like Si (the momenta are different), recombination can only be indirect, requiring a transition to an intermediate trap level. Therefore they are classified as indirect transition or simply *indirect* semiconductors.

Figure 4.4.3a and b symbolically shows the two types of transition. The above classification is of great importance in the technology of LEDs and LASERs.

The discussion of Eqs. (4.4.1) through (4.4.12) implicitly assumed a mechanism of direct recombination. In the case of indirect transitions, for low-level injection, the net recombination rate remains approximately proportional to the excess concentration, so (4.4.7) still holds within reasonable approximation, although with different values of the minority carrier lifetime τ_e.

As in the cases considered the indirect recombination and generation mechanisms rely on the trapping action of impurity-induced states, it stands to reason that the probability of such transitions should increase with the concentration N_t of the trap-generating impurity and with its efficiency in trapping the carriers.

A quantitative analysis [14] confirms that the minority carrier lifetime is inversely proportional to N_t and that the trapping efficiency depends on the energy level E_t of the trap state.

The recombination time constant τ_r, respectively the generation time constant τ_g, is differently related to E_t, as shown in Fig. 4.4.4. It is evident that the maximum trapping efficiency (and so the minimum minority lifetime) is achieved

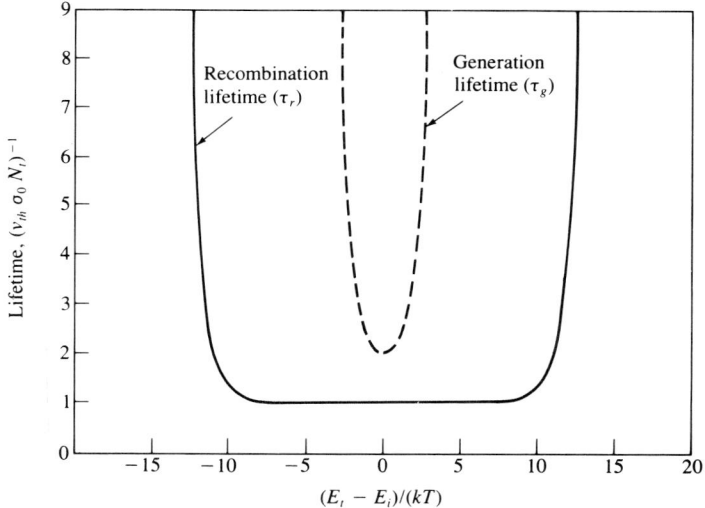

FIGURE 4.4.4
Recombination (solid line) and generation (dashed line) lifetimes as functions of the energy level of the trap state. Both lifetimes display a minimum when the trap state is at the intrinsic Fermi level [15].

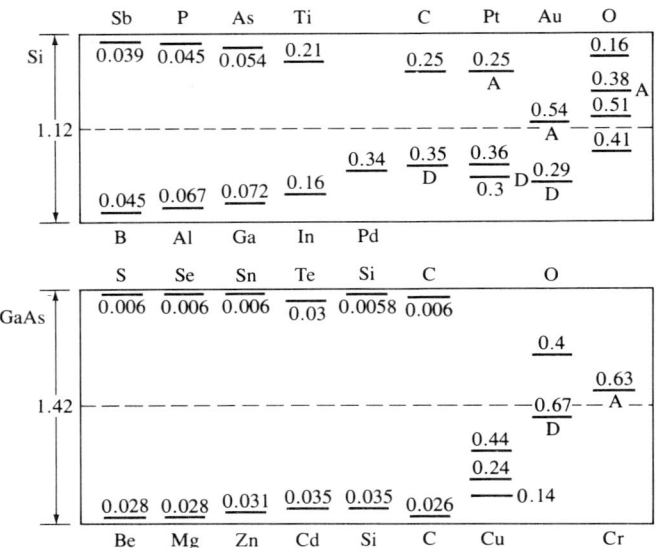

FIGURE 4.4.5
Levels of infragap states induced by different impurities in Si and GaAs. Levels are in electronvolts and are measured from the nearest edge of the gap [14].

by those impurities (such as Au and Cu) that generate a trap level very close to the center of the gap E_i.

The indirect recombination rate is therefore

$$R_{Ai} = \frac{n_{\text{P}} - n_{o\text{P}}}{\tau_r} \tag{4.4.14}$$

and the generation rate is

$$G_{Ai} = \frac{n_i}{\tau_g} \tag{4.4.15}$$

where it is recognized that recombination requires carrier injection ($n_{\text{P}} > n_{o\text{P}}$) and generation carrier depletion ($n_{\text{P}} \approx 0$, $n_{o\text{P}} \approx n_i$).

Relaxation dynamics are also still acceptably described by (4.4.11) and (4.4.12), provided the appropriate value of τ_r is substituted for τ_e. Direct measurement of τ_e and τ_r is often performed on the basis of Eq. (4.4.12) (cf. Chap. 5).

The importance of the above considerations in device design becomes apparent when it is observed that some characteristics of semiconductor devices depend on the time constants of the recombination and generation transitions (e.g., in a PN junction, a fast turnoff time requires a short τ_r and a small leakage current mandates a high τ_g; cf. Sec. 6.9). Having decided on the desired ratio τ_g/τ_r, the designer, using Fig. 4.4.4, can determine a suitable energy level for the trap state and accordingly choose the appropriate trap-inducing impurity, with

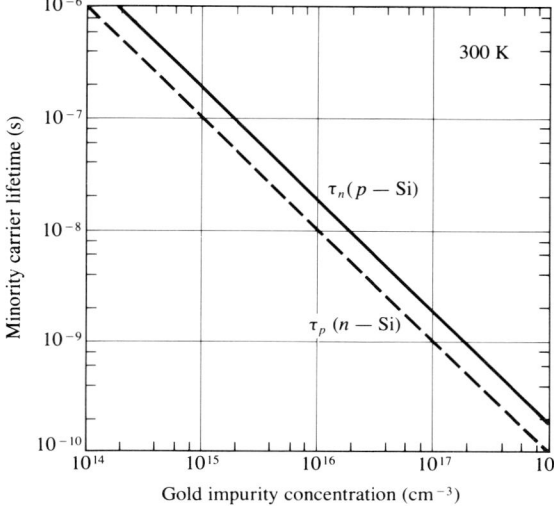

FIGURE 4.4.6
Recombination time constant for minority carriers in Si vs. Au trap impurity concentration [16].

the help of tables or diagrams of the type shown in Fig. 4.4.5. The appropriate trap impurity concentration N_t can then be computed from the above-mentioned inverse proportionality relationship, as shown in Fig. 4.4.6. Further reference to this topic will be made in Chap. 6.

Trap levels within the gap can be induced not only by impurity atoms but also by any other cause of disruption of the regular periodic lattice structure. Disruptions of this type evidently occur at the crystal boundaries, where the lattice is abruptly interrupted. Indirect generation and recombination phenomena similar to those discussed above often occur at the surface of semiconductor crystals. Unless special precautions are taken, they can result in significant surface leakage currents, sometimes large enough to disrupt device performance (cf. Sec. 6.9). A quantitative analysis of the phenomenon is beyond the scope of this text; it shall only be noted here that, just as in bulk phenomena, such analysis leads to the definition of *surface recombination and generation time constants*. These surface parameters can be computed from the characteristics of the device material and structure, but, of course, the mathematical expressions differ from those valid for the bulk case.

High-Level Injection

Under high-level injection conditions, the approximation used to obtain (4.4.7) ($\Delta n \ll n_{oP}$) fails and the net recombination rate is more closely approximated by a square law dependence on the excess population. The attendant mathematics become a little more complicated and the subject is better treated at the graduate level. When high-level injection effects are considered in the following, results will be offered without proof, usually at the qualitative level.

4.5   THE EQUATION OF CONTINUITY

Under nonequilibrium conditions, when carrier transport, generation, and recombination occur, the carrier populations within a given element of volume of the semiconductor keep varying. Consider each type of carrier, for instance, conduction electrons: as they enter this element of volume from the outside, or are generated inside it, new charge appears within the volume. Conversely, as they leave it, or are recombined, charge disappears from it. By the fundamental principle of conservation of charge, the rate at which electron charge appears minus the rate at which it disappears must equal the rate of variation of the electron charge contained in the volume. The same is true for the charge due to holes.

By simply taking inventory of the charge appearing and disappearing within a given volume, one can then obtain a relationship that must be satisfied by all carriers at all times. This is known as the *equation of continuity* and, because of its very general validity, we shall find it a powerful tool for investigating carrier behavior.

Referring to Fig. 4.5.1 and designating the electron current density as $J_e(x, y, z)$, one can easily compute the rate at which electron charge enters and/or leaves the elementary volume $dx\, dy\, dz$ through each of the faces. Through the face perpendicular to the x axis at point $[x, y, z]$, electron charge enters the volume at a rate of

$$J_{ex}(x, y, z)\, dy\, dz \qquad (4.5.1)$$

and the rate at which charge leaves the volume through the face perpendicular to the x axis at $[x + dx, y, z]$ is

$$J_{ex}(x + dx, y, z)\, dy\, dz \qquad (4.5.2)$$

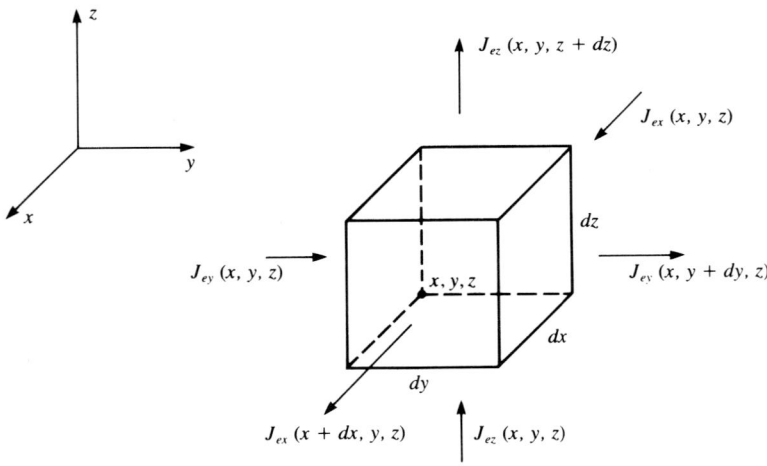

FIGURE 4.5.1
Rate of charge injection into an infinitesimal element of volume.

However, by definition of partial derivatives,

$$\frac{\partial J_x}{\partial x} = \frac{J_x(x + dx, y, z) - J_x(x, y, z)}{dx} \tag{4.5.3}$$

or

$$J_{ex}(x + dx, y, z) = J_{ex}(x, y, z) + \frac{\partial J_{ex}}{\partial x} dx \tag{4.5.4}$$

so that the *net* rate at which electron charge enters the volume through the faces perpendicular to the x axis is

$$J_{ex}(x, y, z) \, dy \, dz - J_{ex}(x + dx, y, z) \, dy \, dz = -\frac{\partial J_{ex}}{\partial x} dx \, dy \, dz \tag{4.5.5}$$

in coulombs per second. Similarly, through the pair of faces perpendicular to the y and z axes, the rate is, respectively,

$$-\frac{\partial J_{ey}}{\partial y} dx \, dy \, dz \tag{4.5.6}$$

$$-\frac{\partial J_{ez}}{\partial z} dx \, dy \, dz \tag{4.5.7}$$

The total rate of electron charge flow into the volume is then [3]

$$-\left(\frac{\partial J_{ex}}{\partial x} + \frac{\partial J_{ey}}{\partial y} + \frac{\partial J_{ez}}{\partial z}\right) dx \, dy \, dz = -\nabla \cdot \mathbf{J}_e \, dv \tag{4.5.8}$$

By definition of the generation rate, the rate of conduction electron charge generation in the volume dv is

$$-qG_e \, dv \tag{4.5.9}$$

Similarly, the rate at which conduction electron charge disappears from the volume by recombination is

$$-qR_e \, dv \tag{4.5.10}$$

The net rate of total free electron charge increment in the volume is therefore

$$\frac{\partial Q_e}{\partial t} = -\nabla \cdot \mathbf{J}_e \, dv - G_e q \, dv + R_e q \, dv \tag{4.5.11}$$

Assuming low-level injection and remembering that $Q_e = -qn \, dv$, $G_e = G_{eth} + G_{eA}$, and $R_e = R_{eth} + R_{eA}$, then using (4.4.2) and (4.4.7) and dividing by $-q \, dv$,

$$\frac{\partial n_P}{\partial t} = \frac{\nabla \cdot \mathbf{J}_{eP}}{q} + G_{eA} - R_{eA} = \frac{\nabla \cdot \mathbf{J}_{eP}}{q} + G_{eA} - \frac{n_P - n_{oP}}{\tau_e} \tag{4.5.12}$$

Only the rate of variation of the minority carrier concentration is computed, because, for low-level injection, variations of majority carrier concentration play a negligible role. For a one-dimensional system, under the same conditions,

$$\frac{\partial n_P}{\partial t} = \frac{1}{q}\frac{\partial J_{eP}}{\partial x} + G_{eA} - \frac{n_P - n_{oP}}{\tau_e} \tag{4.5.13}$$

Similarly, for hole motion,

$$\frac{\partial p_N}{\partial t} = -\frac{1}{q}\frac{\partial J_{hN}}{\partial x} + G_{hA} - \frac{p_N - p_{oN}}{\tau_h} \tag{4.5.14}$$

If the current density is given by (4.3.5), then

$$\frac{\partial n_P}{\partial t} = D_e\frac{\partial^2 n_P}{\partial x^2} + \mu_e\frac{\partial n_P}{\partial x}\mathscr{E} + \mu_e n_P\frac{\partial \mathscr{E}}{\partial x} + G_{eA} - \frac{n_P - n_{oP}}{\tau_e} \tag{4.5.15}$$

for electrons and

$$\frac{\partial p_N}{\partial t} = D_h\frac{\partial^2 p_N}{\partial x^2} - \mu_h\frac{\partial p_N}{\partial x}\mathscr{E} - \mu_h p_N\frac{\partial \mathscr{E}}{\partial x} + G_{hA} - \frac{p_N - p_{oN}}{\tau_h} \tag{4.5.16}$$

for holes. We shall often have occasion to solve the above partial differential equations under various boundary conditions. In order to do so, the relationship of $\mathscr{E}$ to the carrier concentrations must be used. This is given by *Poisson's equation* [5]:

$$\nabla^2\psi = -\frac{\rho_v}{\varepsilon}$$

which, under the assumed one-dimensional conditions, becomes

$$\frac{\partial \mathscr{E}}{\partial x} = \frac{\rho_v}{\varepsilon}$$

where ρ_v must be appropriately related to the carrier density for each case considered.

4.6 SUMMARY

Carriers in a crystal at temperature $T \neq 0$ K are animated by random motion with *zero average velocity* and *rms speed* $v_{th} = (3kT/m)^{1/2}$, colliding with each other and with the lattice atoms. The *average time between collisions* is τ_c.

Under the action of an electric field the carriers *drift*, with *drift velocity* proportional to the electric field $\mathbf{v}_d \pm \mu\mathscr{E}$, where $\mu = q\tau_c/m$ is the *mobility*, varying with semiconductor material, doping, and temperature.

This motion results in a *conduction current density* mediated by both electrons and holes $\mathbf{J} = q\,(n\mu_e + p\mu_h)\,\mathscr{E}$, so that the *conductivity* is $\sigma = q(n\mu_e + p\mu_h)$.

When the carrier concentration is not uniform, *diffusion* occurs: the carriers move from high to low concentration regions. The *diffusion flow-rate density* is given by Eq. (4.3.1), with *diffusivity* D related to the mobility by *Einstein's relation* (4.3.2). This adds a *diffusion component* to the current. The total current density is therefore as given by Eq. (4.3.4).

Under the action of an external source of energy, *high- or low-level injection* may occur, characterized by an *additional generation rate* G_A. This increases the *recombination rate* by an amount R_A, which, for low-level minority carrier injection of electrons, at steady state is given by Eq. (4.4.7), where τ_e is the *minority carrier lifetime*.

Semiconductors are classified as *direct* and *indirect* semiconductors depending on whether direct band to band recombination is possible without contravening the law of conservation of momentum. The two types are easily distinguishable with the help of energy–momentum band diagrams.

Carrier motion obeys the *equation of continuity*, which, for electrons under conduction and diffusion current in a one-dimensional system, is given by (4.5.15).

APPENDIX 4A
DIFFUSION

To determine quantitatively the net flow-rate density, compute the net number of particles that flow through a unit surface during the time τ_c, defined in Sec. 4.2 as the average time interval over which the particles move freely (between two consecutive collisions).

Referring to Fig. 4A.1, consider a unit surface S perpendicular to the x direction. Evidently, only the x component of velocity can cause a flow through this surface. As discussed in Sec. 4.1, from a statistical point of view, it is legitimate to assume that all particles are animated by the same thermal velocity, having components of equal magnitudes [see (4.1.3)]:

$$|v_{thx}| = |v_{thy}| = |v_{thz}| = \sqrt{\frac{kT}{m}} \qquad (4A.1)$$

each component being positive for half of the particles and negative for the other half (as the average must be zero).

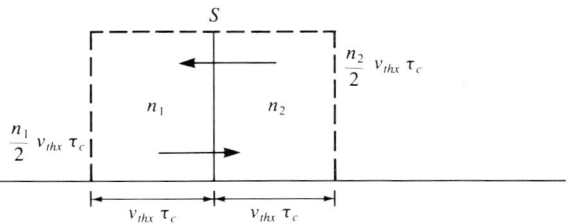

FIGURE 4A.1
Flow rates through a surface S between two regions characterized by different concentrations of mobile particles n_1 and n_2.

Of the particles to the left of S, the only ones that can contribute to the flow during time τ_c are those within a distance $v_{thx}\,\tau_c$ from S (in the x direction), i.e., those that are within the volume ($v_{thx}\,\tau_c \times 1$) and have a positive component of velocity v_{thx} (i.e., half of the particles within the volume, as previously remarked). If the particle density to the left of S is n_1, then this flow in the x direction is $(n_1/2)v_{thx}\,\tau_c$. Similarly, during the same interval of time, the particles to the right contribute a flow $-(n_2/2)v_{thx}\,\tau_c$ (this flow is negative, because it is in the negative x direction). The net flow rate in the x direction is therefore

$$F_x = \frac{n_1 - n_2}{2}\, v_{thx}\,\tau_c\, \frac{1}{\tau_c} \tag{4A.2}$$

where n_1 and n_2 are the particle concentrations to the left, respectively the right, of S. If the distance $v_{thx}\,\tau_c$ is very small and the concentration n is a continuous function of x, then, from McLaurin's theorem,

$$n_1 = n(-v_{thx}\,\tau_c) \approx n(0) - \frac{\partial n}{\partial x}\, v_{thx}\,\tau_c \tag{4A.3}$$

$$n_2 = n(v_{thx}\,\tau_c) \approx n(0) + \frac{\partial n}{\partial x}\, v_{thx}\,\tau_c \tag{4A.4}$$

Introducing (4A.3) and (4A.4) in (4A.2),

$$F_x = -v_{thx}^2\,\tau_c\, \frac{\partial n}{\partial x} = -D\, \frac{\partial n}{\partial x} \tag{4A.5}$$

where $D = v_{thx}^2\,\tau_c$ cm^2/s is a constant to which is given the name of *diffusivity*. Similarly, the flow rate through unit surfaces in the y and z directions is

$$F_y = -v_{thy}^2\,\tau_c\, \frac{\partial n}{\partial y} = -D\, \frac{\partial n}{\partial y} \tag{4A.6}$$

$$F_z = -v_{thz}^2\,\tau_c\, \frac{\partial n}{\partial z} = -D\, \frac{\partial n}{\partial z} \tag{4A.7}$$

where, from (4A.1), the *diffusivity*

$$D = \frac{kT}{m}\,\tau_c \tag{4A.8}$$

is the same in all directions.

A *flow-rate density* can then be defined as a vectorial quantity, characterized by the above components, or, by a definition of the gradient [5],

$$\mathbf{F} = -D\nabla n \qquad \text{particles/(cm}^2 \cdot \text{s)} \tag{4A.9}$$

The diffusion flow rate through any surface S is therefore

$$\phi = \int_S - D\nabla n \cdot d\mathbf{S} \qquad \text{particles/s} \tag{4A.10}$$

In a one-dimensional system (n depends only on one coordinate, for example x), a scalar definition of the flow-rate density is sufficient:

$$F = -D \frac{dn}{dx} \tag{4A.11}$$

Notice that, for the one-dimensional system, the total rather than the partial derivative is in order.

For electrons, respectively for holes, using (4.2.3) and (4.2.4) in (4A.8), the diffusivities can be expressed as

$$D_e = \frac{kT}{q} \mu_e \tag{4A.12}$$

$$D_h = \frac{kT}{q} \mu_h \tag{4A.13}$$

known as the *Einstein relationships*.

PROBLEMS

4.1. Compute the rms thermal speeds and average velocities of electrons and holes in Si in equilibrium at 300, 600, and 900 K in the absence of electric or magnetic fields.

4.2. A sample of extrinsic Si at 300 K has resistivity $\rho = 5 \, \Omega \cdot cm$. Assuming that As is the only dopant, determine the impurity concentration.

4.3. A sample of Si is doped with 10^{15} B atoms/cm³. Compute the concentration of As dopant required to turn the sample into an N-type semiconductor with resistivity $\rho = 5 \, \Omega \cdot cm$.

4.4. Compute the conductivity of an intrinsic Si sample at:
(a) 200 K
(b) 300 K
(c) 400 K

4.5. Compute the resistivity of Si samples with N-type dopings of $N_D = 10^{16}$, $N_D = 5 \times 10^{17}$, and $N_D = 2 \times 10^{19}$ at 200, 300, and 400 K. (Assume that total doping is equal to net doping.)

4.6. A Si sample, doped with 10^{14} As atoms/cm³, has a resistivity of 20.8 $\Omega \cdot cm$. What is the temperature?

4.7. Compute the diffusivity of electrons under the conditions of Prob. 4.6.

4.8. In the region from $x = 0$ to $x = 0.1$ cm of a P-type semiconductor with uniform $N_A = 10^{15}$ cm^{-3} at 300 K there exists a uniform electron diffusion current density of $J_{de} = 1$ mA/cm². If the conventional current density is in the $-x$ direction, and knowing that at $x = 0$ the electron concentration is 1.8949×10^{13} electrons/cm³, compute their concentration at $x = 0.1$ cm.

4.9. Because of some injection process, the minority carrier concentration in a P-type Si semiconductor at 300 K is

$$n_P = 5 \times 10^{13} \, e^{-x/10^{-3}} + 2.25 \times 10^6$$

where the semiconductor surface is at $x = 0$. If the semiconductor is doped with 10^{14} B atoms/cm^3, write an expression for the electron diffusion current density J_e in the semiconductor. In what direction is the electric current?

4.10. Compute the longest wavelength of light that can result in extra carrier generation in intrinsic Si.

4.11. The surface of a Si semiconductor doped with 10^{14} B atoms/cm^3 has been illuminated for a long time when, at $t = 0$, the illumination is stopped. The temperature is 300 K. At $t = 0$ the minority carrier concentration at the surface is 5×10^6 cm^{-3}. After 1 μs the minority carrier concentration at the surface has decreased to 3×10^6 cm^{-3}. Compute the generation rate at the surface Ga and the minority carrier lifetime τ_e.

4.12. Minority carriers have been injected for a long time into a Si sample doped with B and As atoms in the ratio of 3 atoms of B per atom of As. The rate of injection was 10^{14} carriers/(cm$^3 \cdot$ s). At time $t = 0$ the injection is stopped. Measurements of the minority carrier concentration at the surface n_P show it to decrease exponentially with time. At $t = 0$, $n_P(0) = 10^8$ cm^{-3}; after a long time $n_P(\infty) = 5 \times 10^5$ cm^{-3}.
(a) What is the electron diffusivity of the sample?
(b) What is the minority carrier lifetime?
(c) Write an expression for $n_P(t)$.

4.13. What is the lowest ratio between generation and recombination lifetime at 300 K due to:
(a) Acceptor O impurities in Si?
(b) Acceptor Au impurities in Si?

4.14. A minority carrier lifetime of 0.1 μs is desired in N-type Si at 300 K. To this end, recombination traps are to be induced by addition of Au impurities.
(a) What is the desired impurity concentration?
(b) What is the corresponding generation time constant?

REFERENCES

1. Dekker, Adrianus J.: *Solid State Physics*, 3d ed., Prentice Hall, Inc., Englewood Cliffs, N. J., 1959.
2. Kittel, Charles: *Introduction to Solid State Physics*, 2d ed., John Wiley & Sons, Inc., New York, 1959.
3. Adler, R. B., A. C. Smith, and R. L. Longini: *Introduction to Semiconductor Physics*, John Wiley & Sons, Inc., New York, 1964.
4. Haynes, J. R., and W. Shockley: "The Mobility and Life of Injected Holes and Electrons in Germanium," *Phys. Rev.*, vol. 81, p. 835, 1951.
5. Hayt, William H., Jr.: *Engineering Electromagnetics*, 4th ed., McGraw-Hill Book Company, Inc., New York, 1981.
6. Conwell, E. M.: *Proc. IRE*, vol. 46, pp. 1281–1300, June 1958.
7. Moll, J. L.: *Physics of Semiconductors*, McGraw-Hill Book Company, Inc., New York, 1964.
8. Smith, R. A.: *Semiconductors*, Cambridge, London, 1978.
9. Beadle, W. F., J. C. C. Tsai, and R. D. Plummer: *Quick Reference Manual for Semiconductor Engineers*, John Wiley & Sons, Inc., New York, 1985.
10. Yariv, Amnon: *Quantum Electronics*, John Wiley & Sons, Inc., New York, 1975.
11. Yariv, Amnon: *Introduction to Optical Electronics*, Holt, Rinehart and Winston, Inc., New York, 1971.
12. Zambuto, Mauro: "Lasers," in *Handbook of Electrical and Computer Engineering* (ed. Sheldon S. L. Chang), John Wiley & Sons, Inc., New York, 1983.

13. Pantell, Richard H., and H. E. Puthoff: *Fundamentals of Quantum Electronics*, John Wiley & Sons, Inc., New York, 1969.
14. Sze, S. M.: *Physics of Semiconductor Devices*, 2d ed., John Wiley & Sons, Inc., New York, 1981.
15. Sze, S. M.: *Semiconductor Devices, Physics and Technology*, John Wiley & Sons, Inc., New York, 1985.
16. Wolf, H. F.: *Semiconductors*, John Wiley & Sons, Inc., New York, 1971.

ADDITIONAL READING

Halliday, David, and R. Resnik: *Physics for Students of Science and Engineering*, John Wiley & Sons, Inc., New York, 1960.
Holton, Gerald, and H. D. Roller: *Foundations of Modern Physical Science*, Addison-Wesley Publishing Company, Inc., Reading, Mass., 1958.
Plonsey, R., and R. E. Collins: *Principles and Applications of Electromagnetic Fields*, McGraw-Hill Book Company, Inc., New York, 1961.
Ramey, Robert E. L.: *Physical Electronics*, Wadsworth Publishing Company, Belmont, Calif., 1961.
Schroder, D. K.: "The Concept of Generation and Recombination Lifetimes in Semiconductors," *IEEE Trans. Elect. Devices*, vol. ED-29, p. 1336, 1982.
Sears, Francis, M. W. Zemansky, and H. D. Young: *College Physics*, 4th ed., Addison-Wesley Publishing Company, Reading, Mass., 1974.

CHAPTER

5

FABRICATION TECHNOLOGY

The remarkable growth and expanded scope of data processing, communications, controls, etc., that have characterized the last decades have been made possible by the development of the modern technologies of semiconductor device fabrication. The impact of these technologies, not only on the availability of new products and on our lifestyles but also on the very nature of engineering, of scientific research, and even on the pattern of our thought processes, promises to be as far reaching as that of the industrial revolution. This technology is still developing at bewildering speed, tapping every field of scientific knowledge.

Essentially, microelectronics technology consists of a few basic techniques (the fundamental *fabrication steps*). The industrial production of each device usually takes the form of a *sequence* of these basic fabrication steps.

In industrial practice, a wide range of variations is introduced in the implementation of these techniques; indeed, industrial success or failure of a whole process has often depended on apparently trivial details (which are sometimes guarded by the most jealous industrial secrecy). However, this concerns the specialists.

It is not the purpose of this book to train the reader as a specialist in microelectronics, or even to describe with any thoroughness each detail of the processes. Readers seeking such detailed expertise are referred to more specialized and advanced texts and, most of all, to the periodic literature.

However, proficiency in designing modern electronic systems requires some familiarity and understanding of device fabrication techniques. In the following chapters, whenever a new device is introduced, some information about its mode of fabrication will be provided. This will prove useful in understanding the geometrical and physical properties of the device and in analyzing its operation and determining its limitations. Adequately to profit by such a description, the reader should first gain some familiarity with the basic fabrication steps. This chapter is meant as an introduction to these fundamental processes, briefly outlining the principles involved and the attendant nomenclature, indicating some approaches to the control of product characteristics with consequent advantages and disadvantages.

5.1 WAFER PREPARATION

The substrate material on which devices are fabricated is usually a monocrystalline wafer of appropriately doped Si, GaAs, or Ge.

Wafer preparation has become a very specialized technology and integrated circuit manufacturers often rely on specialized silicon foundries to provide them with wafers made to their specifications.

Wafer manufacturing includes the operations of crystal growing, wafer slicing, shaping, identification, and finishing.

Crystal Growing

The operation of growing the crystals from which the semiconductor wafers will ultimately be obtained must meet very strict industrial standards. The size and shape of these very large single crystals must permit efficient fabrication of wafers of the desired dimensions and characteristics, their crystalline lattice must be free from flaws, and the orientation of their crystallographic axes must be accurately determined and maintained.

The purity of the crystalline material must be extremely high and the desired doping must be achieved within narrow limits of accuracy. In industrial practice the doping impurity concentrations (net doping) are specified and controlled in terms of the extrinsic semiconductor resistivity (cf. Sec. 4.2).

Before the crystal is grown, the host semiconductor material, in polycrystalline form, must be produced and refined by chemical means. The degree of purity required depends on the type of device to be fabricated. Particularly low impurity levels are demanded for high-voltage (low-doping) devices. In the case of GaAs, this chemical process entails a careful synthesis operation. From this material a single crystal with the proper orientation and doping must be grown. In the most popular crystal-growing processes the refined material is first melted and then brought back to solid state temperatures in the presence of a properly oriented *seed crystal*. During this process, the appropriate amount of dopant is added to the high-purity semiconductor material. Several techniques are in common use:

1. In the *Czochralski technique*, shown in schematic form in Fig. 5.1.1, the refined material is melted in a slowly rotating silicon crucible, enclosed in a conducting graphite *susceptor* heated either by rf energy from an induction coil or by resistive heating elements. An appropriately oriented seed crystal is immersed at the surface of the melt and slowly pulled out. The melted material sticks to the seed and, solidifying at the lower temperature as the seed is withdrawn, forms a large, single-crystal *boule* with the same crystallographic parameters as the seed. The *puller assembly* operates in a clean, inert gas environment. During the pulling operation the seed is rotated by the seed holder in the direction opposite to the crucible rotation, obtaining a cylindrical boule of the proper shape and size.

   Precisely measured quantities of dopant are added to the melt to obtain an extrinsic semiconductor of the required resistivity. The computation of the quantity of dopant for each melt is complicated by the fact that the dopant

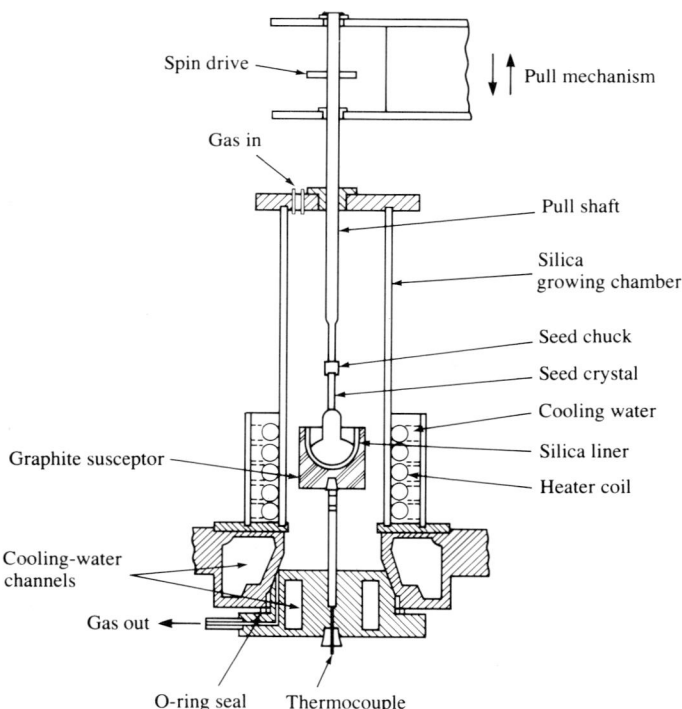

FIGURE 5.1.1
Czochralski crystal pulling apparatus [1].

concentration in the solid phase equals that of the liquid phase times an *equilibrium segregation coefficient* depending on both dopant and host materials. Following are some of the most common dopants and their equilibrium segregation coefficients (in parentheses):

Host	N-type	P-type
Si	P (0.35); As (0.3)	B (0.8)
GaAs	Se (0.1); Si (2)	Zn (0.42)

The fact that most segregation coefficients are <1 indicates that the concentration of the impurity in the solid is lower than in the liquid phase (for instance, to obtain a concentration of 10^{15} As atoms/cm^3 in a Si crystal, the melt must contain $10^{15}/0.3 = 3.3 \times 10^{15}$ As atoms/cm^3). It follows that, as the crystal is pulled, the percentage of atoms leaving the melt is higher for the host material than for the impurity, so that, during the process of solidification, the impurity concentration in the bath tends to increase. As the crystal develops, *the doping in the most recently solidified layers increases.* This effect can be minimized by proper choice of the speeds of rotation and pulling.

In the Czochralski technique, spurious impurities tend to enter the crystal due to some crucible fusion. This can become a serious drawback when low-doping (high-resistivity) semiconductors are required, as in high-voltage device fabrication.

2. Less contaminated boules are obtained by the *float zone process*, in which the refined polycrystalline material is originally made into a rod, rather than melted in a crucible. A small region of the rod is then melted by localized heating through a small rf coil, while the rod is slowly rotated, as shown in Fig. 5.1.2. Such *narrow zone melting* starts at the bottom of the rod, where a seed crystal is located to determine the crystallographic orientation. The rf coil is then slowly raised, so that the molten zone moves toward the top of the rod, while the lower layers, no longer under rf radiation, solidify into a single crystal. There is no crucible and so no contamination source. Good doping uniformity is achieved in the upper crystal layers, especially for thin melt zones.

Very low levels of contamination can be obtained by the float zone process, especially by repeated recrystallizations through several passes, so the process is often used for high-resistivity devices.

3. In the *Bridgman technique*, mostly used for GaAs materials, two adjacent regions of the melting oven are kept at two different temperatures (Fig. 5.1.3). The chemically purified material is melted at about 1250°C in the high-temperature zone and slowly passes into the lower temperature zone, where it gradually solidifies from one end up at about 620°C. This temperature is chosen so that a small amount of As, always kept in this zone, will provide an As atmosphere in which crystallization of the GaAs occurs. This process is slowly being replaced by the Czochralski technique.

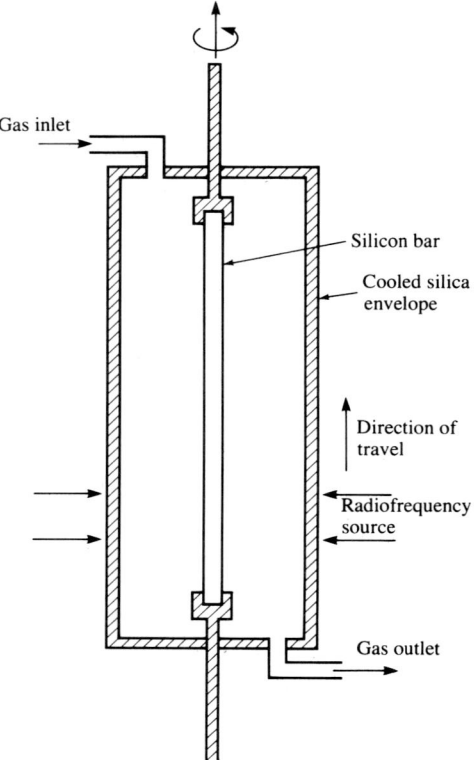

FIGURE 5.1.2
Float zone apparatus schematic. The narrow
zone of the polysilicon bar under rf radiation
melts and recrystallizes into a single crystal
as the rf source is raised [2].

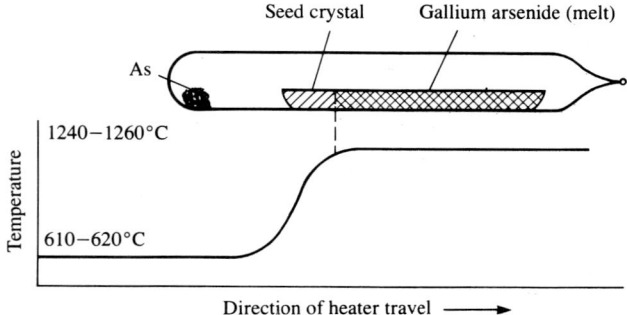

FIGURE 5.1.3
Bridgman apparatus schematic. The graph shows the temperature "step" existing in the oven and
indicates the location of crystallization. The source of the As atmosphere is shown at the extreme left
in the 620°C zone [2].

Wafer Slicing and Shaping

Once a boule with appropriate chemical and crystalline characteristics is obtained, it is sliced into wafers, the physical characteristics of which must be suitable for the subsequent processing, so that wafer dimensions and their strict tolerances are dictated by the requirements of the automatic equipment used in device processing.

At the present writing, wafer dimensions vary from about 50 to 150 mm in diameter with thicknesses of about 0.5 to 0.7 mm, depending on the diameter.

The boule is accordingly ground to obtain a cylinder of the desired diameter and then sliced to the appropriate thickness, usually by high-speed diamond saws.

Before the slicing operation, reference *flats* are ground on the cylindrical surface for reference and identification, as shown in Fig. 5.1.4. A large primary flat is accurately oriented with respect to the crystallographic axes to permit proper automatic orientation of the crystal during the subsequent fabrication processes. Crystal orientation at the surface is usually maintained within 1° tolerance.

Small secondary flats identify other crystal characteristics. The angular position α of the secondary flats relative to the primary flat indicates, by a standard code, crystal type (P or N), conductivity (indicating the net impurity concentration), and surface orientation in terms of the surface Miller indices

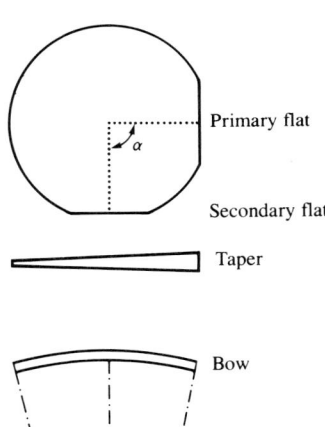

α	Semiconductor type	Orientation
45°	N	111
90°	P	100
180°	N	100
0°	P	111

FIGURE 5.1.4
Wafer with identification flats. The angle α indicates several characteristics of the wafer material.

($\langle 100 \rangle$, $\langle 111 \rangle$, etc.). Strict tolerance limits are set not only on the absolute value of the wafer thickness but also on its uniformity (taper) and flatness (maximum bow).

Finally, the wafer surfaces are lapped and mirror polished to permit the necessary subsequent high-precision lithographic processes.

5.2 EPITAXY

Epitaxy is used to grow precisely calibrated thin single-crystal semiconductor layers, usually with doping different from that of the substrate. The technique, commonly used in bipolar device fabrication, is now gaining popularity in MOSFETs. It is essentially a single-crystal growth technique, differing from the previously described ones in that the function of the seed crystal is taken over by the substrate and crystal formation occurs without first reaching the melting point, usually by incorporation of adsorbed atoms in the crystal lattice.

Typical epitaxial techniques are *vapor phase epitaxy* (VPE), *liquid phase epitaxy* (LPE), and *molecular beam epitaxy* (MBE), depending on the conditions under which atoms are made available.

Vapor Phase Epitaxy

The wafers to be epitaxially coated, supported by a graphite susceptor, are introduced into an induction oven, where the susceptor is heated by rf energy to a temperature below the substrate melting point (about 1200°C for Si). Susceptor and oven geometries vary widely. Figure 5.2.1 indicates some common types.

The oven atmosphere is a gas mixture, pumped as a continuous flow by a system of pipes which permit a precise control of its composition, rate of flow, temperature, etc. Epitaxy on Si substrates is based on the *reversible reaction*

$$2H_2 + SiCl_4 \rightleftharpoons 4HCl + Si \tag{5.2.1}$$

where, at the oven temperature, all components are gaseous except the silicon. If a mixture of hydrogen and silicon tetrachloride is pumped into the oven, the reaction proceeds from left to right and the silicon produced is deposited on the substrate as an extension of the substrate crystal lattice: *epitaxial growth* occurs. The hydrochloric acid produced is washed away in the flow. If instead a hydrochloric acid atmosphere is pumped into the system, then by mass law the reaction proceeds from right to left, the HCl attacks the silicon surface, producing gaseous silicon tetrachloride, which is blown away by the flow, and *etching* of the substrate surface occurs. This process is often used to perform a final cleaning of the wafer surface.

If the atmosphere is a mixture of H_2 and $SiCl_4$, then the rate of epitaxial growth at a given temperature depends on the composition of the mixture, as shown in Fig. 5.2.2 where the rate of growth at 1270°C is plotted as a function of

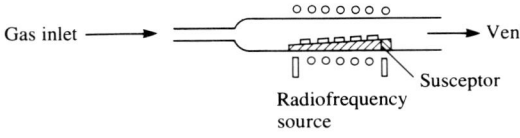

(a)

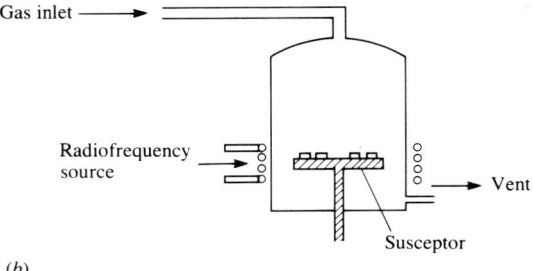

(b)

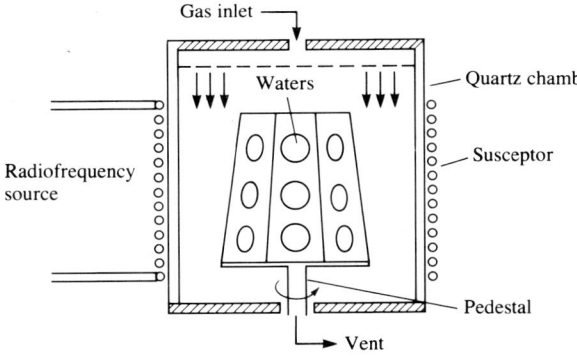

(c)

FIGURE 5.2.1
Some commonly used configurations for vapor phase epitaxy [2].

the ratio y of the $SiCl_4$ concentration to the total molecular concentration of the gas atmosphere (mole fraction).

At low values of y the rate of growth is seen to increase linearly with y; then the plot begins to deviate from linearity, reaching a maximum around $y \approx 0.11$. For $y > 0.28$ the rate of growth actually becomes negative, resulting in etching instead of epitaxial growth. This behavior is due to the reaction

$$SiCl_4 + Si \rightleftharpoons 2SiCl_2 \qquad (5.2.2)$$

competing with (5.2.1) and becoming prevalent at high concentrations of $SiCl_4$. The $SiCl_2$ is washed away by the gas flow.

Normally, epitaxy proceeds at growth rates of about 1 μm/min and correspondingly low values of y. Small quantities of dopant impurites are added to the

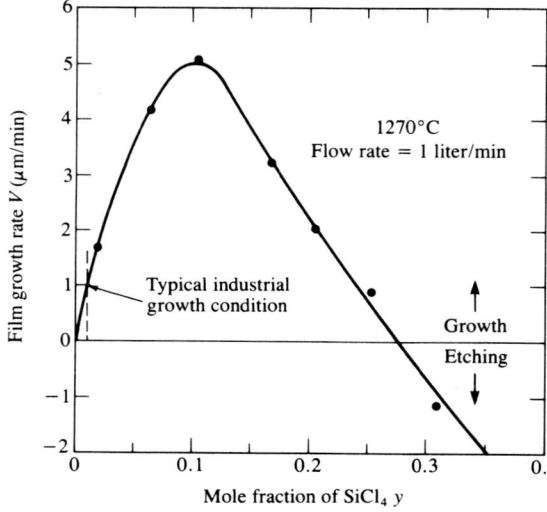

FIGURE 5.2.2
Rate of growth of epitaxial layers vs. mole fraction of $SiCl_4$ in gas atmosphere at 1270°C. Notice negative rate of growth (indicating etching) for $y > 0.28$ [3].

mixture to obtain the desired doping of the epitaxially grown film. Most frequently arsine (AsH_3) and phosphine (PH_3) are used for N-type and diborane (B_2H_6) for P-type semiconductor layers.

Example 5.2.1. A layer 10 μm thick is to be epitaxially grown on a substrate wafer. For how many seconds must the wafer remain in an oven at 1270°C if the atmosphere is a mixture of gases with the following concentrations in atoms per cubic centimeter:

$$H_2 \Rightarrow 2.94 \times 10^{19}; \qquad SiCl_4 \Rightarrow 6 \times 10^{17}$$

Solution. The mole fraction of SiC_4 in the atmosphere is

$$y = \frac{6 \times 10^{17}}{2.94 \times 10^{19} + 6 \times 10^{17}} = 0.02$$

From Fig. 5.2.2 the epitaxial growth rate is 1.67 μm/min, so the duration of the epitaxial growth should be

$$\frac{10}{1.67} = 6 \text{ min} \equiv 360 \text{ s}$$

It is assumed that the wafer is properly preheated to 1270°C.

Liquid Phase Epitaxy

In this technique the material that will form the epitaxial layer is dissolved in a solvent. The temperature is then slowly dropped until the solution is slightly supersaturated. When the substrate is put in contact with this solution, the material precipitates forming the epitaxial layer.

Growth rate is slow, but the process is well suited to GaAs epitaxy and it is

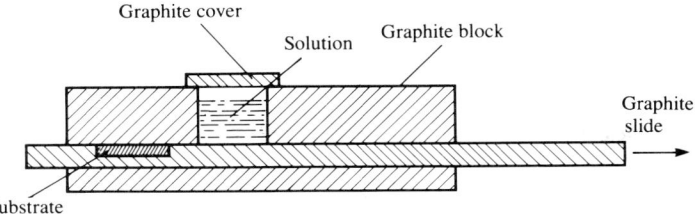

FIGURE 5.2.3
Graphite block and slide for liquid phase epitaxy. When the slide is moved to the right, the substrate material becomes the bottom of the well and so comes in contact with the solution. At the proper temperature, crystal growth is obtained by precipitation from the supersaturated solution [2].

possible to produce well-controlled growth (several layers of different composition) with one oven pass.

The basic system geometry is shown in Fig. 5.2.3. The substrates are housed in a graphite slider, precisely fitting in a groove machined in a graphite block (or boat). The top of the boat has one or more wells in which the desired solution will be poured. The wells have no bottom of their own, but, when the boat is assembled and the slider inserted in the groove, the slider acts as the bottom of the wells, preventing the solution from escaping.

The assembly, with the appropriate solutions, is inserted in the oven and the slider is positioned so that the wafers do not coincide with the wells, the bottoms of which are therefore made of the slider graphite material. The oven temperature is now changed, until, when the proper temperature for precipitation is reached, the slider is repositioned, bringing the substrate at the bottom of the appropriate well and so in contact with the desired solution. Epitaxial growth now occurs under precise small temperature variations and for a precisely controlled time to obtain the desired layer thickness. At this moment the slider is again repositioned to move the wafers away from the solutions. The cycle can be repeated, moving the wafer from one well to another to obtain multilayered epitaxial films, if so desired. Several wafers can be processed simultaneously.

Molecular Beam Epitaxy

The different constituents of the layer to be grown (including the impurities) are evaporated, each in its individual effusion oven. The ovens are arranged inside a high vacuum chamber housing the substrate on its rotating support, as shown in Fig. 5.2.4. Each oven is calibrated to obtain a specific evaporation rate of the component in it, yielding a gas that leaves the effusion oven in the form of a *thermal molecular beam* of the desired properties for optimal deposition of that constituent.

This means that each beam, when it finally reaches the substrate, is endowed with statistical molecular characteristics, such as temperature, mean free path, density, viscosity, diffusivity, etc., determining the conditions under which the corresponding constituent is deposited on the substrate. By controlling indi-

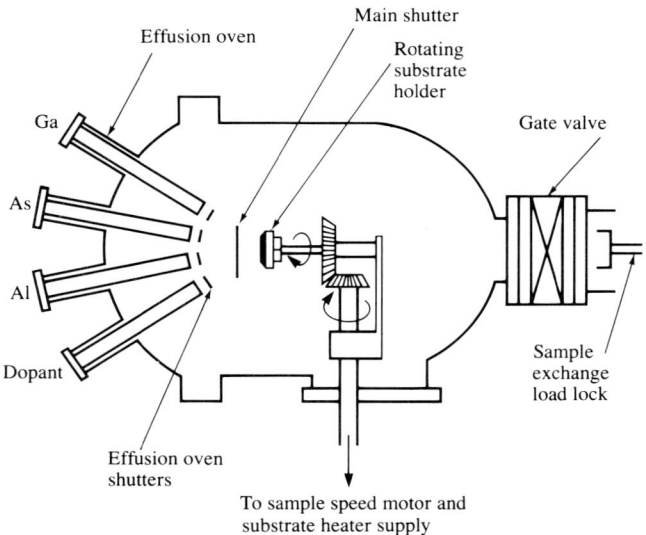

FIGURE 5.2.4
Schematic arrangement for molecular beam epitaxy, permitting individual adjustment of the conditions of deposition from each effusion oven [4].

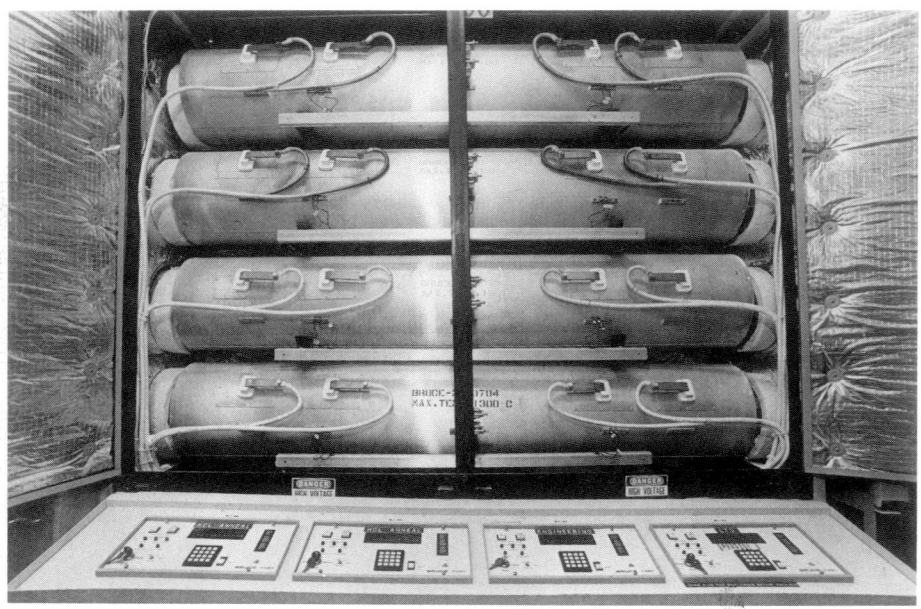

FIGURE 5.2.5
Processing ovens. Four furnaces with separate controls for sequential processing. (*Courtesy NJIT Microelectronics Laboratories.*)

FIGURE 5.2.6
Gas cabinet for the control of the oven atmosphere. (*Courtesy NJIT Microelectronics Laboratories.*)

vidual oven temperatures and effusion times it is possible to grow epitaxially single crystals while accurately calibrating the properties of each of their many component layers, down to a resolution of single-molecule layers. This technique can be used, for instance, to obtain extremely accurate doping profiles.[1]

The technique is particularly well suited for GaAs epitaxy, but is also used on silicon devices, notwithstanding the extremely slow growth rates (as low as 1 nm/min).

5.3 OXIDATION AND CHEMICAL VAPOR DEPOSITION (CVD)

Oxidation

Layers of SiO_2 (and some other materials) are used as insulators, dielectrics, protective films, and, at several fabrication stages, as masks, passivators, inhibitors,

---

[1] The enhanced sophistication of modern devices mandates greater accuracy in controlling fabrication parameters. In response to these demands and other circumstances, there is evidence of a trend in the industry toward high vacuum fabrication techniques, of which molecular beam epitaxy is an example. Some sources predict that such technologies will be a mainstay of the space laboratories of the future.

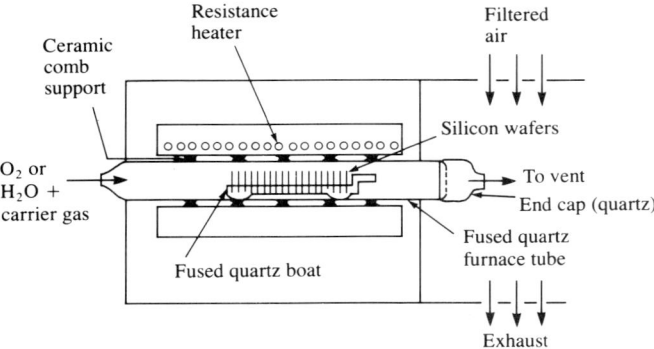

FIGURE 5.3.1
Schematic of one of the many possible configurations of an oxidation apparatus. Gas atmosphere composition and flow rate in the furnace tube are accurately controlled [5].

etc. A wide variety of technologies can be employed. In silicon-based devices, SiO_2 layers can efficiently be obtained by direct oxidation of the substrate. In GaAs the oxides produced by such means are often poorly suited for the intended purpose and other technologies are employed.

Silicon substrate oxidation is performed in oxidation ovens very similar to those used in epitaxy (cf. Fig. 5.3.1). Several wafers, held in a quartz *boat* are placed in a quartz tube, which is slowly inserted into a furnace, heated by resistive elements to temperatures usually between 900 and 1200°C. A controlled flow of an oxidant atmosphere is maintained through the tube. Flow rate of the oxidant atmosphere, time of exposure, entry, and withdrawal rate into and from the furnace, and gas composition are all automatically and carefully controlled. Two types of process are commonly used:

1. *Dry oxidation* occurs in an oxygen atmosphere according to the reaction

$$Si + O_2 = SiO_2 \tag{5.3.1}$$

The rate of growth is slow, but the oxide layers are very uniform, relatively free from defects, and electrically very reliable. The process is used for thin dielectric layers, such as required in MOSFET fabrication and often referred to as *gate oxides* (cf. Chap. 9).

2. *Wet oxidation* uses a steam atmosphere. The reaction

$$Si + 2H_2O = SiO_2 + 2H_2 \tag{5.3.2}$$

yields much higher rates of growth with layers well suited for insulation, protection, and masking applications, but not as free of microscopic faults as the dry oxidation products. Such layers are often referred to as *field oxides*.

As the wafer surface is oxidized, the next Si molecules to react with the oxidizing agent are those in the next deeper lattice layer of the crystal, so the silicon interface moves deeper and deeper into the crystal. This mechanism has two consequences:

1. Taking the density and molecular weight of Si and SiO_2 into account, it is easy to see [5, p. 343] that, when a SiO_2 layer of thickness x is formed, the Si interface moves into the crystal by only $0.4x$, so the insulating layer grows partly into and partly above the original Si crystal.
2. The molecules of the oxidizing agent, in order to react with more Si, must first reach the Si interface, crossing the already generated SiO_2 layer. This penetration occurs by solid state *diffusion*, so the rate at which new SiO_2 is produced (the *rate of growth*) is influenced both by the *diffusion rate* (the time required to diffuse through the oxide, depending on the diffusion coefficient D) and by the speed at which the atoms react (the *reaction rate* coefficient r_c).

At first, when the oxide layer is thin, the reaction rate plays the most important role and the growth is *reaction controlled*. Later on, when the oxide layer to cross is thicker, it is the diffusion process that mainly determines the rate of growth and the process is *diffusion controlled*.

A simple mathematical analysis (cf. App. 5A) shows that, in the reaction controlled domain, growth is essentially linear in time. In the diffusion controlled domain, growth follows a parabolic time variation.

The thickness of the oxide layer after oxidation of duration t is expressed by

$$x = \frac{A}{2}\left[-1 + \sqrt{1 + \frac{4B(t + \tau)}{A^2}} \right] \tag{5.3.3}$$

where A, B, and τ are constants related to the characteristics of the oxidation process and are defined in App. 5A.

For reaction controlled oxidation (thin oxide layers: $x \ll A/2$, short duration oxidation), $4B(t + \tau)/A^2 \ll 1$ and (5.3.3) can be approximated by

$$x \approx \frac{B}{A}(t + \tau) \tag{5.3.4}$$

This is the linear law of growth and it is seen to depend on the *linear rate constant* B/A.

For diffusion controlled oxidation (thick oxide layers: $x \gg A/2$, long duration of oxidation), $4B(t + \tau) \gg 1$ and (5.3.3) becomes

$$x \approx \sqrt{B(t + \tau)} \tag{5.3.5}$$

showing that growth is parabolic with *parabolic rate constant* B.

At atmospheric pressure, the linear rate constant A/B is plotted vs. temperature in Fig. 5.3.2 and the parabolic rate constant in Fig. 5.3.3.

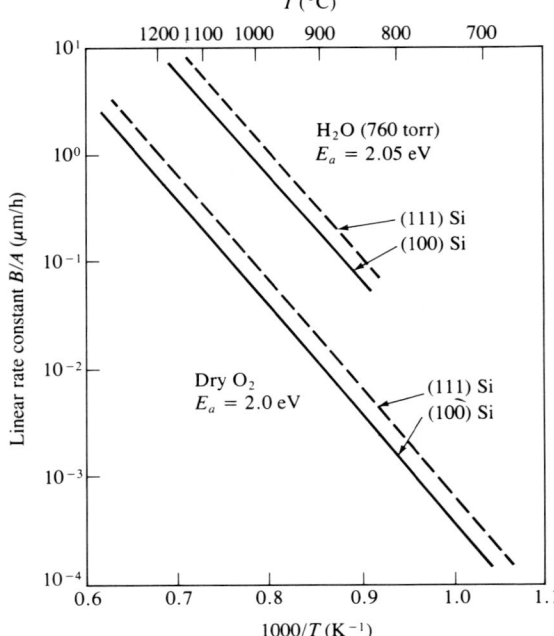

FIGURE 5.3.2
Temperature dependence of the linear rate constant in oxidation (notice the dependence on the crystallographic surface orientation) [6].

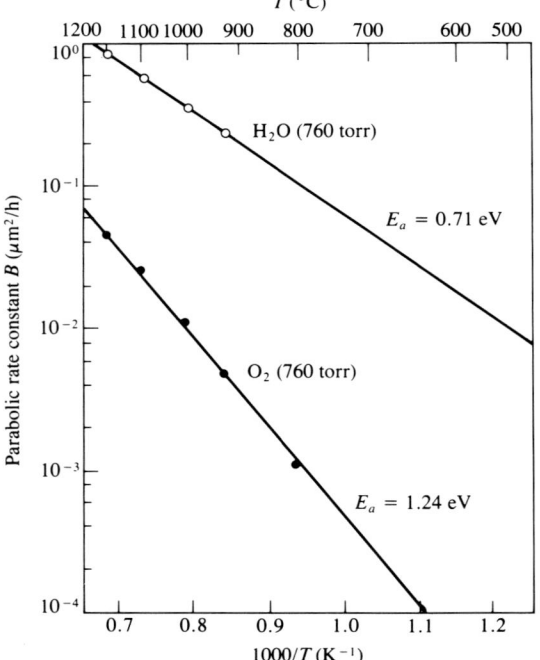

FIGURE 5.3.3
Temperature dependence of the parabolic rate constant in oxidation [6].

For computer calculations, the rate constants can be obtained analytically from

$$\text{Rate constant} = R_\infty \, e^{-E_a/(kT)} \tag{5.3.6}$$

in which E_a (known as the activation energy) is indicated for each case in Figs. 5.3.2 and 5.3.3, $k = 8.625 \times 10^{-5}$ eV/K is Boltzmann's constant, and, for process design, the following values of R_∞ yield satisfactory results:

Linear rate: R_∞ depends on type of process and on the crystal's surface orientation and has the following values in micrometers per hour:

Orientation	Dry	Wet
100	4.1×10^6	1.07×10^8
111	7.5×10^6	1.8×10^8

Parabolic rate:

Dry oxidation: $R_\infty = 810 \ \mu m^2/h$

Wet oxidation: $R_\infty = 243 \ \mu m^2/h$

Process design also requires computation of the constant τ appearing in (5.3.4) and (5.3.5). As shown in App. 5A,

$$\tau = \frac{x_0^2 + A x_0}{B} \tag{5.3.7}$$

where A and B can be computed from the linear and parabolic rate constants and x_0 is the oxide thickness at the beginning of the process, so that Eq. (5.3.7) implies that an oxide layer of thickness x_0 is already present on the substrate before the oxidation process begins. If oxidation is performed on an uncoated Si surface, this initial oxide layer may be due to some protracted room temperature oxidation of the sample, occurring before it is introduced in the furnace (e.g. during storage). Usually such oxidation reaches an asymptotic value of 30 Å and so it is negligible for most practical cases. On the other hand, for the dry process (in O_2), a layer of about 200 Å is grown in a very short time as soon as the sample enters the oven. This occurs at such a fast rate that it cannot be accounted for by Eq. (5.3.3).

In conclusion, for the purposes of practical computations, for dry oxidation, x_0 in (5.3.7) can be assumed to be 200 Å. For wet oxidation instead, x_0 (and so τ) can be assumed to be 0.

In accordance with Eq. (5.3.3), the plot of the thickness x of the deposited oxide layer vs. the duration t of the oxidation process is shown in normalized form in Fig. 5.3.4, which clearly shows the asymptotic conditions of reaction and diffusion controlled oxidation. Notice the transition region for $x_0 \approx A/2$.

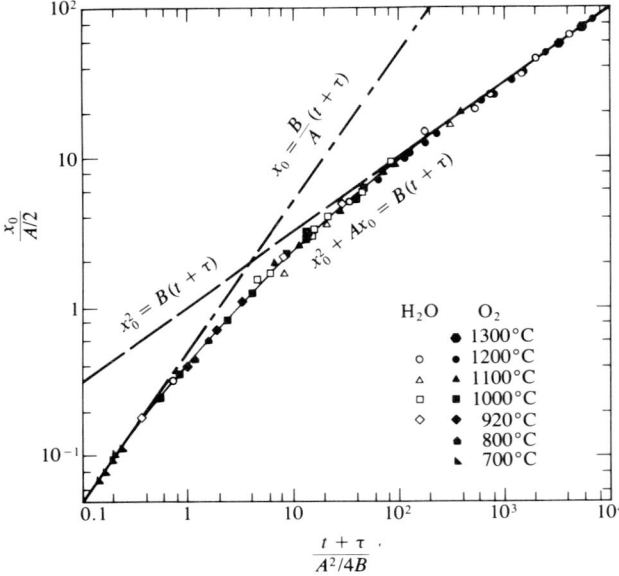

FIGURE 5.3.4
Normalized graph of SiO_2 layer thickness vs. oxidation time. Notice the linear and parabolic nature of reaction, respectively diffusion, controlled processes [6].

This plot implies a technologically important phenomenon. Consider the oxidation of a Si surface, part of which, at the beginning of the oxidation, is already covered by an oxide layer of thickness x_0, while the remaining surface is exposed, as indicated in Fig. 5.3.5a. Initially, the oxidation of the exposed area is reaction limited and proceeds linearly at the fast rate B/A; however, on the precoated area, if x_0 is large enough ($x_0 \gg A/2$), the oxidation is diffusion controlled

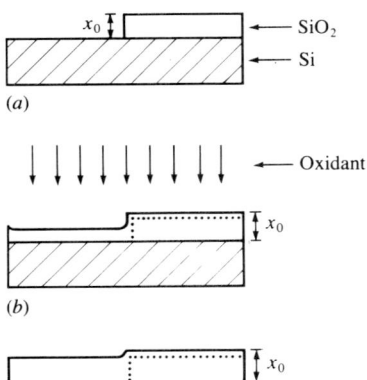

FIGURE 5.3.5
Oxidation of a Si surface already partially covered by an oxide layer: (a) initial condition, (b) after comparatively short oxidation time, (c) after prolonged oxidation. The dotted lines in (b) and (c) indicate initial oxide layer.

and proceeds parabolically at a much lower rate. Consequently, the oxide coating will grow fast over the exposed Si, while the existing coating over the protected area grows by an almost negligible amount, as shown in Fig. 5.3.5b.

It is evident that prolonged oxidation will soon result in a coating of essentially uniform thickness over the whole face of the specimen, as in Fig. 5.3.5c, practically eliminating the initial oxide step at the surface. Several fabrication sequences take advantage of this phenomenon to produce essentially uniform oxide coatings without having to strip previous coatings from the specimen's surface (cf. Chaps. 6, 8, 9, etc.).

Example 5.3.1. Oxidation of a Si crystal with surface orientation $\langle 100 \rangle$ is performed at atmospheric pressure and 1373 K for a duration of 1 h. Compute the thickness of the oxide layer for (a) dry oxidation in O_2 atmosphere; (b) wet oxidation.

Solution

(a) From (5.3.6), using the appropriate values of E_a and R_∞:

$$\frac{B}{A} = 4.1 \times 10^6 \, e^{-23188.4/1373} = 0.19 \; \mu m/h$$

The same result can be obtained from Fig. 5.3.2. Analogously:

$$B = 810 \, e^{-14377/1373} = 0.023 \; \mu m^2/h$$

The same result can be obtained from Fig. 5.3.3. From the above,

$$A = \frac{0.023}{0.19} = 0.12 \; \mu m$$

Using these values in (5.3.7) and also setting $x_0 = 200 \, \text{Å} \equiv 0.02 \; \mu m$:

$$\tau = \frac{0.02^2 + 0.12 \times 0.02}{0.023} = 0.121 \; h$$

so that

$$\frac{t + \tau}{A^2} \times 4B = 7.2$$

and finally, from (5.3.3) (or Fig. 5.3.4),

$$x = \frac{0.12}{2} (-1 + \sqrt{8.2}) = 0.112 \; \mu m$$

(b) From (5.3.6) (or Figs. 5.3.2 and 5.3.3),

$$\frac{B}{A} = 3.24 \; \mu m/h$$

$$B = 0.6 \; \mu m^2/h$$

so that

$$A = 0.18 \; \mu m$$

Letting $\tau = 0$ (wet oxidation),

$$t \times \left(\frac{4B}{A^2}\right) = 66.5$$

and finally, from (5.3.3) (or Fig. 5.3.4),

$$x = 0.65 \ \mu m$$

Notice the much thicker layer obtained by wet oxidation, as compared to dry oxidation.

The above data refer to the oxidation of essentially pure Si at atmospheric pressure. The oxidation process, however, is influenced by both the presence of impurities and by the operating pressure. High levels of P or B doping substantially increase the rate constants. Oxidation of thick layers is speeded up by operating at high pressure. These modes of operation are often used to grow thick oxide layers at lower temperatures, minimizing the perturbing effects of high temperature on previously generated device structures.

The recessed oxide process is often used, especially in MOSFET fabrication (cf. Chap. 9) to isolate active device regions.

Before oxidation, a portion of the Si surface is protected (by chemical deposition followed by a photo step, as discussed later) by a layer of Si_3N_4 as shown in Fig. 5.3.6a. If this sample is subjected to an oxidation process, the portion of the Si surface under the Si_3N_4 is not oxidized at all, as the oxidizing agent cannot diffuse through the Si_3N_4 to reach the Si surface. Oxidation occurs only on the exposed Si surface and there, as already noted, 40 percent of the oxide thickness grows into the specimen below the surface. This oxidation also spreads laterally for a small distance below the edges of the Si_3N_4 coating, as shown in Fig. 5.3.6b, and, if the oxide layer is sufficiently thick, the edges of the nitride film are pushed up and away from the Si surface.

The final result, after the Si_3N_4 layer is removed (Fig. 5.3.6c), is that the nitride has acted as an efficient mask preventing oxidation of the protected

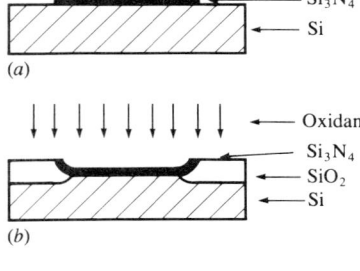

(a)

(b)

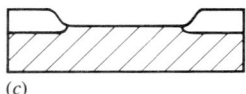

(c)

FIGURE 5.3.6
The recessed oxide process: (a) Si_3N_4 mask in place (sometimes a thin oxide layer is interposed between the Si_3N_4 and the Si to improve adherence); (b) during oxidation, the SiO_2 spreads laterally lifting the edges of the Si_3N_4 mask; (c) after removal of the Si_3N_4 mask.

surface. Sometimes, to improve adherence, a thin oxide coating is interposed between the protective Si_3N_4 and the Si surface (cf. Chap. 9).

Chemical Vapor Deposition (CVD)

When the device is not Si based, or when insulating layers other than SiO_2 are required, chemical vapor deposition (CVD) is most often used.

CVD ovens are similar to oxidation furnaces and so are the physical processing steps. Chemically, however, oven atmospheres are different and the insulating layer is deposited from the vapor phase *above* the surface of the substrate. Chemical composition of the vapor phase, temperature, and pressure vary widely with the chemical and physical characteristics of the desired dielectric layer.

Energy-enhanced CVD, in which deposition occurs in the presence of an electric field or other source of energy (such as ultraviolet radiation, electron beam, or plasma energy), often occurs at low temperature and deposition can be limited to precisely localized areas. As the technology is refined and stricter requirements are imposed, new details are introduced in the techniques. These are sometimes vital in making industrial fabrication of new and improved devices possible, especially in VLSI technology.

SiO_2 layers can also be obtained by CVD processes. These oxides are usually inferior to those obtained by oxidation. In Si-based devices they are used for special purposes, for instance when low-temperature processing is mandated.

Silicon nitride, deposited by CVD, is often used for its high dielectric constant and its chemical and physicochemical properties.

Strongly doped polysilicon layers are often obtained by CVD. They are used as electrodes, because of their superior resistance to high temperatures and because of other favorable properties (cf. Chap. 7). Their introduction has been of great importance, especially in MOS technology (cf. Chap. 9).

5.4 THE PHOTOLITHOGRAPHIC STEP

Often, certain fabrication processes must be limited to well-defined regions of the device. Then these regions must be precisely delimited and the rest of the device must be protected, so as not to be affected by the action of the fabrication process. This is most often done by a photolithographic step, which creates a protective mask capable of selectively inhibiting the specific process in certain regions while leaving other parts of the device open to its action. In practice, to achieve this end, a whole sequence of masks is usually created. In a typical sequence, a photomask is used to create a photoresist mask, which is in turn used to produce a SiO_2 mask.

The Photomask

Once the designer has determined which regions of the device must be exposed to a subsequent process (such as etching) and which parts must be protected from it, an appropriate photomask is made. This is a pattern in which the regions to be exposed are opaque and the ones to be protected are transparent (or vice versa, depending on the subsequent processing). A typical photomask is shown in Fig. 5.4.1. A whole variety of photomask production techniques are used.

Artwork,[2] often thousands of times larger than the final mask, can be hand-produced. The size of this pattern can then be scaled down by a succession of photographic steps with a reduction camera. During this photoreduction, the pattern can also be duplicated many times, resulting in an array of identical, scaled-down patterns, arranged in adjacent rows and columns like elements in a matrix, as in Fig. 5.4.1. This permits simultaneous transfer to a wafer of many identical patterns for chip mass-production.

Conversely, the pattern features may be appropriately programmed on a computer, which produces the mask by driving a scribing beam of electrons to expose the photographic plate directly (Fig. 5.4.6).

In all cases, the final product is a plate, similar to a photographic transparency. The minimum detail size to be reproduced determines the required *resolution* (often of the order of 1 μm or even less). This requirement is very stringent, mandating the use of short-wavelength radiation (uv and, recently, x-ray) and special optical techniques and materials. For instance, high-resolution photographic emulsions generally cannot resolve detail smaller than about 4 μm, so special, hard (metallic) photosensitive coatings may be required.

Many technological problems are connected with the need for cleanliness and registration in the preparation and use of these photomasks.

The Photoresist Mask

The pattern is transferred from the optical mask to a photoresist coating on the wafer surface by optical printing.

A typical photoresist film consists of an organic material soluble in some solvent. If a region of this film is exposed to light (usually ultraviolet radiation), it polymerizes, so that the molecules in the exposed region coalesce into very large macromolecular aggregates. When developed, i.e. treated with the solvent, the unexposed regions of the film are washed away in the bath, but the exposed ones remain, forming an image. The portions of the wafer under the exposed regions are covered by a polymerized film; the rest of the wafer surface is left unprotected and can be affected by some subsequent fabrication process. This image constitutes the photoresist mask.

The photoresist just described is of the *negative type* (regions exposed to light become insoluble). A *positive photoresist* is originally insoluble, but regions

---

[2] As the devices produced become more and more complex, hand-drawn artwork becomes unmanageable. At the present writing hand-drawn artwork is industrially obsolete and has exclusively pedagogical and historical interest.

(a)

(b)

FIGURE 5.4.1

Photomask used in the production of power transistors. (a) The entire photomask, reproducing the individual device pattern many times in a matrix-like array, for the simultaneous fabrication of many identical devices on one wafer. (b) Two individual BJTs of different power rating showing how the same pattern can be used at different magnifications.

133

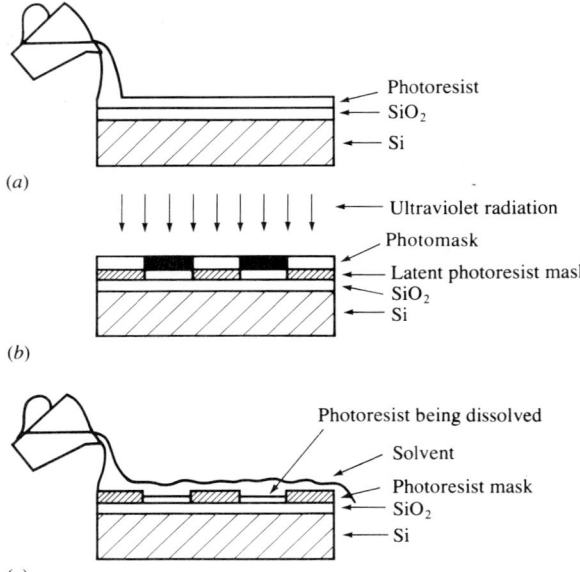

(a)

(b)

(c)

FIGURE 5.4.2

Photostep: exposure and development of photoresist mask. (a) Pouring the photoresist layer. (b) Contact exposure of the photoresist through a photomask. Notice the polymerization of the exposed regions of the photoresist. (c) Developing the photoresist mask by dissolving the unpolymerized photoresist with a solvent. The polymerized portions are insoluble.

exposed to light become soluble. If a positive and a negative photoresist are exposed using the same photomask, then, after development, they yield images related to each other (and to the photomask) exactly as positive and negative photographic plates—hence the names.

The photoresist is originally in the liquid state. It is spread over the surface to be processed and dries to a thin film adhering to the SiO_2 (cf. Fig. 5.4.2a where the surface to be protected is assumed to be SiO_2). Exposure of the photoresist is a photographic printing process. The light used is commonly ultraviolet.

Contact, proximity, and *projection* printing techniques are used (Fig. 5.4.2b shows contact printing). Many precautions and specialized techniques are necessary to meet the special requirements of the process, especially with regard to resolution, registration, and cleanliness.

After exposure the photoresist mask is developed by the removal of the soluble portion by washing with a solvent (cf. Fig. 5.4.2c).

Silicon Dioxide and Other Insulator Masks

Once the photoresist mask is in place, the next localized fabrication process can be performed. This is very often the etching of an insulating layer to produce an insulator mask (most often consisting of SiO_2). With the photoresist mask in place, the surface of the oxide is exposed to the action of an *etching agent*, which attacks the SiO_2 without affecting the photoresist mask (Fig. 5.4.3a). Most often the *wet etching* technique is used: the etching agent is a water (or acetic acid)

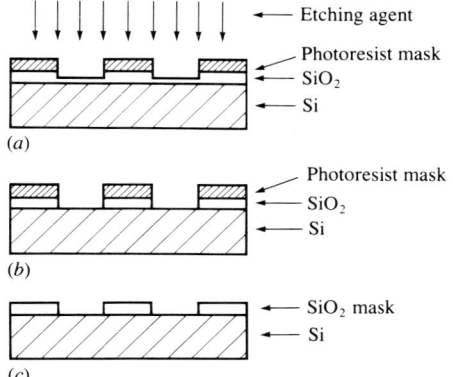

FIGURE 5.4.3
Production of a SiO$_2$ mask. (a) Etching of the SiO$_2$ by HF. In the region where the SiO$_2$ is covered by the photoresist, the acid does not reach the SiO$_2$ and etching does not take place. (b) The SiO$_2$ under the openings in the photoresist mask has been etched away. (c) SiO$_2$ mask after stripping away the photoresist.

solution of hydrofluoric acid (often buffered with an oxidizing agent such as nitric acid). The etching reaction is

$$SiO_2 + 6HF \rightarrow H_2SiF_6 + 2H_2O \tag{5.4.1}$$

and the soluble H$_2$SiF$_6$ is washed away in the bath.

Where the SiO$_2$ layer comes in contact with the solution it is etched away, leaving the underlying semiconductor surface exposed, but where the SiO$_2$ is protected by the photoresist mask, it remains intact. The final result is therefore a SiO$_2$ mask, exactly reproducing the pattern of the photoresist mask (Fig. 5.4.3b).

The photoresist is then stripped away by oxidizing or dissolving it, leaving only the SiO$_2$ mask, as shown in Fig. 5.4.3c.

In practice, the wet etching action does not occur only in the direction perpendicular to the SiO$_2$ surface. Some etching spreads parallel to this surface under the photoresist, eating away some SiO$_2$ near the edges of the photoresist mask, as shown in Fig. 5.4.4. This limits the resolution of SiO$_2$ masks produced by the wet etching technique.

More accurate masks can be obtained by *dry etching*, in which the etching agent is a gas mixture, usually ionized by an electric field (plasma activated). This technique yields SiO$_2$ masks with sharp, well-defined edges with no appreciable tangential spread.

Etching is not used exclusively on SiO$_2$ layers; nor is it limited to the production of insulating masks. A wide variety of etching techniques are in common use and are continuously being developed, as evidenced by the copious literature on the subject.

The photolithographic step (or photostep, for short) is only schematically outlined in the sequence of Figs. 5.4.1 through 5.4.3. Many technological variants

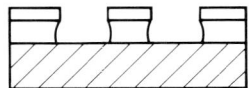

FIGURE 5.4.4
Effect of lateral spreading of liquid phase etching solution: image spread in the SiO$_2$ mask, reducing the resolution.

are often introduced. The photostep is a powerful tool to limit the action of subsequent processing to precisely localized, limited regions of the device under fabrication. Several photosteps are usually employed at different stages of the fabrication sequence, using different mask patterns to yield devices with rather complex structures.

Economically, the already mentioned problem of registration of the relative positioning of subsequent masks is critically important.

The final cost of fabrication and the percentage yield of acceptable devices are strongly influenced by the number of photolithographic steps required during the complete fabrication cycle, so that the minimization of such steps is one important criterion in device fabrication design. Indeed, fabrication sequences are often characterized and referred to by the number of masks used.

Because of its frequent use and critical impact on device cost and reliability, photolithography has become a highly specialized art, the technological details have been extensively analyzed, and many variations and refinements are constantly being introduced. The state of the art is abundantly reported in the literature. Figure 5.4.5 shows a registration station used to align masks.

In the following chapters, whenever a new device is introduced, examples of fabrication cycles will be offered. In the description of these sequences, reference

FIGURE 5.4.5
Photomask projection and alignment station. (*Courtesy NJIT Microelectronics Laboratories.*)

FIGURE 5.4.6
Station for photomask production by computer. (*Courtesy NJIT Microelectronics Laboratories.*)

will often be made to the photolithographic step. In these references the step will be treated as a block of operations, without repeating a detailed description of the different processes that compose the step.

5.5   LOCALIZED DOPING—DIFFUSION AND ION IMPLANTATION

As stated in Sec. 5.1, accurately controlled concentrations of impurities can be obtained during wafer preparation by appropriately doping the melt. By such methods, however, it is extremely difficult to produce precisely localized variations in doping, such as those required in PN junctions (cf. Chap. 6). This is generally achieved by introducing the impurity atoms into the semiconductor crystal in the solid state, i.e. operating at temperatures below the melting point. The process is localized by means of suitable protective masks. The two main methods used are: solid state *diffusion* (a high-temperature process) and *ion implantation* (a low-temperature process).

Diffusion

Selected regions of the semiconductor crystal surface are exposed to high concentrations of dopant atoms under conditions assuring reasonably high diffusivity (e.g., high temperature). The impurity atoms then enter into *solid solution* at the crystal surface and from there diffuse into the body of the crystal.

Diffusion ovens are similar to oxidation furnaces, but the gas atmosphere consists of an appropriate chemical compound of the desired impurity. The dopant source can be a solid, liquid, or gas.

Some of the most frequently used compounds for dopant sources are:

	B (P-type)	As (N-type)	P (N-type)
Solid state	BN	As_2O_3	P_2O_5
Liquid state	BBr_3	$AsCl_3$	$POCl_3$
Gaseous state	B_2H_6	AsH_3	PH_3

The source is exposed to the flow of an inert gas, which flushes away with it a controlled amount of the source's molecules. Figure 5.5.1 is an example of a liquid source diffusion apparatus.

The resulting dopant-enriched gas, mixed, if necessary, with other active and/or inert gases, is pumped into the oven containing the wafers. At the oven temperature, the gases in the mixture react with each other and with the wafer material, wherever this is left exposed by the SiO_2 mask. A typical sequence of reactions may be

$$4BBr_3 + 3O_2 \rightarrow 6Br_2 + 2B_2O_3 \qquad (5.5.1)$$

followed, at the Si surface, by

$$3Si + 2B_2O_3 \rightleftharpoons 3SiO_2 + 4B \qquad (5.5.2)$$

Different reactions occur, of course, with different gas mixtures, but they all result in the production of a high concentration of dopant gas [boron in the example of

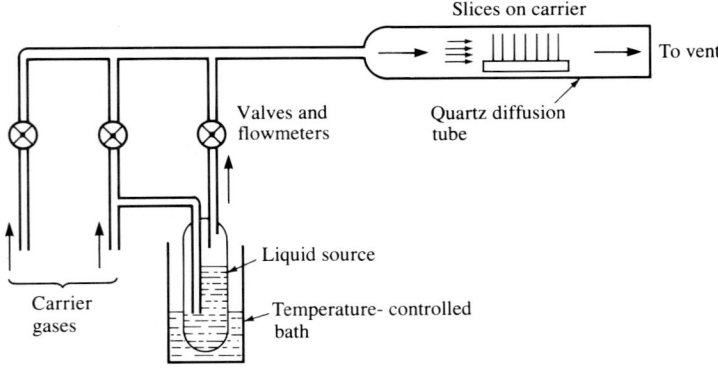

FIGURE 5.5.1
Diffusion system illustrating a liquid source diffusion apparatus. Solid and gas source systems differ essentially only in the atmosphere control system [2].

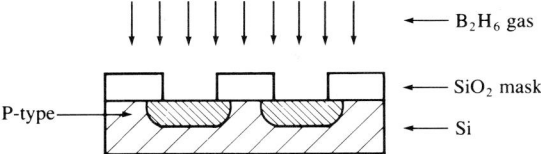

FIGURE 5.5.2
Schematization of diffusion process: the dopant species diffuses into the crystalline silicon, where its surface is exposed. Notice the limited amount of lateral spreading.

(5.5.2)] at the exposed semiconductor surface. There the dopant is dissolved into the semiconductor, as symbolized in Fig. 5.5.2.

The surface concentration of the dopant after dissolution depends on the dopant and semiconductor species (e.g., B and Si), on the oven temperature and on the partial pressure of the doping gas. If the latter is high enough, the dopant concentration equals the *solid solubility* N_0. Solid solubility curves are shown in Fig. 5.5.3 for several dopants in Si.

The high concentration of dopant atoms in the solid solution at the surface now creates a concentration gradient, which in turn results in diffusion of the dopant molecules into the body of the semiconductor in accordance with the theory of Sec. 4.3.

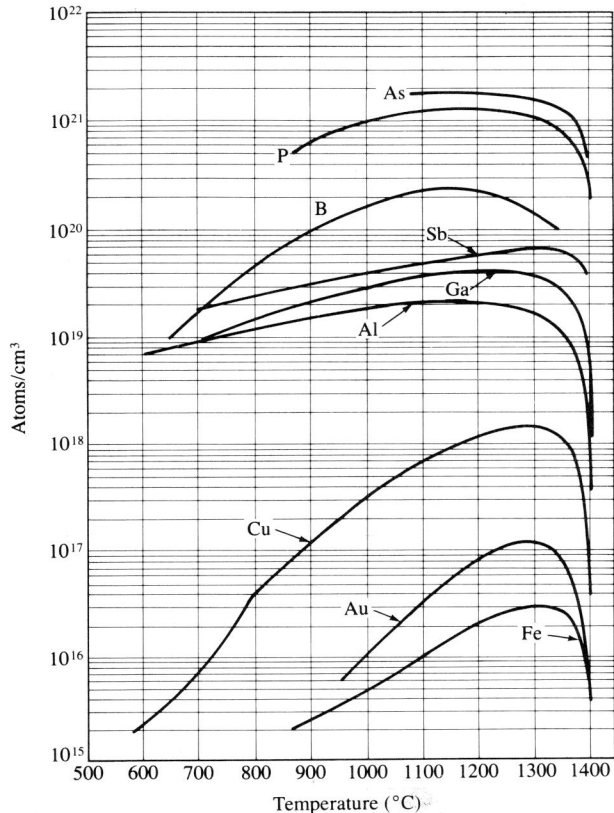

FIGURE 5.5.3
Solid solubility of various species in Si vs. temperature [2].

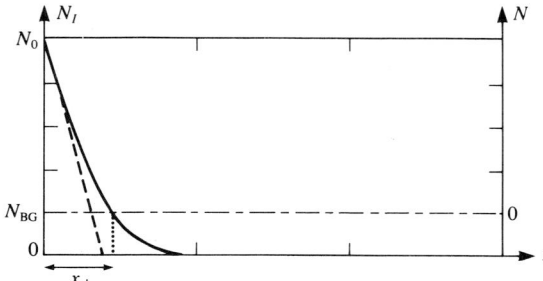

FIGURE 5.5.4
Impurity concentration vs. depth of penetration in typical diffusion process. The background doping N_{BG} of the extrinsic semiconductor is indicated by the dash-dot line. For background doping of type opposite to that of the diffused impurity, the junction depth is x_d.

As time elapses, the dopant molecules penetrate deeper and deeper into the semiconductor and the extrinsic layer generated becomes thicker. Notice that the surface concentration remains constant because the atoms migrating by diffusion are replaced as new atoms enter the surface solution from the external dopant atmosphere. The impurity concentration $N_I(x)$ decreases from the surface into the crystal body to maintain diffusion and so a *doping profile* is created, as shown in Fig. 5.5.4.

Notice that, if the substrate is already extrinsic, with a uniform dopant *background* concentration N_{BG}, then two possibilities arise:

1. If the background and diffusing doping agents are of the same type (i.e., both donors or both acceptors) then the resulting net concentration is $N = N_I(x) + N_{BG}$.

2. However, if they are of opposite type, then the net concentration is $N = N_I(x) - N_{BG}$. In this case, if N is positive, the resultant semiconductor is of the same type as N_I; if negative, it is of the background type.

In Fig. 5.5.4, N_{BG} is represented by the dash-dot line in the concentration diagram. Let x_d be the abscissa of the point where this line meets the N_I graph. Then, for all points to the left of x_d, $N_I > N_{BG}$ and the semiconductor is of the same type as the diffusing dopant. To the right of x_d, $N_I < N_{BG}$ and the semiconductor is of the background type.

An N- and a P-type semiconductor now exist *one next to the other in the same crystal*. This structure constitutes a *PN junction*; its properties are of fundamental importance in semiconductor device theory and will be discussed in some detail in Chap. 6.

In conclusion, in a diffused PN junction, the surface layer $(x < x_d)$ of the extrinsic semiconductor has changed type (after diffusion it has become of the type of the diffusing impurity); deeper in the body of the crystal $(x > x_d)$ it remains of background type. The depth x_d, where $N_I = N_{BG}$ and so the semiconductor changes type, is the *junction depth*.

Diffusion occurring under the condition described above is known as *predeposition*. For given conditions the doping profile (and with it the junction depth)

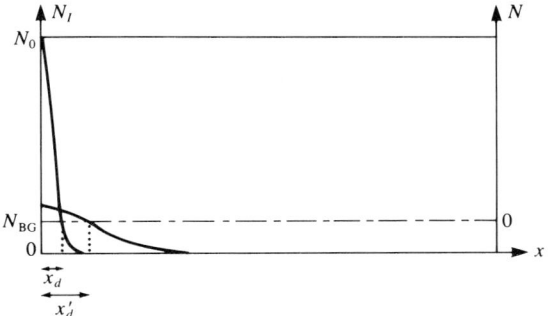

FIGURE 5.5.5

Predeposition vs. drive-in diffusion profiles.

extends deeper into the crystal the longer the duration of the diffusion process, but the surface doping N_0 is independent of this duration.

It is often desirable to alter the doping profile, making it smoother (cf. Fig. 5.5.5); then a second diffusion process is applied: *drive-in diffusion*. Technologically this process is identical to predeposition, except that the gas atmosphere in the oven now consists of inert gases exclusively and does not contain any dopant. As surface dopant molecules migrate by diffusion, they are no longer replaced from the outside. Therefore, as the process progresses, the surface impurity concentration decreases, the doped layer is driven deeper into the crystal and so, in most cases, is the junction depth.

Quantitative Analysis

For quantitative computations, a mathematical simulation of the diffusion process uses the diffusion law (4.3.3). For a one-dimensional system, the rate of flow of diffusing atoms is

$$F = -D\frac{\partial N}{\partial x} \tag{5.5.3}$$

where N is the impurity concentration and D the diffusivity of the impurity atoms in the semiconductor crystal.

Using (5.5.3) in the continuity equation (4.5.15), under the assumption that diffusion is the only source of impurity atoms ($G = R = \mathscr{E} = 0$),

$$\frac{\partial N}{\partial t} = -\frac{\partial}{\partial x}\left(-D\frac{\partial N}{\partial x}\right) \tag{5.5.4}$$

Solution of this differential equation yields the concentration N of impurity atoms at any depth x after drive-in diffusion of duration t. The doping profile so obtained depends on the diffusivity D and on the boundary conditions.

Diffusion coefficients vs. temperature for several impurities in Si and GaAs are shown in Fig. 5.5.6. For computer calculations, the diffusivity can be obtained

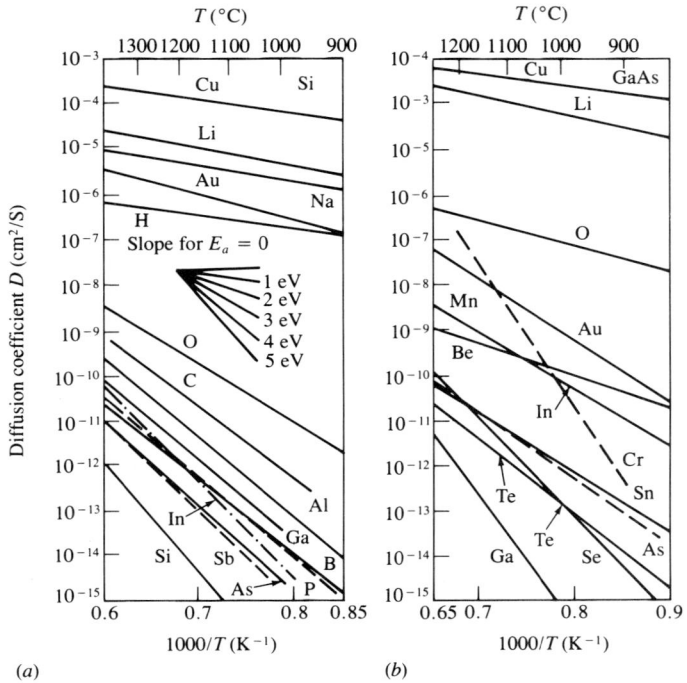

FIGURE 5.5.6
Diffusivities of several species in Si and GaAs vs. temperature [5].

from

$$D = D_\infty \, e^{-E_a/(kT)} \tag{5.5.5}$$

where E_a is the activation energy in electronvolts and $k = 8.625 \times 10^{-5}$ eV/K is Boltzmann's constant. For practical computations, D_∞ and E_a can be obtained from Fig. 5.5.6 (cf. Prob. 5.10).

Under the assumption of *concentration independent diffusivity*, from (5.5.4),

$$\frac{\partial N}{\partial t} = D \, \frac{\partial^2 N}{\partial x^2} \tag{5.5.6}$$

In setting the boundary conditions for the solution of this equation, two important cases must be distinguished:

PREDEPOSITION. The boundary conditions are $N(0) = N_0$; $N(\infty) = 0$, where N_0 can usually be assumed to equal the solid state solubility of Fig. 5.5.3. The solution of the partial differential equation (5.5.6) becomes

$$N(x, t) = N_0 \, \text{erfc} \left(\frac{x}{2\sqrt{Dt}} \right) \tag{5.5.7}$$

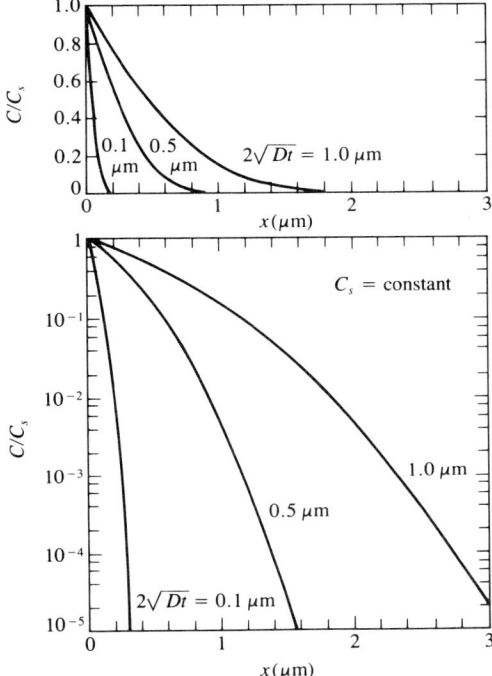

FIGURE 5.5.7
Normalized impurity concentration profiles for predeposition diffusion [3].

where the *error function complementary* is a transcendental function often encountered in probability and stochastic process theory [7]. A plot of erfc(x) is presented in Fig. 5.5.7 (also see Example 5.5.1) and in Fig. 5.5.8.

An important characteristic of the predeposition process is the impurity dose:

$$Q(t) = \int_0^\infty N(x, t) \, dx = \frac{2}{\sqrt{\pi}} N_0 \sqrt{Dt} \tag{5.5.8}$$

in which, in performing the integration, use has been made of Eq. (5.5.7) [7, p. 299].

The impurity dose represents the total number of impurity atoms that have been injected into the crystal per unit exposed surface area over a predeposition duration t. This quantity is important in computing drive in diffusion profiles.

Notice that, for predeposition into an extrinsic semiconductor with a background density N_{BG} of a dopant of type opposite to that of the diffusing impurity, a PN junction is formed with net impurity concentration distribution (doping profile):

$$N(x, t) = N_0 \, \text{erfc} \left(\frac{x}{2 \sqrt{Dt}} \right) - N_{BG} \tag{5.5.9}$$

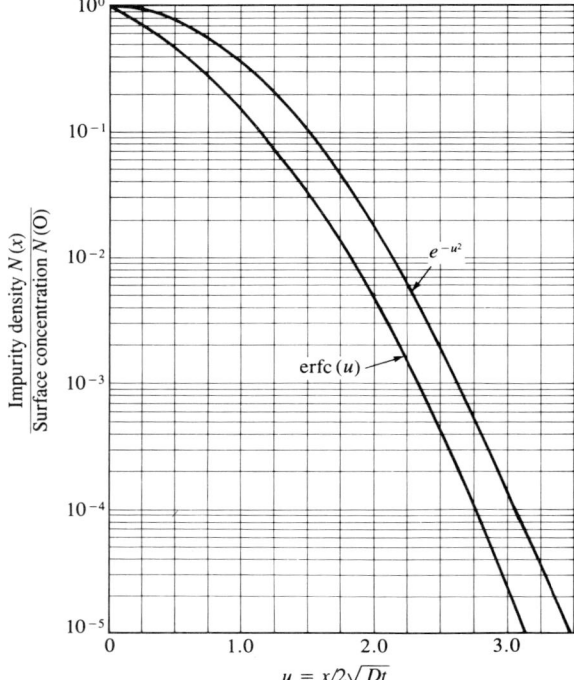

FIGURE 5.5.8
Plots of erfc (u) and e^{-u^2} (related to the gaussian function) [8].

and the junction depth x_d, as computed by setting $N = 0$ in (5.5.9), is

$$x_d = 2\sqrt{Dt}\ \text{erfc}^{-1}\left(\frac{N_{\text{BG}}}{N_0}\right)$$ (5.5.10)

where the inverse error function complementary can easily be computed from Fig. 5.5.7 (cf. Example 5.5.1) or from Fig. 5.5.8.

The time-dependent quantity

$$L_D = \sqrt{Dt}$$ (5.5.11)

prominently appearing in (5.5.8) through (5.5.10), has the dimensions of a length and is known as the *diffusion length* of the process. Predeposition diffusion profiles are shown for different values of L_D in Fig. 5.5.7, where $N(x, t)/N_0$ is plotted vs. depth x from the crystal surface. The top graph shows N/N_0 on a linear scale, the bottom graph on a logarithmic scale for more precise computation. For the case in which a PN junction is formed, a horizontal line at N_{BG}/N_0 can be used to determine the junction depth x_d (cf. Fig. 5.5.4). The linear scale (top graph) predeposition profiles of Fig. 5.5.7 are sometimes approximated by a *linear profile* (cf. linearly graded junctions in Sec. 6.3). This defines a triangle of base $2L_D$ and height $N/N_0 = 1$ (see dotted line in Fig. 5.5.4).

Example 5.5.1. Predeposition of B in a Si sample doped with 10^{17} As atoms/cm^3 has a duration of 3 min at 1200°C. Compute (a) the net doping at 0.8 μm from the surface; (b) the junction depth.

Solution

(a) From Fig. 5.5.6, $D = 10^{-12}$ cm^2/s. From Fig. 5.5.3, $N_0 = 2.3 \times 10^{20}$ cm^{-3}. From (5.5.9), therefore,

$$N = 2.3 \times 10^{20} \, \text{erfc} \left(\frac{8 \times 10^{-5}}{2\sqrt{10^{-12} \times 180}} \right) - 10^{17} = -9.4 \times 10^{16} \, \text{cm}^{-3}$$

where the value of erfc (2.98) $= 0.25 \times 10^{-4}$ has been obtained from Fig. 5.5.7.[2]

As the result of a negative net concentration, at this depth the semiconductor is of the N-type (we are deeper than the junction) with net impurity concentration $N_D = 9.4 \times 10^{16}$ cm^{-3}.

(b) From (5.5.10),

$$x_d = 2.68 \times 10^{-5} \, \text{erfc}^{-1} \left(\frac{10^{17}}{2.3 \times 10^{20}} \right)$$

From Fig. 5.5.7, $\text{erfc}^{-1} (4.3 \times 10^{-4}) = 2.4$, so

$$x_d = 0.64 \, \mu\text{m}$$

As 0.64 μm < 0.8 μm, this confirms the result of part (a).

DRIVE-IN DIFFUSION. In this case what remains constant is not the impurity concentration at the surface but the total number of impurity molecules injected in the semiconductor. From (5.5.8), therefore,

$$\int_0^\infty N(x, t_P) \, dx = Q(t_P) = \text{constant} \tag{5.5.12}$$

which constitutes a boundary condition for the solution of Eq. (5.5.6) for drive-in diffusion. The other boundary condition is $N(\infty, t) = 0$. Notice that, as implied in the first boundary condition, the variable t in (5.5.6) indicates the duration of the drive-in diffusion process, while the duration of the preceding predeposition process is indicated as t_P. The solution of (5.5.6) now becomes

$$N(x, t) = \frac{Q(t_P)}{\sqrt{\pi D t}} e^{-x^2/(4Dt)} \tag{5.5.13}$$

a *gaussian profile*. The gaussian distribution also prominently appears in probability and stochastic process theory and it is abundantly discussed and tabulated in the literature [7, p. 351]. A normalized plot of the gaussian function is presented in Fig. 5.5.8 for ease of computation. Drive-in profiles are shown in

---

[2] In accordance with Eq. (5.5.7), Fig. 5.5.7 plots erfc $(x/2L_D)$, so the plot for $2L_D = 1$ is actually a plot of erfc (x). Inverting the axes, this plot can be used to compute $x = \text{erfc}^{-1}(y)$.

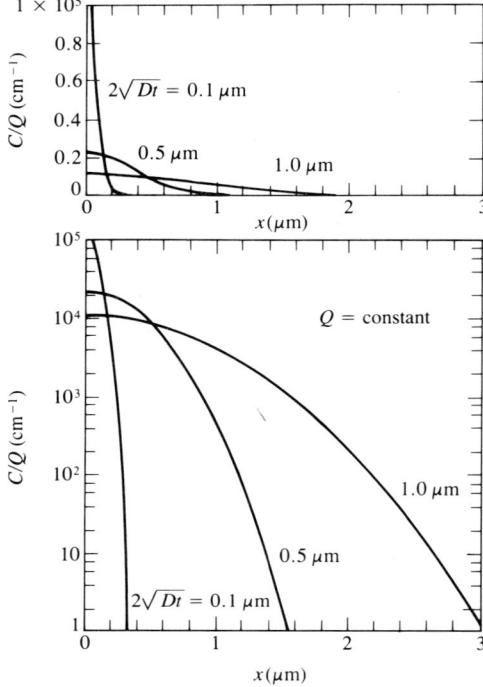

FIGURE 5.5.9
Normalized impurity concentration profiles for drive-in diffusion [3].

Fig. 5.5.9, where it is easy to recognize the fact that the surface impurity density $N(0, t)$ does not remain constant during the drive-in process but decreases with process duration. This has already been qualitatively predicted and is in accord with (5.5.13), from which

$$N(0, t) = \frac{Q(t_P)}{\sqrt{\pi Dt}} \qquad \text{for } t > 0 \qquad (5.5.14)$$

At $t = 0 \rightarrow N(0, 0) = N_0$ and the surface impurity concentration is seen to be inversely proportional to the diffusion length L_D.

Example 5.5.2. After predeposition, the specimen of Example 5.5.1 undergoes drive-in diffusion at 1100°C for 4 h. Compute (a) surface concentration of the diffused species, (b) impurity dose, (c) junction depth after drive-in diffusion, (d) the background doping that would be required to drive the junction depth to 2 μm after drive-in diffusion.

Solution. From (5.5.8), with the data of Example 5.5.1, the total number of impurity atoms diffused per square centimeter of surface during predeposition is

$$Q(t_P) = \frac{2}{\sqrt{\pi}} \, 2.3 \times 10^{20} \sqrt{10^{-12} \times 180} = 3.48 \times 10^{15} \text{ cm}^{-2}$$

(a) From Fig. 5.5.6 at $1100°C$, $D = 1.7 \times 10^{-13}$, so from (5.5.14),

$$N(0, t) = \frac{3.48 \times 10^{15}}{\sqrt{\pi 1.7 \times 10^{-13} \times 14{,}400}} = 0.4 \times 10^{20} \text{ cm}^{-3}$$

Notice that this is much less than the predeposition surface concentration (2.3×10^{20}).

(b) The drive-in diffusion does not change the total number of atoms diffused, so

$$Q(t) = Q(t_P) = 3.48 \times 10^{15} \text{ cm}^{-2}$$

(c) From (5.5.13),

$$10^{17} = 0.4 \times 10^{20} \, e^{-x_d^2/(9.79 \times 10^{-9})}$$

or

$$x_d = \sqrt{9.79 \times 10^{-9} \ln\left(\frac{0.4 \times 10^{20}}{10^{17}}\right)} = 2.42 \ \mu m$$

much larger than the predeposition junction depth of $0.64 \ \mu m$.

(d) From (5.5.13), the concentration of the diffused species at a depth $x = 2 \ \mu m$ is

$$N(2 \ \mu m, 3600 \text{ s}) = 0.4 \times 10^{20} \, e^{-(2 \times 10^{-4})^2/(4 \times 1.7 \times 10^{-13} \times 14{,}400)}$$

$$= 6.73 \times 10^{17} \text{ cm}^{-3}$$

To set the junction to $2 \ \mu m$ from the surface, the background doping should have this same value.

It has been stated (Sec. 4.2) that, in integrated circuits, resistors are often fabricated as thin sheets of extrinsic semiconductor material of appropriate thickness and shape. These are usually thin diffused layers, so that the dopant concentration is not uniform but varies with depth in accordance with (5.5.9) or (5.5.13), depending on the process (one- or two-stage diffusion).

The computation of the sheet resistance can be done on the basis of Eq. (4.2.13), using the average conductivity of Eq. (4.2.15), which, in the notation used in this chapter, becomes

$$\bar{\sigma} = \frac{q}{x_d} \int_0^{x_d} N_D(x, t)\mu_e \, dx \tag{5.5.15}$$

for N-type semiconductor and analogous for P-type. Notice that, as the total impurity concentration N_t varies with x, then the mobility is a complicated function of the depth x. Evaluation of the integral may consequently become very difficult, so, in practice, average conductivity calculations are made with the aid of graphs such as that of Fig. 5.5.10.

Example 5.5.3. A resistor consists of a sheet of P-type semiconductor material obtained by the two-stage diffusion process of Examples 5.5.1 and 5.5.2. If the resistor's length equals 10 times its width, compute the resistance.

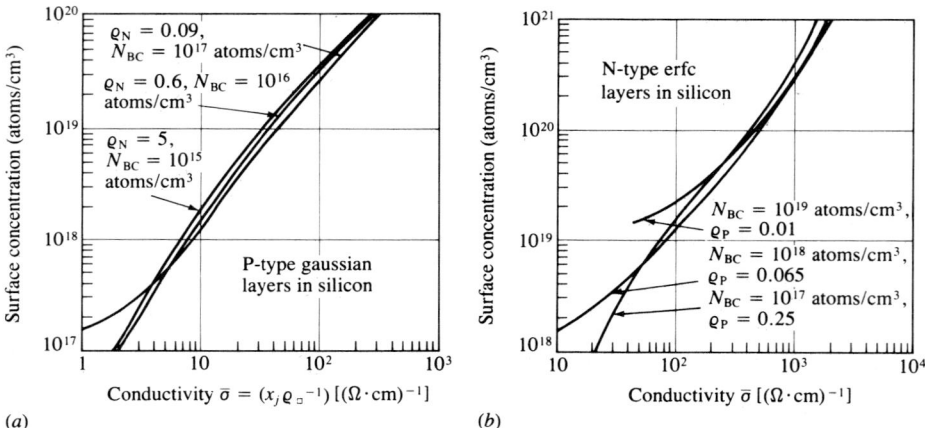

FIGURE 5.5.10
Average conductivity of sheet resistors vs. surface concentration for different background dopings: (a) drive-in diffusion, (b) predeposition diffusion [9].

Solution. From Fig. 5.5.10a, for the background doping $N_{BG} = 10^{17}$ cm^{-3}, gaussian distribution, and surface concentration $N(0, t) = 4 \times 10^{19}$ cm^{-3} (cf. Example 5.5.2), the average conductivity is read as

$$\bar{\sigma} = 150 \text{ mho/cm}$$

This value could also have been computed using Eq. (5.5.15). To simplify the integration, assume the mobility to be constant and to correspond to a total average impurity concentration of 10^{19} cm^{-3}, or, from Table 4.2.1, $\mu_h = 60$ cm^2/(V · s). Then, from (5.5.15), using the results of Example 5.5.2,

$$\bar{\sigma} = \frac{1.6 \times 10^{-19}}{2.2 \times 10^{-4}} 60 \int_0^{2.42 \times 10^{-4}} (e^{-x^2/(9.79 \times 10^{-9})} - 10^{17}) \, dx$$

From standard tables of integrals [7, p. 297, eqs. 7.1.1, 7.1.2] and by proper variable substitution,

$$\bar{\sigma} = 3.97 \times 10^{-14} \left(\sqrt{9.79 \times 10^{-9}} \, 4 \times 10^{19} \int_0^{2.446} e^{y^2} \, dy - 2.42 \times 10^{13} \right)$$

and reading from Fig. 5.5.8, erfc (2.446) $\approx 3 \times 10^{-4}$, finally,

$$\bar{\sigma} = 140 \text{ mho/cm}$$

Considering the limited accuracy of both Fig. 5.5.11 and of the semiarbitrary choice of the approximate average value of μ_h used in the integration, this is in excellent agreement with the previous result.

Assuming $\bar{\sigma} = 150$ mho/cm, then with the above values, from Eq. (4.2.13), the sheet resistance becomes

$$R_\square = \frac{10^4}{150 \times 2.42} = 27.5 \ \Omega_\square$$

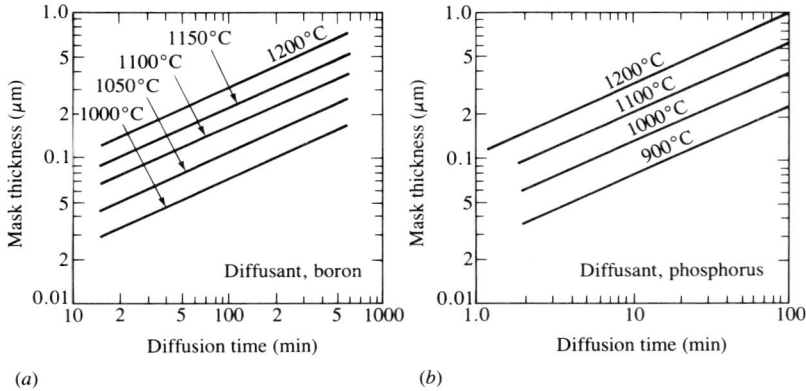

FIGURE 5.5.11
Minimum SiO$_2$ mask thickness required for effective masking of diffusion vs. diffusion time for several diffusion temperatures: (a) diffusion species: B; (b) diffusion species: P [10].

As the resistance consists of 10 squares, the total resistance is

$$R = 275 \ \Omega$$

In the above, the diffusivity D has been assumed to be independent of the impurity concentration; this is not true at high doping, requiring some correction of the above results.

It has been mentioned that limitation of the diffusion process to precisely localized areas of the device is obtained by means of SiO$_2$ masks. The protective qualities of SiO$_2$ are due to the very low diffusivity of dopant molecules in oxide layers. To be effective, however, such layers must be thick enough.

The minimum oxide thickness required for effective masking depends on the duration t and temperature T of the diffusing process and on the diffusion species. For B and P in SiO$_2$ it can be estimated from Fig. 5.5.11.

It has also been assumed that the system is unidimensional, i.e., that diffusion occurs only in the direction perpendicular to the surface. This assumption is satisfactory only comparatively far from the edges of the protective mask. Near these edges, there exists a noticeable tangential component, so that impurities also diffuse parallel to the surface, doping a small region of the semiconductor under the edges of the mask, as shown in Fig. 5.5.2.

Lateral diffusion usually does not penetrate as deeply into the crystal as the normal component. Lateral depths of penetration range from about 60 to 80 percent of corresponding normal depths.

Ion Implantation

Considerably improved accuracy in controlling the doping profiles, lateral diffusion, etc., can be achieved by aiming a beam of high-velocity impurity ions at the target window to be doped.

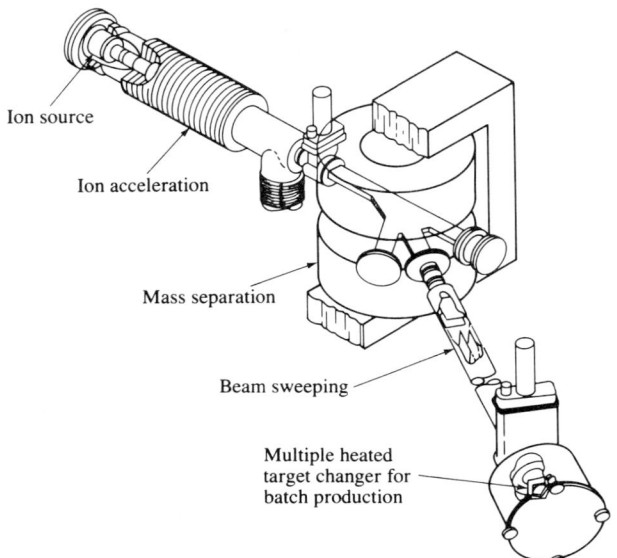

Ion source

Ion acceleration

Mass separation

Beam sweeping

Multiple heated
target changer for
batch production

FIGURE 5.5.12
Ion implantation apparatus.
Mass separation, particle
speed, beam focusing, and
deflection are provided by
conventional electrical and
magnetic means, permitting
accurate control of implantation parameters [11].

A typical ion implantation system is depicted in Fig. 5.5.12. Mass separation is obtained by an analyzer magnet focusing only ions of the desired
mass/charge ratio into the beam, thereby limiting possible unwanted impurities.
The desired beam energy is controlled by the voltage applied to the ion acceleration tube and the beam is aimed by the vertical and horizontal scanner electrodes. Beam energies commonly vary between tens and hundreds of
kiloelectronvolts.

The ions penetrate the surface (i.e., are implanted in the crystal) because of
their high initial energy and come to rest only when this energy has been dissipated by collisions with the crystal elements. By controlling the ion beam energy,
the depth of penetration can be controlled. Because of the randomness of the
collisions with crystal atoms, the range of penetration (or projected range) is
subject to statistical variations and so the implanted ion population is not all
concentrated at the projected range, but is distributed in a gaussian manner
around it over a range of depths corresponding to a statistical *projected straggle.*
Similarly, the ions are spread laterally from the direction in which the ion beam
is originally aimed. This spread is also gaussian and is measured by a statistical
parameter called *lateral straggle.*

Such spreads notwithstanding, ion implantation can afford doping control
orders of magnitude more accurate than thermal diffusion, both vertically and
laterally, improving control, not only of doping profiles (and junction depth,
where applicable) and of lateral spread under the mask edges but also of the total
number of impurity ions injected per unit target surface (the impurity dose).

A major advantage of ion implantation is the fact that it is a low-
temperature process, so that it can be introduced in the fabrication sequence at

moments in which the high temperature involved in diffusion would damage already fabricated structures. This permits greater flexibility in designing the sequence of some critically important operations.

Ion implantation is a high precision tool, becoming increasingly popular with the more stringent requirements of advanced technology. Its use is not limited to semiconductor doping but extends to a range of other applications, some of which will be mentioned in the following chapters as each type of device is introduced.

One important drawback of ion implantation is crystal lattice damage. The energetic ions, colliding with crystal atoms, may displace some of them from their lattice positions, creating electron-hole pairs and interfering with lattice regularity. This may even cause localized regions of the crystal to lose their crystalline properties and convert to an amorphous state, adversely affecting important material characteristics (lifetime, mobility, etc.). To help restore desired crystal properties, *annealing* processes are used, generally under special precautions to minimize doping profile degradation by diffusion.

FIGURE 5.5.13
Ion implanter. (*Courtesy NJIT Microelectronics Laboratories.*)

FIGURE 5.5.14
Vacuum processing apparatus. (*Courtesy NJIT Microelectronics Laboratories.*)

5.6 ELECTRODE DEPOSITION

For use in circuits, electrical access to the devices requires appropriate electrodes, which must make low-resistance, ohmic contact with the various portions of the device. Electrode characteristics are often the limiting factor in determining device performance and reliability.

The *electrode material* must be of low resistance, especially in fast devices, where RC time constants can play a critical role. Because of this requirement, electrodes are usually metallic. Several types of metals are used, the most common at the present writing being Al and the silicides, such as $TiSi_2$, $CoSi_2$, $TaSi_2$, Pd_sSi, $NiSi_2$, etc. An increasingly important role in electrode production is being played by heavily doped polysilicon films, especially in MOSFET technologies.

The *metal to semiconductor contact* must be ohmic, so appropriate care must be taken to avoid rectifying contacts (cf. Sec. 7.8). The contact itself must be of low resistance and of excellent mechanical characteristics, not only during use in circuits but also during subsequent processing steps, which often require comparatively high temperatures.

A particular problem in this regard arises from the fact that most of these metals form eutectic alloys with Si, i.e., the system formed by the addition of metal and silicon melts at a temperature lower than the melting point of either of the two components, so they tend to dissolve into each other altering the geometry of the structure. This is particularly damaging when the phenomenon does not occur uniformly over the contact surface but is concentrated at a few points where the metal penetrates the Si, forming *spikes,* as often happens with Al. Techniques to improve contact characteristics include the interposition between electrode and semiconductor of thin layers of other metals or, especially with silicides, of heavily doped polysilicon films. This is often done for the gate contact in MOSFETS (cf. Chap. 9).

Metal electrodes are most often deposited on the semiconductor surface from the vapor phase, in a process called *metallization.* This is usually obtained by either physical or chemical means.

Chemical vapor deposition (CVD) technologies for metal deposition are quite similar to those used for dielectric deposition (Sec. 5.3), except, of course, for the materials used and the chemical reactions involved.

In *physical vapor deposition* (PVD) the wafers are placed in a vacuum chamber, often under low pressures of inert gases. Here they are exposed to the metal vapors produced by heating solid metal sources. The rate of deposition depends on the shape of the source and its position relative to the wafer, so proper positioning is required to assure uniform deposition, especially because usually several wafers are coated simultaneously.

The whole surface of the wafers may be coated and the metal film dry-etched in a photolithographic step to obtain the desired electrode configuration. Conversely, the photoresist pattern may be created on the substrate before the metal deposition, so that both the exposed regions and the photoresist mask are coated with the metallic film. Then the photoresist mask is dissolved by a suitable solvent. When this happens, the portions of the metallic film formed on the photoresist are lifted away, while the portions adhering to the substrate remain in place, again yielding the desired electrode configuration.

This second method, known as the *liftoff technique,* yields remarkably precise patterns, provided the photoresist is much thicker than the metal film, thus making the metal discontinuous at the edges of the pattern (cf. Fig. 5.6.1).

Sputtering

In the sputtering technique, a solid piece of the material to be deposited, called the *target,* is placed in the vicinity of the substrate, in a high-vacuum container. An electric field is established between an appropriately located anode and the

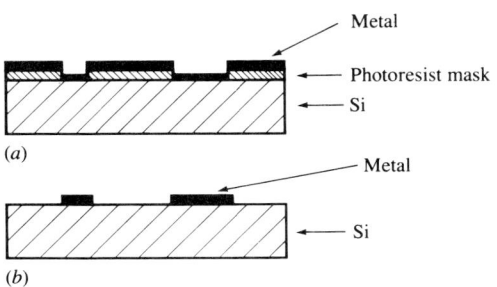

(a)

(b)

FIGURE 5.6.1

Liftoff technique of electrode generation. (*a*) Both the exposed semiconductor surface and the photoresist mask are coated with the metallic film. Notice that the photoresist mask is much thicker than the metal film, which is therefore discontinuous. (*b*) When the photoresist material is dissolved, the metal deposited on it is lifted off. The metal film formed on the semiconductor surface adheres to it and remains, constituting the electrodes.

target, which thus constitutes the cathode of the system. Ions, usually of an inert gas, are present in the vacuum chamber and are accelerated to the target, where, by impact, they extract some target surface atoms. A gas of target atoms is therefore generated by impact, rather than by heat, as in the evaporation technique. Notice that the atoms of this gas are neutral, i.e., not charged, so that they are deposited on the substrate as a neutral, low-temperature, low-kinetic-energy vapor.

By the proper choice of impact ion energy, system geometry, process duration, etc., the technique permits excellent control of the characteristics of the deposited films with minimum damaging of the substrate and its use is becoming increasingly popular, replacing less sophisticated technologies, not only for the production of device electrodes but also as a general thin-film deposition technique.

5.7 INTEGRATED CIRCUIT FABRICATION— AN INTRODUCTION

It has repeatedly been mentioned that, by operating on a wafer with an appropriate sequence of the steps described in this chapter, it is possible to fabricate a large variety of devices even if their internal structure is quite complex. Several examples of fabrication sequences will be offered in the rest of this text, when a new device is introduced. To find out how each device operates, we shall then analyze its internal structure, which, in turn, will best be described by outlining the sequence of operations used in its fabrication.

One important characteristic of the technologies described so far is that they make it easy to fabricate a large number of identical devices on one wafer. Indeed, each component of the device structure can be fabricated simultaneously on any number of devices on the same wafer by one and the same operation, so that many devices are produced "in parallel" as it were.

The economic advantages of this mode of operation are evident: many devices can be produced for essentially the price of one. However, another technical advantage can also be derived: all devices fabricated simultaneously enjoy the best probability of being well matched, i.e., of having essentially identical characteristics, so that they can be trusted to operate in a well-balanced fashion.

The next step in this line of reasoning is quite natural. Is it possible to produce simultaneously on the same wafer not only identical devices but also a variety of different circuit components? If so, then a single chip can contain not just a circuit component, but the entire circuit, automatically assembled with the attendant advantages on production time and cost, size, and uniformity of product. This can be done economically by properly scheduling the sequence of operations and by the use of masks, limiting the action of each fabrication step to selected regions of the wafer. The result is the integrated circuit (IC)—the key to the enormous success of modern semiconductor technology.

IC fabrication is based on the fundamental steps described in this chapter. However, efficient IC production also requires several fabrication artifices, and, in the final analysis, it has also profoundly changed the criteria and techniques of electronic circuit design. We shall offer here an introduction to some of the most important and general techniques and concepts.

Device Isolation

When two devices are fabricated on the same support they are also electrically connected by that support. If the devices are meant to be separate individual circuit components, then they are physically separated by cutting them apart, forming two separate chips. However, if they must operate on the same chip, then some other form of isolation must be provided.

Consider an example. In Sec. 5.5 it was seen that diffusing donor impurities into a P-type semiconductor (or acceptor impurities in an N-type) results in the formation of a PN junction. Figure 5.7.1 shows how two such structures can be simultaneously formed on the same support (shown in Fig. 5.7.1a), by a sequence

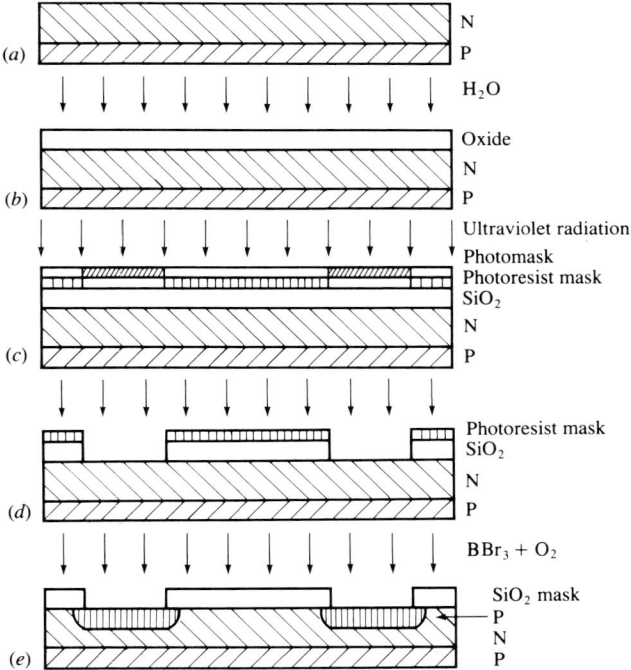

FIGURE 5.7.1

Example of the production of two identical PN junctions on the same wafer. (a) The support consists of a P-type wafer covered by an epitaxially grown N-type layer (Sec. 5.2). This layer may be typically 10 to 2 μm thick. (b) By wet oxidation, a SiO_2 layer is grown on the Si surface (Sec. 5.3). (c) A photoresist coating is exposed to uv radiation through a photomask in a photostep (Sec. 5.4). (d) After developing the photoresist mask, the SiO_2 layer is etched, obtaining a SiO_2 mask with two wells exposing the underlying Si surface (Sec. 5.4). (e) A donor impurity (B) is diffused in the exposed region of the N-type semiconductor surface to form the PN junction (Sec. 5.5).

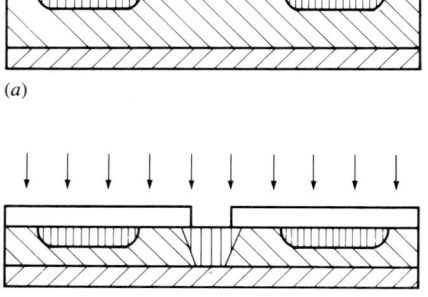

(a)

(b)

FIGURE 5.7.2
Isolation of the devices fabricated in Fig. 5.7.1. (a) New SiO$_2$ mask protecting the devices, but leaving exposed a region of the N-type silicon between them. (b) Isolation ring (p^+), obtained by reach-through diffusion.

of oxidation (b), photostep (c), etching (d), and diffusion (e). However, as evidenced by Fig. 5.7.1e, the resulting devices, having the N-type semiconductor in common, are electrically connected to each other. To make them independent some isolation must be provided.

One way of obtaining device isolation is shown in Fig. 5.7.2. By a sequence of oxidation, photostep, and etching a SiO$_2$ mask is produced, protecting the devices but exposing the silicon surface between them, as in Fig. 5.7.2a. Remember that, when the junctions were first formed, the support used consisted of an N-type layer (such as might have been epitaxially grown) over a p^+ wafer. Figure 5.7.2b indicates that, using the SiO$_2$ mask, a deep p^+ layer is now diffused between the two devices. This layer reaches all the way through the N-type semiconductor to the underlying p^+ substrate (*reach-through* diffusion).

When many structures must be isolated from one another, each device is encircled by an isolation ring, obtained by reach-through diffusion. Each device is now enclosed by a continuous p^+ shell, forming a PN junction with the N-type semiconductor.

As shall be discussed in Chap. 6, a PN junction acts similarly to an open switch, whenever the P-type semiconductor is maintained at a negative voltage relative to the N-type. Consequently, in Fig. 5.7.2b, it is sufficient to connect the p^+ substrate to a large enough negative voltage to assure that no electrical signal can be transmitted across the p^+ shell, which, therefore, effectively isolates one device from the other.

In usual practice, the isolation rings are obtained by reach-through diffusion as soon as the N-type layer is grown, at the very beginning of the fabrication sequence, delimiting device regions isolated from each other. The different devices are then fabricated simultaneously, but each within its proper region.

This often-used artifice, although effective, takes up a large portion of the wafer surface. When space is at a premium, other techniques (such as chanstop) are available.

To understand the vital importance of conserving space, consider that the number of devices fabricated on a single chip has increased exponentially since the invention of the IC. This has led to a classification of ICs: small-scale integration (SSI) for ICs with 100 or less devices, such as the early ICs, MSI (medium-scale integration) for 100 to 1000 devices, LSI (large-scale integration) for 1000 to 100,000 devices, and VLSI (very large scale integration) for more than 100,000 devices on a single chip. At the present writing, several commercially available memory chips contain in excess of one million devices per chip.

Circuit Design for ICs

The advent of IC technology has profoundly affected the criteria on which circuit design is based. Before ICs, the cost of a circuit was mainly determined by the number of active devices it contained. In ICs, the active device is usually the least expensive circuit component; capacitors and resistors take up much more of the wafer surface and the basic wired connection has become the most expensive of all circuit elements. Consequently, whenever possible, capacitors and even resistors are simulated using transistors, while inductors have virtually disappeared as circuit elements.

To assure the frequent use and so the commercial success of chips, they are often endowed with such a large gamut of capabilities that usually only a small portion of them is actually used in each application.

Furthermore, the process of design has become a strict collaboration among at least four types of specialists: the process, device, circuit, and system designers repeatedly interface with each other to achieve the most efficient, reliable, and inexpensive system. If this strict collaboration is to be successful, each of these specialists must be at least acquainted with the techniques used by the others, easily recognizing, and possibly anticipating, their specific problems, design requirements, and limitations and mastering their technical jargon and point of view.

Finally, the importance of the computer as a means of simulation, analysis, and design, and, in many cases, as a direct tool for manufacturing, is paramount in IC technology and some familiarity with the software available or at least with the approaches to problem solution by computer is a desirable, often a necessary, condition for effective membership in the design team.

The main purpose of this textbook is to provide prospective designers with at least the minimum understanding and information about the principles of device operation and fabrication required of each and every engineer involved in the design cycle.

5.8 MEASUREMENT AND TEST TECHNIQUES

The characteristics of the materials and structures fabricated by the above-described technology must, of course, be measured and verified. Following is an introductory description of some of the measurements and tests most frequently used to characterize materials and devices.

Characterization and testing constitutes a very sophisticated and specialized art, playing a fundamentally important role, not only at the research and development stage but also at the industrial production level, where statistical quality control and failure mechanism determination play a vital role in achieving reliability standards, which in turn may be the determinant factor in the commercial and industrial success or failure of a product.

The techniques described are classified on the basis of the property or physical quantity measured.

In the following descriptions of measurement procedures, reference is repeatedly made to the pertinent theory and concepts outlined in the text. The student should review this material, making sure that the theoretical background is clearly understood and mastered so that not only the measurement techniques but also the meaning of the characteristics measured, their implications, their limitations, and their roles in the process of device design and fabrication and in device utilization in circuits can be adequately appreciated.

Resistivity

As stated in Sec. 5.1, the doping of extrinsic semiconductor materials is characterized in terms of their resistivity. The relationship of resistivity to dopant concentration is given by (4.2.8), also remembering (3.3.13) and (3.3.14).

Measurements of resistivity of homogeneous materials are usually performed by the four-contact probe method of Fig. 5.8.1.

The four contacts in the probe are located along a straight line at a constant distance s between any two adjacent contacts and the probe is touched to the sample. A constant current generator injects a current I into the external contacts AA and a voltmeter V measures the difference of potential between the internal contacts BB.

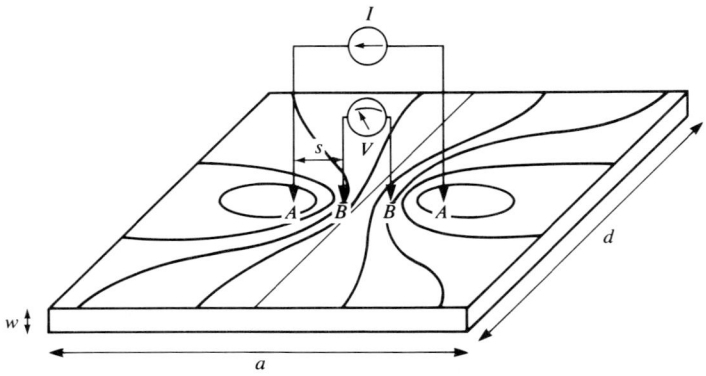

FIGURE 5.8.1

Schematic of the four-probe method of resistivity measurement. For reliable results $w \ll s$. The distribution of the equipotential lines shows that the measurement can be influenced by the dimensions of the sample.

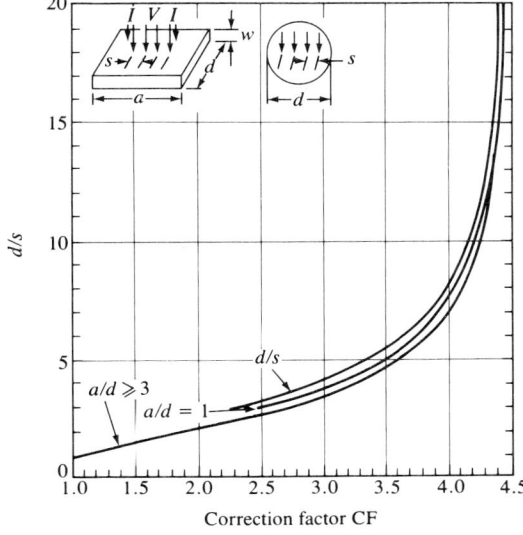

FIGURE 5.8.2
Correction coefficient for Eq. (5.8.1) for samples of dimensions comparable to the interprobe distance s [12].

The figure shows qualitatively the configuration of the equipotential lines generated in the sample. Notice that this configuration is influenced by the finite dimensions of the sample. For thin resistive layers ($w \ll s$), it is easily shown [12] that

$$\rho = \frac{V}{I} w K_F \qquad (5.8.1)$$

where K_F is a shape coefficient plotted vs. d/s in Fig. 5.8.2 and tends to 4.532 for $d \gg s$.

The condition that the sample must be thinner than the interelectrode distance ($w \ll s$) is usually amply met by both wafers and diffused resistors. For the latter, however, the condition $d \gg s$ is often not valid and the correction coefficient must be determined from Fig. 5.8.2 or by computing the distribution of the equipotential curves for the specific resistor shape.

Resistivity measurements are routinely performed on wafers to verify their doping, and, together with measurements of junction depth and of surface impurity concentration, can be used to check the type of doping profile in diffused layers.

Doping Profile

The theoretical doping profiles for diffused layers are discussed in Sec. 5.5 for both predeposition and drive-in (two-stage) diffusion. Different, precisely controlled doping profiles can be obtained by ion implantation.

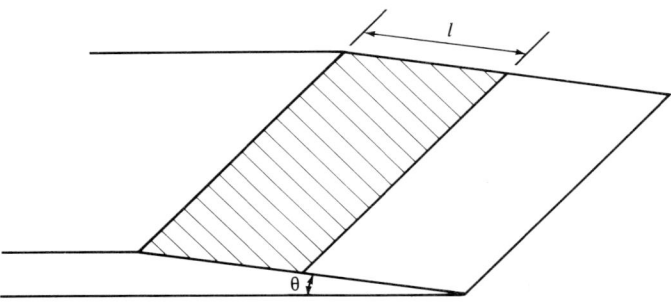

FIGURE 5.8.3
Schematic of microscopic measurement of junction depth. The P-type semiconductor is chemically stained darker than the N-type.

Doping profiles can be evaluated by reverse-biased capacitance measurements. These techniques will be discussed in Chaps. 6 and 9. Electrical measurements and predictions made from them are often compared with optical microscopic measurements of diffused layer parameters.

Junction Depth

This parameter can be measured by optical inspection. A wedge-shaped cross section of the junction is obtained by lapping the sample at a small angle θ from the surface, as shown in Fig. 5.8.3. The exposed section of the junction is then treated with a chemical which stains the P-type semiconductor darker than the N-type. Observation under a microscope permits measurement of the length l of the section of the top region. The junction depth is then

$$x_d = l \sin \theta \approx l\theta \qquad (5.8.2)$$

for $\theta \leq 1°$. There are several variants to this measurement. Often a groove of known shape is ground in the junction. After staining, microscopic measurements are made and the junction depth is determined by appropriate formulas depending on the groove shape. Microscopic measurements of junction depth are usually compared with theoretical computations and with the results of profile measurement.

Electron Microscope and Optical Measurements

The characteristics of layers and surfaces and the dynamics of fabrication processes are often checked by electron microscope observations. These observations yield extremely representative pictures containing a very large amount of detailed information. Often these techniques also involve rather delicate mechanical or chemical specimen preparation and may require skilled operators for their interpretation, especially when quantitative results are desired. Electron microscope observation is a powerful tool, used mainly in the research and development stage of device technology.

Other optical measurements, such as laser and holographic interferometry, are also in use and some of them have been automated (cf. Fig. 5.8.4) [13].

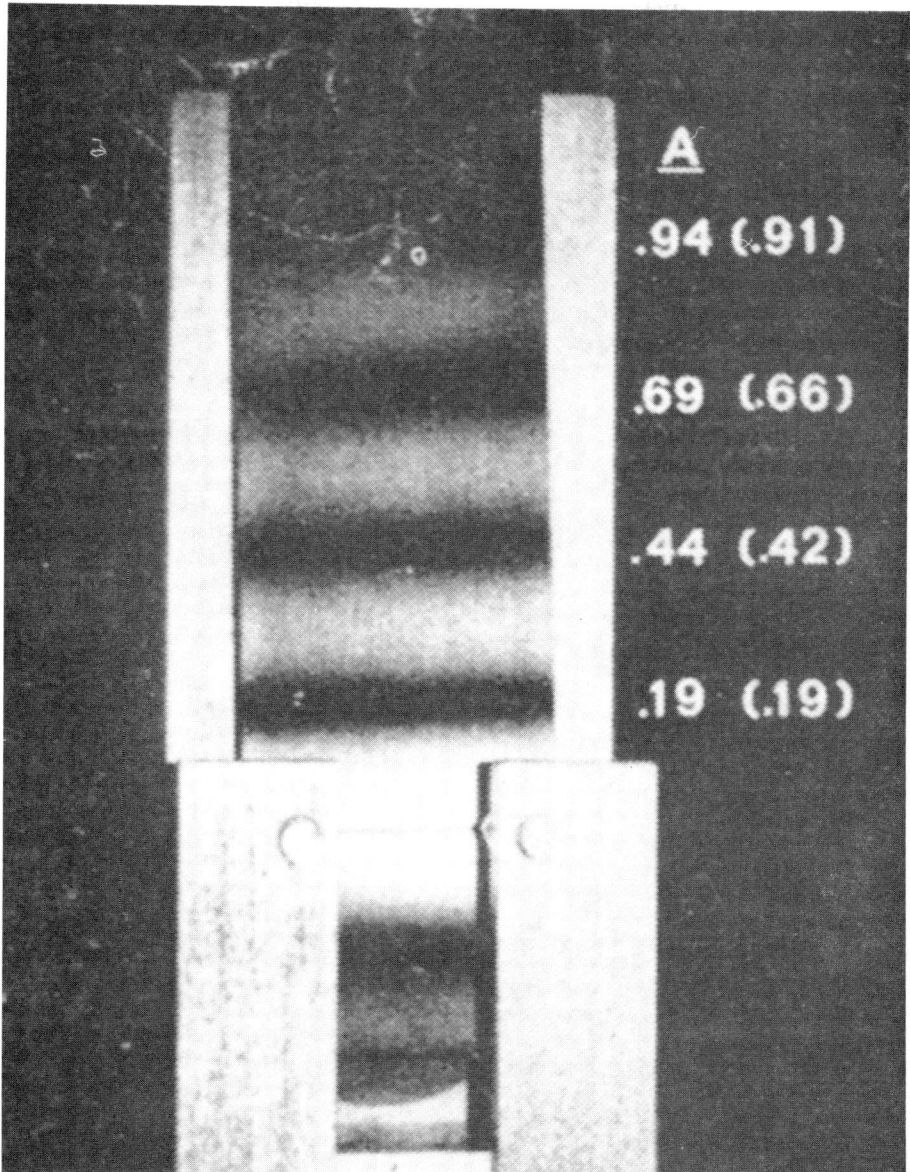

FIGURE 5.8.4
Typical interferometric hologram reconstruction showing distribution of surface displacement [13].

Hall Effect

Consider the specimen in Fig. 5.8.5. It is subjected to a magnetic field **B** (in the $-z$ direction in the figure) while a current density **J** is made to flow through it in a direction perpendicular to the magnetic field (in the **y** direction in the figure). The Lorentz force acting on each current-carrying charge is

$$\mathbf{F} = q(\mathscr{E} + \mathbf{B} \times \mathbf{v}) \tag{5.8.3}$$

where **v** is the charge velocity. However, $\mathbf{J} = \rho_v \mathbf{v} = Nq\mathbf{v}$, where N is the current carrier concentration and q the charge of the individual current carrier. At equilibrium $\mathbf{F} = 0$ and so

$$\mathbf{B} \times \mathbf{v} = \frac{1}{qN} \mathbf{B} \times \mathbf{J} = -\mathscr{E} = \frac{BJ}{qN} \mathbf{u}_x \tag{5.8.4}$$

in which $\mathbf{u}_x$ is the unit vector of the x axis and the cross-product has been computed remembering the orientations of the several vectors of Fig. 5.8.5. In conclusion,

$$\mathscr{E}_x = -\frac{BJ}{qN} = -\frac{V}{l} \tag{5.8.5}$$

from which

$$V = \frac{BJl}{qN} \tag{5.8.6}$$

showing that, under the conditions of the experiment, a voltage V is generated across the specimen.

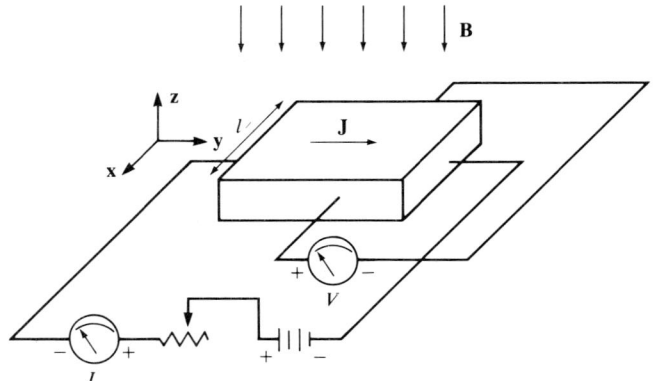

FIGURE 5.8.5
Schematic of Hall effect measurement. V is the Hall effect voltage developed in the specimen under the magnetic flux density **B** and the current density **J**. Orientations and meter polarities correspond to the values and sign conventions of Eqs. (5.8.3) through (5.8.6).

Some thought should be given to the polarity of the Hall effect voltage V. With the orientations of the magnetic flux density **B** and of the current density **J** shown in the figure, the voltage V is developed across the x dimension of the specimen, as shown. As for its polarity, from Eq. (5.8.5), the x component of $\mathscr{E}$ is oriented in the direction opposite to the quantity BJ/qN, and so in the $-\mathbf{u}_x$ direction if $q > 0$ and in the $+\mathbf{u}_x$ direction if $q < 0$. Remembering that $\mathscr{E}$ always points from high toward low potential, it can be concluded that the voltmeter polarity shown in the figure is such that, if $q > 0$ (carriers are holes), then $V > 0$, and if $q < 0$ (carriers are electrons), then $V < 0$.

This conclusion can be confirmed by considering charge displacements in the specimen for the two possible cases:

1. Hole carriers ($q > 0$). The velocity **v** is in the $\mathbf{u}_y$ direction (same direction as **J**); therefore $\mathbf{B} \times \mathbf{v} = Bv(-\mathbf{z} \times \mathbf{y}) = Bv\mathbf{u}_x$ is in the $+\mathbf{u}_x$ direction and the force acting on the holes is $\mathbf{F} = qBv\mathbf{u}_x$, also in the $\mathbf{u}_x$ direction, because q is positive. We conclude that holes are pushed toward the front face of the specimen and accumulate there, making it positive and generating a positive voltage as measured by a voltmeter with the polarity shown in the figure.
2. Electron carriers ($q < 0$). The velocity **v** is in the $-\mathbf{u}_y$ direction (opposite to the current flow), so that $\mathbf{B} \times \mathbf{v} = Bv(-\mathbf{u}_z) \times (-\mathbf{u}_y) = -Bv\mathbf{u}_x$ is in the negative $\mathbf{u}_x$ direction. The force acting on the electrons if $\mathbf{F} = -qBv\mathbf{u}_x$ and, as q is negative, is in the $+\mathbf{u}_x$ direction, so that electrons are pushed toward the front face of the specimen and accumulate there, making this face negative (because their charge is negative). A voltmeter connected with the polarity shown in the figure will therefore register a negative voltage.

If B, J, and l are known, measurement of the transverse Hall effect voltage V permits the computation of qN in *magnitude and sign*, showing that Hall effect measurements permit determination of the net concentration of dopant impurities in semiconductor specimen and also detect the type of the majority carriers (electrons or holes).

Hall effect measurements are not limited to the determination of extrinsic semiconductor type and doping concentration: several other properties of conducting materials, not only in the solid state but also in liquid or plasma conditions, can be detected and measured by this technique, which constitutes a widely used means of research and development.

Carrier Lifetime

In Sec. 4.4 it was shown that the action of an external source of energy on an extrinsic semiconductor may raise the carrier generation rate above the thermal value, finally increasing the minority carrier concentration at steady state. If the external source is abruptly removed, recombination becomes prevalent and the minority carrier concentration relaxes exponentially to the thermal equilibrium

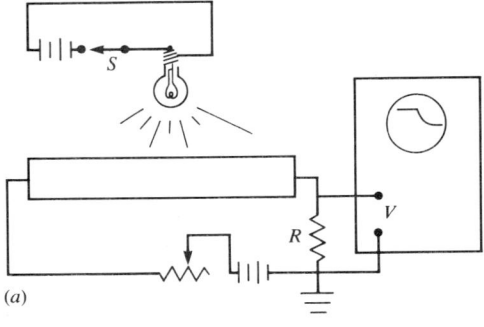

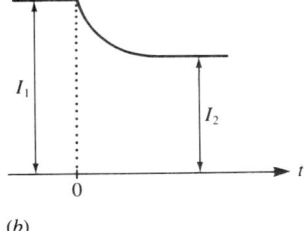

(a)

(b)

FIGURE 5.8.6
Measurement of carrier lifetime. (a) Experimental arrangement. At $t = 0$ switch S is opened and the source of additional carrier generation is extinguished. (b) Oscilloscope display. I_1 indicates enhanced conduction, I_2 dark conduction.

value, with a time constant τ defined as the carrier lifetime. This important characteristic can be measured by simply reproducing under laboratory conditions the phenomenon of relaxation to thermal equilibrium described above.

A current is made to flow in the semiconductor sample under test and is displayed on a scope, as shown schematically in Fig. 5.8.6a.

If the sample is flooded with light, then, after steady state is reached, the current shown by the scope is constant and proportional to the enhanced carrier concentrations given for P-type semiconductors by (4.4.10) (similar expressions hold for N-type semiconductors). If the light source is suddenly extinguished, the scope displays a time-varying current, in accordance with (4.4.12), as shown in Fig. 5.8.6b. The minority carrier lifetime can then be measured as the time constant of the displayed decay.

Notice that, for accurate measurements, the current must at all times remain proportional to the carrier concentration. Remembering that the conductivity is expressed by (4.2.7), it is seen that, to ensure this proportionality, some care must be taken to keep the voltage applied to the sample constant. In most cases it is sufficient to assure that the voltage variation across the resistor R is much smaller than the voltage drop across the sample.

In practice, for measurements of intrinsic semiconductors, this technique may require some experimental skill. Indeed, (4.2.7) shows that the conductivity is determined primarily by the majority carrier concentration. Under conditions of low-level injection, by definition the majority carrier concentration remains essentially constant, so that the difference between I_1 and I_2 in the figure may be very small, making observation difficult and computations imprecise. Differential test techniques may help the measurement, but may require some careful experimental adjustments.

The Haynes-Shockley Experiment

This technique can be used to measure several important parameters, such as mobility, diffusivity, and even minority carrier lifetime. Figure 5.8.7 shows the experimental arrangement for an N-type semiconductor (variations for P-type measurements are self-evident).

At time $t = 0$, the pulse generator produced a positive pulse of voltage $v(t)$, triggering the oscilloscope time axis. At the same time the pulse forward-biases the metal to the semiconductor point contact junction A, which injects into the semiconductor a pulse of minority carriers at the position x_0 of the point contact. The mechanism causing this injection will be analyzed in some detail in Chap. 7, when metal to semiconductor contacts will be investigated. For the moment the student is asked to accept the fact that, under the conditions described, a spurt of minority carriers is injected into the body of the semiconductor. We can then proceed to analyze the behavior of these injected charges.

Let N be the total number of minority carriers injected per unit contact area. The enhanced hole population must satisfy the continuity equation (4.5.16). Two cases will be considered:

1. If switch S is open, then there is no significant electric field in the semiconductor, $\mathscr{E} = 0$, there is no drift current, and the cloud of injected holes shown

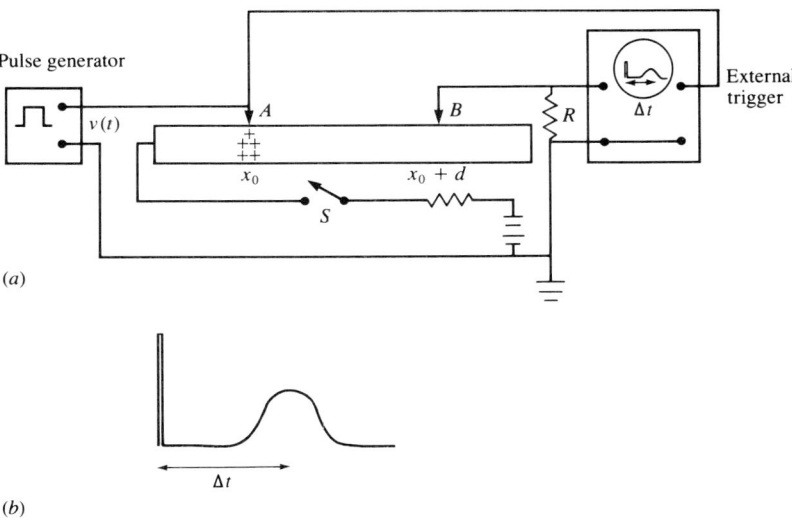

(a)

(b)

FIGURE 5.8.7
Haynes-Shockley experiment: (a) experimental arrangement, (b) oscilloscope display.

in Fig. 5.8.7 at position x_0 moves exclusively by diffusion. It can therefore be expected that the localized high concentration of holes will spread along the length of the semiconductor, recombining as it goes; as times goes by it will "flatten out" and disperse.

Mathematically, as $\mathscr{E} = 0$ and $G_{hA} = 0$ (there is no extra injection after the pulse is over), Eq. (4.5.16) becomes

$$\frac{\partial p_N}{\partial t} = D_h \frac{\partial^2 p_N}{\partial x^2} - \frac{p_N - p_{oN}}{\tau_h} \tag{5.8.7}$$

Assuming as boundary conditions that at $t = 0$ the holes injected are all concentrated at one point, so that the hole concentration at $t = 0$ is an impulse (δ function) and that, after a long time, the concentration relaxes to the thermal equilibrium value p_{oN}, then the solution to this partial differential equation becomes

$$p_N(x, t) = p_{oN} + \frac{N}{\sqrt{4D_h t}} \exp -\left[\frac{(x - x_0)^2}{4D_h t} + \frac{t}{\tau_h}\right] \tag{5.8.8}$$

where, as already stated, N is the number of holes per square centimeter of contact surface injected at the point contact A of abscissa x_0. It is seen that the initial impulse function spreads, as expected, into a gaussian distribution in space, while its amplitude decreases with time constant τ_h, i.e., the distribution of the concentration spreads and flattens out.

2. If switch S is closed, then a uniform field $\mathscr{E}$ in the $\mathbf{u}_x$ direction is present in the semiconductor, resulting in a small current proportional to the carrier concentration, flowing through the specimen into resistor R. We can expect this field to add a drift component to the motion of the cloud of injected holes, so that, while spreading by diffusion, as in case (1), the cloud also moves to the right at a constant drift speed.

Mathematically, with $\mathscr{E} = \mathscr{E}\mathbf{u}_x = $ constant (so that $\partial\mathscr{E}/\partial x = 0$), Eq. (4.5.16) becomes

$$\frac{\partial p_N}{\partial t} = D_h \frac{\partial^2 p_N}{\partial x^2} - \mu_h \mathscr{E} \frac{\partial p_N}{\partial x} - \frac{p_N - p_{oN}}{\tau_h} \tag{5.8.9}$$

the solution to which becomes

$$p_N(x, t) = p_{oN} + \frac{N}{\sqrt{4D_h t}} \exp -\left[\frac{(x - x_0 - \mu_h \mathscr{E}t)^2}{4D_h t} + \frac{t}{\tau_h}\right] \tag{5.8.10}$$

which indicates that the cloud of injected holes spreads into a gaussian space distribution and flattens out, just as in case (1), but it also drifts to the right with a drift velocity:

$$v_d = \mu_h \mathscr{E} \tag{5.8.11}$$

as expected from the previous qualitative analysis.

Notice that in this drift motion, the cloud will eventually reach the contact B, where its localized higher carrier concentration will generate an increment in the drift current normally flowing through R and produce a peak of voltage at the oscilloscope input. The scope display will then look as in Fig. 5.8.7b. If the distance between A and B is d, as in the figure, then, remembering Eq. (5.8.11), the delay with which the peak of the cloud reaches B and is detected by the scope is

$$\Delta t = \frac{d}{\mu_h \mathscr{E}} \qquad (5.8.12)$$

As both $\mathscr{E}$ and d are known and Δt can be measured from the oscilloscope trace, μ_h can easily be computed.

Once μ_h is known, D_h can be computed from the Einstein relations (4.3.2) and a comparison of the shape and amplitude of the transmitted vs. the original pulse permits computation of τ_h from Eq. (5.8.10). This last determination is usually rather difficult and much experimental skill is required in order to obtain more than order-of-magnitude indications.

5.9 SUMMARY

Fabrication techniques are based on a few basic steps.

Wafer preparation grows large single-crystal wafers. The most usual techniques are the Czochralski, the zone melting, and the Bridgman techniques. Wafers are then sliced, brought to specific dimensions, finished, and coded.

Doping is characterized by the material's conductivity.

Precisely controlled layers of single-crystal semiconductor with the desired doping and the same crystallographic characteristics as the support are grown by epitaxy, which can be vapor phase, liquid phase, or molecular beam. The epitaxy process, under proper conditions, can result in etching.

Oxidation, dry or wet, is often used to produce SiO_2 layers on Si semiconductors. These layers can be used as masks, or passivation or insulator layers. In some cases their quality and dimensions must be controlled with great accuracy.

The photolithographic step is used to limit the action of subsequent procedures to selected regions of the semiconductor. It is a crucial and usually expensive step, requiring high-precision registration. It usually consists of the production of a photomask, which is then transferred to a layer of photoresist poured and dried on the surface being processed by a printing and developing procedure to form the photoresist mask. This in turn is usually used to control the etching of a SiO_2 mask which exposes selected regions of the Si to further processing (such as diffusion, ion implantation, etc.).

Diffusion of impurities permits the doping of selected portions of the device under fabrication to be varied following a desired doping profile. The predeposition profile, of the erfc type, can be modified and smoothed into a gaussian profile by the drive-in diffusion process. Ion implantation is a higher precision process to perform a similar task.

Electrodes and other materials can be deposited on the surface being processed by metallization and other deposition processes, such as chemical and physical deposition and sputtering.

The above steps permit the simultaneous fabrication of large numbers of identical devices on one wafer, affording economic production of devices with consistent characteristics. The simultaneous production of different types of devices on one wafer has resulted in the techniques of integrated circuit (IC) fabrication, developing by subsequent steps into MSI, LSI, and VLSI, in which circuits containing in excess of a million devices are produced on one chip by a sequence of the fundamental steps outlined.

Integrated circuit fabrication requires special technologies, including ways to isolate the different devices from one another.

Testing and verification of device properties is often referred to as characterization. Some important techniques are: resistivity measurements (the four-probe technique), doping profile measurements to be described later, junction depth measurements and other optical measurements, mobility, diffusivity, and other measurements that can be performed using the Hall effect or the Haynes and Shockley experiment procedure, and minority carrier lifetime measurements, often performed through oscillographic observation of relaxation to thermal equilibrium.

APPENDIX 5A
OXIDATION DYNAMICS

Determination of the appropriate parameters (temperature, processing duration, etc.) of an oxidation step requires a quantitative description of the process. To this end, let N be the concentration distribution of the diffusing oxidizing species. The diffusion flux rate density is (cf. Chap. 4)

$$F_D = -D \frac{dN}{dx} \qquad \text{cm}^{-2} \text{ s}^{-1} \qquad (5A.1)$$

The oxidation reaction rate F_r is proportional to the reagent concentration and the reaction rate coefficient γ:

$$F_r = \gamma N \qquad \text{cm}^{-2} \text{ s}^{-1} \qquad (5A.2)$$

As F_r is, by definition, the rate at which the oxidizing molecules are incorporated into the newly formed SiO_2, then, at steady state, $F_D = F_r$ and

$$-D \frac{dN}{dx} = \gamma N \qquad (5A.3)$$

with solution:

$$N = N(0) \, e^{-(\gamma/D)x} \tag{5A.4}$$

where $N(0)$ is the concentration of the reagent at the free surface. Then, by (5A.1),

$$F = -D \frac{dN}{dx} = \gamma N(0) \, e^{-(\gamma/D)x} \tag{5A.5}$$

For thin layers ($x \ll D/\gamma$), substituting the exponential with the first two terms of its McLaurin series expansion:

$$F \approx \frac{N(0)D}{x + D/\gamma} \tag{5A.6}$$

The rate of growth of the oxide equals the rate of flux F of the molecules of the oxidizing species divided by the number N_{ox} of oxidant molecules required to build a unit volume of SiO_2:

$$\frac{dx}{dt} = \frac{F}{N_{ox}} \approx \frac{[N(0)/N_{ox}]D}{x + D/\gamma} \tag{5A.7}$$

where x is the oxide thickness and N_{ox} equals the concentration of SiO_2 molecules in the oxide crystal (or 2.2×10^{22} in SiO_2) times the number of oxidant molecules required to generated one SiO_2 molecule (one for O_2, two for H_2O). This means that, for the SiO_2 case, $N_{ox} = 2.2 \times 10^{22}$ cm^{-3} for dry oxidation and $N_{ox} = 4.4 \times 10^{22}$ cm^{-3} for wet processing. Separating the variables in (5A.7),

$$\left(x + \frac{D}{\gamma}\right) dx = \frac{N(0)}{N_{ox}} D \, dt \tag{5A.8}$$

Integrating under the assumption that, when the oxidation process begins, there already exists an initial oxide layer of thickness x_0:

$$\frac{x^2}{2} + \frac{D}{\gamma} x = \frac{N(0)}{N_{ox}} Dt + \left(\frac{x_0^2}{2} + \frac{D}{\gamma} x_0\right) \tag{5A.9}$$

and, introducing the parameters,

$$A = \frac{2D}{\gamma}; \qquad B = \frac{2N(0)}{N_{ox}} D; \qquad \tau = \frac{x_0^2 + A x_0}{B} \tag{5A.10}$$

finally,

$$x^2 + Ax = B(t + \tau) \tag{5A.11}$$

Therefore, after an oxidation processing of duration t, the thickness of the oxide layer produced is

$$x = \frac{A}{2}\left(-1 + \sqrt{1 + \frac{4B(t - \tau)}{A^2}}\right) \tag{5A.12}$$

For short duration processes (thin oxide layers) in which $A \gg x$, the approximate thickness becomes

$$x \approx \frac{B}{A}(t + \tau) \qquad (5A.13)$$

For long oxidation with $A \ll x$, (5.3.13) yields the approximate oxide thickness

$$x \approx \sqrt{B(t + \tau)} \qquad (5A.14)$$

The approximate expressions (5A.13) and (5A.14) indicate that, in the early stages of thermal oxidation (reaction controlled), the oxide layer builds up linearly with time, with a *linear rate constant* B/A, while as oxidation proceeds and becomes diffusion controlled, the rate of growth tends to a square law, with a *parabolic rate constant* B.

In practice, oxide layers of thickness much larger than $A/2 = D/\gamma$ are often grown, although this is contrary to the assumption on which (5A.6) is based, and so, rigorously, the whole of the above discussion is invalid. Yet, in most cases, (5A.11) through (5A.14) predict actual measured values with acceptable accuracy.

PROBLEMS

5.1. A Si semiconductor boule with acceptor concentration $N_A = 2 \times 10^6$ cm^{-3} must be obtained by the Czochralski technique. What weight of B must be added to the melt if it contains 10 kg of Si?

5.2. An epitaxial layer 12 μm thick is to be grown on a specimen. The duration of the process should be 720 seconds. The gas mixture constituting the oven atmosphere is obtained by mixing the following volumes of gases at the same pressure at a temperature of 1270°C: 1050 cm^3 of H_2 and a volume V_x of $SiCl_4$. What is the required volume V_x?

5.3. The atmosphere in an oven at 1270°C contains 3.856 g of H_2 and 10 g of $SiCl_4$. If a 0.1-mm thick specimen is processed in the oven for 10 minutes, what is the thickness after the process is over?

5.4. A specimen is processed in an epitaxy oven at 1270°C for 10 minutes. It is desired to etch 8.5 μm from the surface of the specimen. What weight of $SiCl_4$ must be added to 4 g of H_2 to obtain the proper gas mixture for the oven atmosphere?

5.5. Obtain Eq. (5.3.4) from Eq. (5.3.3).

5.6. A Si surface, partially covered by a SiO_2 layer of 0.5 μm thickness, is further oxidized by the wet process at 1373 K for 2 hours. Supposing the surface cut has Miller index $\langle 100 \rangle$, compute the thickness of the resulting oxide layer (*a*) over the exposed Si area and (*b*) over the area originally covered by the 0.5-μm oxide layer. (*c*) Comment on the result.

5.7. A layer of gate oxide of very high quality and thickness 0.5 μm is to be grown on a Si surface cut at Miller index $\langle 100 \rangle$. Oxidation is to occur at 1200°C.
(*a*) Is dry or wet oxidation preferable?
(*b*) How deeply will the oxide penetrate below the original Si surface?
(*c*) What is the duration of the oxidation?

5.8. Two square wells with sides of 1 μm must be etched at a distance of 9 μm from each other, centered along the median line of a rectangular SiO_2 surface of dimensions 3×15 μm. Draw artwork for a photomask suitable for photoreduction with a reduction ratio of 5000 to 1 for contact printing on (a) positive photoresist and (b) negative photoresist.

5.9. Repeat Prob. 5.8 for negative photoresist and projection printing with a 2 to 1 linear image ratio.

5.10. (a) Write a formula suitable for the computer calculation of the diffusivity of B in Si as a function of temperature.

  Hint: Find the appropriate coefficients from Fig. 5.5.6.

  (b) Find the activation energy of B in Si.

  Hint: Use Fig. 5.5.3.

5.11. Compute the junction depth resulting from predeposition of B in Si doped with $N_D = 4 \times 10^{15}$ cm^{-3}, if predeposition occurs at 1050°C for 10 h.

5.12. For the conditions described in Prob. 5.11, how long should the predeposition be to set the junction depth at 0.2 μm?

5.13. The resistivity of a square sample of N-type Si semiconductor of 5 cm side and thickness 0.5 mm is measured by means of a four-contact probe with interelectrode distance constant $s = 1$ cm. The constant current generator provides a 100-mA current and the voltmeter reading is 2.8 V. The temperature is 300 K.

  (a) What is resistivity?

  (b) If only one impurity species is present, what is the majority carrier mobility?

  (c) What is the doping?

5.14. Four-contact probe measurements on a sample of N-type Si semiconductor yield a resistivity of 0.55 $\Omega \cdot$ cm. The Haynes-Shockley experiment, performed on the same sample, using a voltage of 20 V across a 2-cm length and with a distance between the probes A and B (Fig. 5.8.7) of 1 cm, yields a delay time between sent and received pulses of 0.435 ms. Compute the net and total doping of the sample.

REFERENCES

1. Runyan, W. R.: *Silicon Semiconductor Technology*, McGraw-Hill Book Company, New York, 1965.
2. Ghandi, S. K.: *VLSI Fabrication Principles*, John Wiley & Sons, New York, 1983.
3. Grove, A. S.: *Physics and Technology of Semiconductor Devices*, John Wiley & Sons, New York, 1967.
4. Cho, A. Y.: "Growth of III–V Semiconductors by Molecular Beam Epitaxy and Their Properties," *Thin Solid Films*, vol. 100, p. 291, 1983.
5. Sze, S. M.: *Semiconductor Devices, Physics and Technology*, John Wiley & Sons, New York, 1985.
6. Deal, B. E., and A. S. Grove: "General Relationships for the Thermal Oxidation of Silicon," *J. Appl. Phys.*, vol. 36, p. 3770, 1965.
7. Abramowitz, M., and I. A. Stegun: *Handbook of Mathematical Functions*, Dover Publications Inc., New York, 1964.
8. Streetman, B. G.: *Solid State Electronic Devices*, Prentice-Hall, Inc., Englewood Cliffs, N.J., 1972.
9. Irvin, J. C.: "Resistivity of Bulk Silicon and of Diffused Layers in Silicon," *BSTJ*, vol. 41, p. 387, March 1962.
10. Glasser, A., and G. Subak-Sharpe: *Integrated Circuit Engineering*, Addison-Wesley, Reading, Mass., 1977.
11. Mayer, J. W., L. Erikcson, and J. A. Davies: *Ion Implantation in Semiconductors*, Academic Press, New York, 1971.

12. Beadle, W. F., J. C. C. Tsai, and R. D. Plummer (eds.): *Quick Reference Manual for Semiconductor Engineers*, John Wiley & Sons, New York, 1985.
13. Zambuto, M., and M. Lurie: "Holographic Measurement of General Laws of Motion," *Appl. Optics*, vol. 9, p. 2066, September 1970.

ADDITIONAL READING

Casey, H. C., Jr., and G. L., Pearson: "Diffusion in Semiconductors," in *Point Defects in Solids* (eds. J. H. Crawford, Jr. and L. M. Slifkin), vol. 2, Plenum Press, New York, 1975.
Fortino, A. G.: *Fundamentals of Integrated Circuit Technology*, Reston Publishing Company, Inc., Reston, Va., 1984.
Keller, S. P. (ed.): *Handbook on Semiconductors*, North Holland, Amsterdam, 1980.
Sze, S. M. (ed.): *VLSI Technology*, McGraw-Hill Book Company, Inc., New York, 1983.

CHAPTER
6

PN JUNCTION

The PN junction plays a fundamental role in almost all semiconductor devices. In its basic form—the PN diode—it is used in a large number of applications with which the reader is certainly familiar from circuit courses. The following discussion provides a basis for the understanding of and an introduction to the quantitative analysis of a variety of diode circuits including not only rectifiers (Secs. 6.6, 6.5, Probs. 6.33, 6.34), switches (Sec. 6.8), and detectors, but also reference voltage sources (Sec. 6.5, App. 6A, Prob. 6.32), voltage controlled capacitors (Secs. 6.5, 6.8, Prob. 6.19, 6.20, 6.21), current controlled resistors (Sec. 6.8, Prob. 6.31), and negative resistance applications (Sec. 6.9).

6.1 PN DIODE STRUCTURE

Figure 6.1.1a shows the circuit symbol of a PN diode identifying its *cathode* (K) and *anode* (A) terminals. The physical structure is shown in principle in Fig. 6.1.1b. Moving along the device from anode to cathode the electrical path between the metallic terminals consists first of a P-type and then of an N-type semiconductor. The two crystals are so intimately joined at the interface that, theoretically, their lattice structure is continuous and unperturbed; thus, at least as far as their lattice is concerned, they constitute one continuous crystal. This structure is a *PN junction*.

In real diodes, the PN junction of Fig. 6.1.1b is implemented in a large variety of structures, to improve performance and/or optimize fabrication condi-

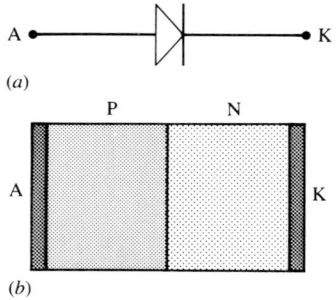

(a)

(b)

FIGURE 6.1.1
The PN diode. Electrodes are anode (A) and cathode (K). (a) Circuit symbol. (b) Physical structure in principle.

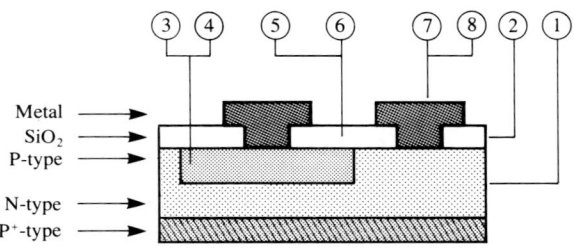

FIGURE 6.1.2
Physical structure of a possible planar implementation of the PN diode. The numbers within circles refer to the fabrication steps listed in the text.

tions. One such structure, of the planar type, is shown in Fig. 6.1.2. The basic steps of the fabrication are listed in the following description. In Fig. 6.1.2 the portions of the structure resulting from each step are indicated by the corresponding step number enclosed in a circle. This figure should be referenced while reading the fabrication step sequence. The masks used in the photolithographic steps are outlined in simplified form in Fig. 6.1.7, assuming the photoresist to be of the negative type (cf. Sec. 5.4).

1. An N-type semiconductor layer is epitaxially grown on a p^+ substrate. This p^+ layer can be used, under proper bias, for device isolation, as discussed in Sec. 5.7. If isolation is not required, this substrate can be of the n^+ type.

2. A thick masking oxide layer is grown on the whole face of the wafer, usually by the wet process.

3. Using a mask (#1) and a photostep, a well is etched in the thick oxide to expose the surface of the underlying N-type semiconductor, where the P-type anode region will be fabricated.

4. By a diffusion process, using the thick oxide as a mask, a P-type layer is diffused into the anode region. This forms the junction. Figure 6.1.3 shows the result of steps 1 through 4. At this stage of the fabrication, to make it possible to inspect the inner structure of the device, we shall display a cross section of the specimen along two perpendicular planes, indicated by the cleavage lines in the figure. This cross-section is shown in Fig. 6.1.4, where the different materials and semiconductor types are identified by different shadings and indicated by the arrows.

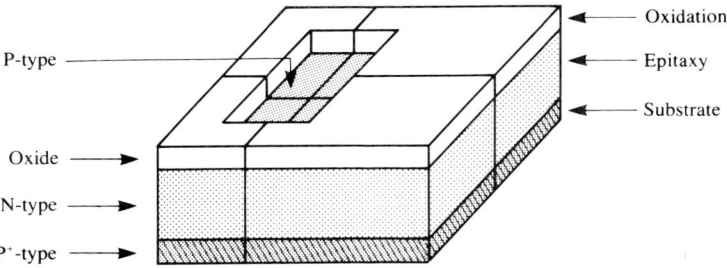

FIGURE 6.1.3
State of the device after step 4 of the fabrication sequence described in the text. The cleavage lines indicated on the surface refer to a possible cross section for demonstration purposes.

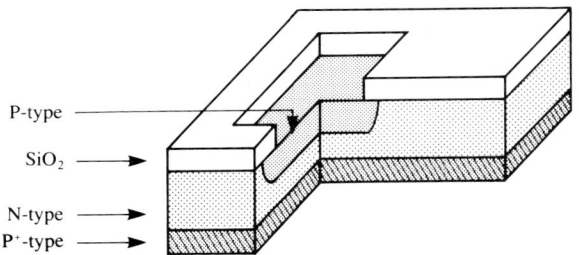

FIGURE 6.1.4
Cross section of Fig. 6.1.3 to show the internal structure of the device at this stage of fabrication.

5. The thick oxide is now etched off and the surface of the wafer is again oxidized. For reliability reasons, this insulating oxide layer is made thinner than the previous mask oxide.

6. Using a mask (#2) and a photostep, two wells are etched in the oxide, to expose a region of the P-type and a region of the N-type semiconductor.

7. By a metallization process, a metal layer is deposited on the whole wafer face.

8. Using a mask (#3) and a photostep, the metal is etched to isolate the two electrodes from each other and give them the desired shape.

The final result of this sequence of operations is shown in Fig. 6.1.5 and, in cross section, in Fig. 6.1.6. Figures 6.1.3 through 6.1.7 should be compared with Fig. 6.1.2.

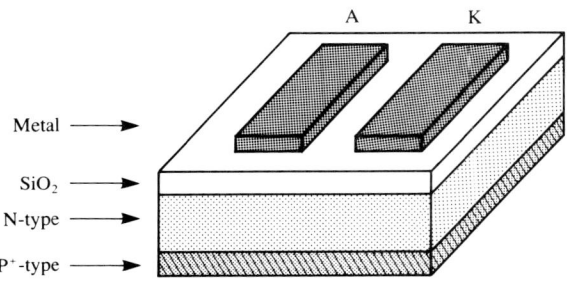

FIGURE 6.1.5
Complete planar PN diode: external view showing anode and cathode terminals.

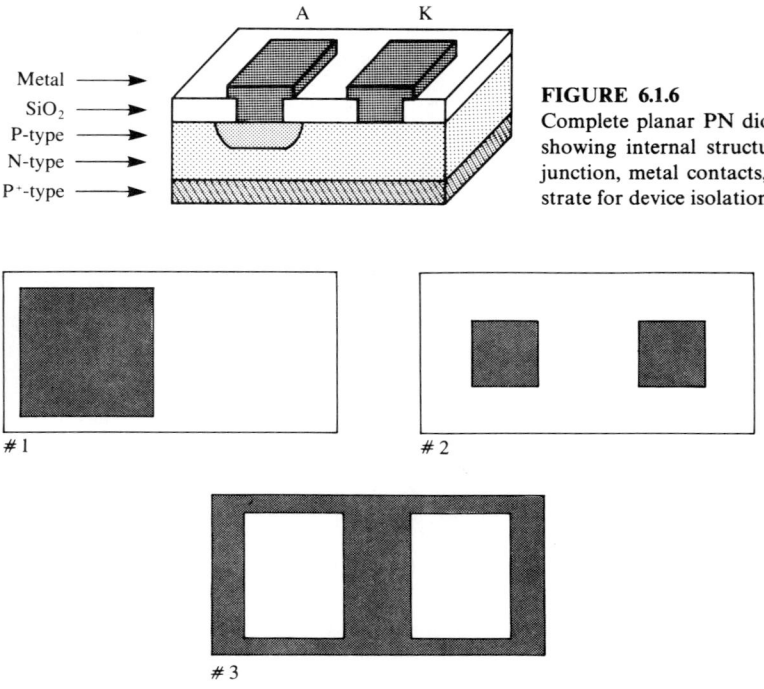

FIGURE 6.1.6
Complete planar PN diode: cross section showing internal structure including PN junction, metal contacts, and P-type substrate for device isolation.

FIGURE 6.1.7
Masks for the diode fabrication sequence outlined in the text. These masks are purely indicative; actual mask shapes may be very different. Negative photoresist is assumed.

This is one of many possible fabrication sequences. Many widely different procedures are employed in actual practice, depending on the use to which the device is destined.

For instance, in step 5, instead of stripping away the masking oxide and substituting it with a new layer of insulating oxide, the device is often subjected to a second oxidation process *without previously stripping the masking oxide*. During this process, oxidation proceeds at a fast rate over the exposed semiconductor areas, covering them with a well-developed insulation layer (reaction limited oxidation, cf. Sec. 5.3). Over the surface already covered by the masking oxide, oxidation is much slower (diffusion limited) and results in only a minor increment of the oxide thickness, as discussed in Chap. 5 (cf. Figs. 5.3.4 and 5.3.5), so that the final oxide coating will be of essentially uniform thickness.

In this case, the masking oxide layer deposited in step 2 will remain on the final device as part of the surface insulating layer. As dictated by reliability considerations, therefore, this layer cannot be very thick, else, being too rigid, it might produce stresses due to possible thermal expansion, severely damaging the device. On the other hand, the oxide layer must be thick enough to constitute an efficient mask for the diffusion process of step 4 (cf. Fig. 5.5.11). The actual thickness is chosen by compromise between these two requirements. Substitution of an implantation process, instead of the diffusion of step 4, may be part of the solution.

Particularly dramatic variations in procedure are adopted when the diode is an element of an integrated chip, rather than an isolated device (cf. Sec. 5.7).

6.2   THE CONTACT POTENTIAL

Figure 6.2.1a shows the energy-position diagrams of a P-type and an N-type semiconductor. The familiar notation is used, but an extra subscript is added: P for P-type, N for N-type semiconductor parameters. The picture also shows schematically the location of the current carriers and of the impurity ions in the

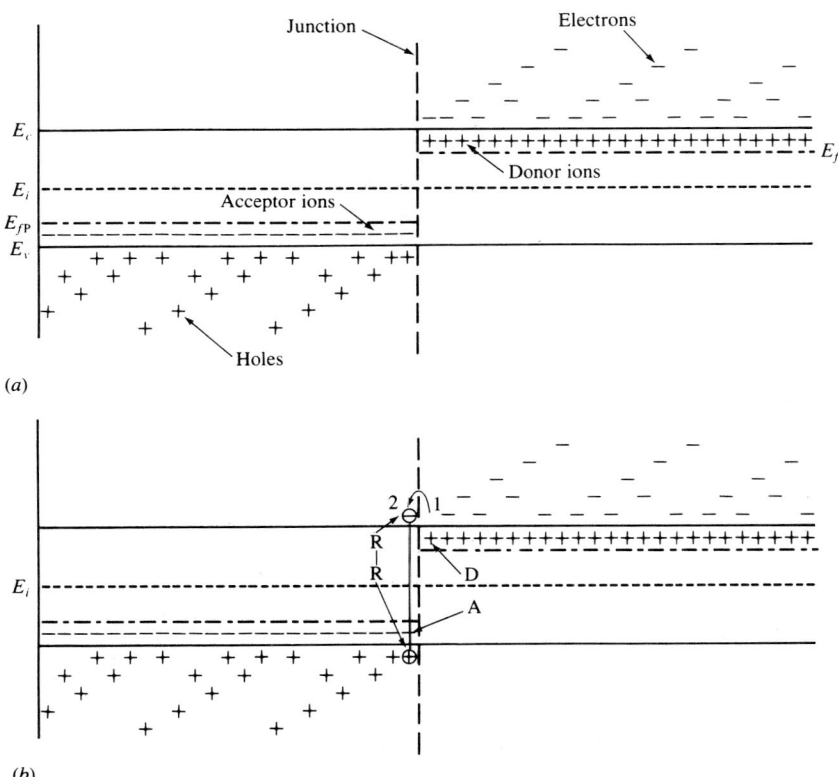

(a)

(b)

FIGURE 6.2.1
PN junction band diagram symbolically showing the conduction electron and hole populations and the charges of the ionized impurity atoms. (a) Just before the junction is established. Notice that each ionic charge is compensated by an equal and opposite carrier charge. (b) Immediately after the junction is made. Migration of an electron (arrow from 1 to 2) leaves a donor charge D uncovered. Recombination with a majority carrier in the P semiconductor (R–R) leaves uncovered the acceptor ion charge A.

energy-position diagram. It is assumed that the host material is the same for both semiconductors and that they are at the same potential.

Under these conditions, all characteristic levels in Fig. 6.2.1a are the same for the two semiconductors, with the exception of the Fermi levels: E_{fP} is below the intrinsic Fermi level, E_{fN} is above it. It is important to remember that (cf. Sec. 1.4) if, for instance, the N-type semiconductor were brought to a potential higher than that of the P-type, then the whole right-hand side of Fig. 6.2.1a would be displaced downwards.

Now let a junction be formed by intimately joining the two semiconductors along one face. The junction interface is now no longer an external surface delimiting the crystal, because the junction has restored crystal continuity over this surface, effectively merging the two semiconductor crystals into one, larger, crystal structure. As a result, the surface potential barrier that originally prevented the current carriers from leaving each semiconductor is now lowered and the current carriers are free to move across the junction interface.

At this junction there exist high gradients of concentration of both holes and electrons, so that *both types of current carriers diffuse* across it. In Fig. 6.2.1b the diffusion of an electron from position 1 to position 2 is indicated by the arrow.

As shown by a comparison of Fig. 6.2.1b with Fig. 6.2.1a, when an electron moves into the P-type semiconductor, the N-type is depleted of one of its negative current carriers, so that one of its positive donor ions (such as D in Fig. 6.2.1b) loses its negative compensating charge. As a result *the N region acquires a net positive charge.* At the same time, the electron, entering the P region, has now become a minority carrier. Within a short time (cf. Secs. 2.4 and 4.4) it recombines with one of the many holes, neutralizing its positive charge. In the figure, to symbolize the recombination process, the minority carrier electron and the hole recombining with it are enclosed in circles connected by a line and identified by the notation R–R (for recombination). The electron and hole, recombining, disappear as net charges, so that the final result of the electron migration into the P-type semiconductor is to eliminate one of its holes.

The P semiconductor, depleted of one of its positive charges, *has now acquired a net negative charge.* In Fig. 6.2.1b this can be identified as the negative charge of acceptor ion A, which is no longer compensated due to the disappearance of the corresponding hole.

Notice that these charges are stored in the impurity ions. They have no mobility and are therefore localized and "frozen" in the lattice, forming a *depletion region* in which impurity ion charges are no longer compensated because the corresponding current carrier charges have vanished from this region (depletion).

Hole migration, similarly depleting both semiconductors, also contributes to the depletion region formation, further charging the P region negatively and the N region positively.

This net charge distribution, in turn, generates an electrostatic potential distribution $\psi(x)$, varying from point to point along the device and tending to make the N-type semiconductor positive with respect to the P-type. As previously discussed (Sec. 1.4) this results in a distortion of the energy diagram. Each point is displaced by $-q\psi(x)$ along the electron energy axis so that the characteristic energy levels also vary from point to point. For instance, the intrinsic Fermi level becomes

$$E_i = E_{i0} - q\psi(x) \tag{6.2.1}$$

where E_{i0} is the intrinsic Fermi level before the junction is made. E_i has now become a function of position and a *potential barrier begins to be raised* across the junction, in the depletion region, as qualitatively shown in Fig. 6.2.2a, where one electron and one hole have migrated across the junction and recombined. Four impurity ions (two donors and two acceptors) have lost their compensating charge, the P semiconductor has acquired a net negative charge of two electronic units, and the N semiconductor an equal, but positive, net charge. Notice that the total charge in the entire device is still zero.

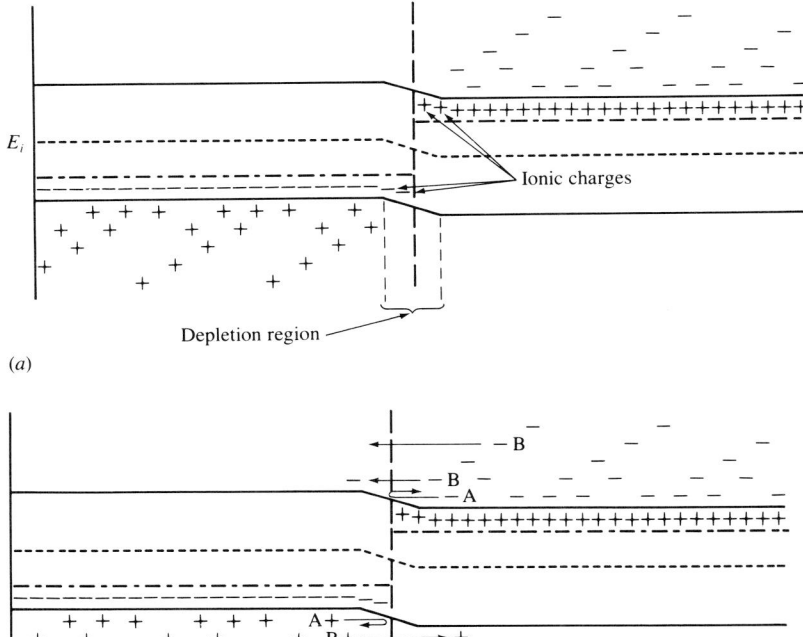

FIGURE 6.2.2
PN junction diagrams and charge distribution during the formation of the depletion region. (*a*) The uncovered ionic charges in the depletion region generate a potential distribution that distorts the energy band diagram, raising a potential barrier. (*b*) Carrier motion. The lower energy carriers (A) cannot diffuse into the body of the adjacent semiconductor, but instead rebound at the barrier. Higher energy carriers (B) diffuse above the barrier, further extending the depletion region.

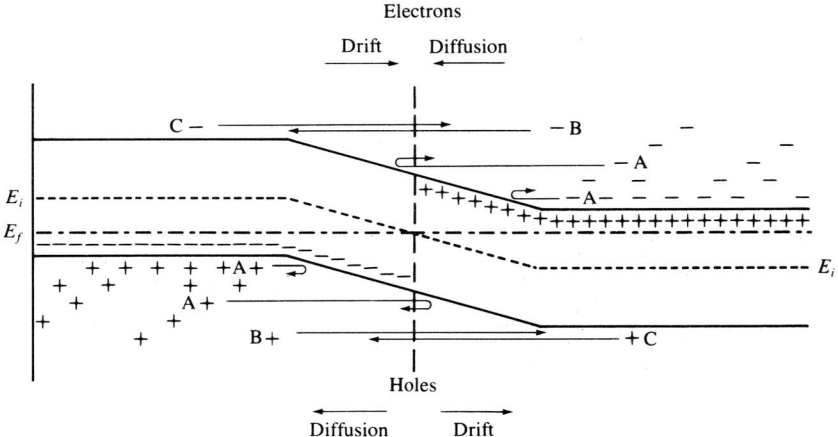

FIGURE 6.2.3
PN junction energy band diagram at equilibrium. Most majority carriers (A) cannot overcome the potential barrier. They diffuse into the depletion region only to drift back. The minority carrier (C) drift current is balanced by a trickle of high-energy majority carrier (B) diffusion across the depletion region: no net current flows. Notice the direction of drift and diffusion motion for electrons and holes.

Now the lower energy current carriers (such as A in Fig. 6.2.2*b*), trying to diffuse into the adjacent semiconductor, do not have enough energy to overcome the potential barrier and so rebound back. This is symbolized by the arrows suggesting their motion. Only the higher energy carriers (such as B in the figure) migrate and, through recombination, *further extend the depletion region*, thereby raising the barrier height.

As time goes by, the potential difference opposing diffusion keeps increasing until it reaches a value high enough to stop current flow completely: this difference of potential is known as the built-in (or *contact*) *potential* V_{bi}. Now no net charge transfer occurs across the junction and the depletion region and the barrier height do not change any more: the system has reached the condition of *dynamic equilibrium* depicted in Fig. 6.2.3.

Two forces act on the current carriers: the push of diffusion and the potential gradient. They generate two current components: a *diffusion* and, respectively, a *drift (or conduction)* current component. The diffusion current is directed from high to low concentration (right to left for electrons, left to right for holes) and so it tends to charge the P semiconductor negatively and the N semiconductor positively, while the drift current, in the direction of the electrostatic force, transports charge in the opposite direction. Figure 6.2.3 explicitly indicates the directions of these different motions for electrons and for holes.

Initially, when the barrier, and so the electrostatic force, is still small, diffusion is much larger than drift, so diffusion results in a pileup of charge and the potential barrier keeps growing higher. The raising of the barrier, in turn, increases the drift current and opposes diffusion, slowing down the barrier buildup process. The condition of dynamic equilibrium is reached when finally the drift current equals the diffusion current so that the rate of transport of charge is the same in the two opposite directions—there is no *net* charge displacement and the barrier stops growing.

In Fig. 6.2.3, the potential barrier is so high that most of the majority carriers rebound from it and return to their original semiconductor. In the figure these carriers are indicated by A and their motion is symbolized by the corresponding arrows. The barrier effectively prevents them from diffusing across the depletion region.

However, considering the minority carriers (designated by C in the figure), it can be noticed that the barrier polarity is such that it actually tends to push them across the depletion region (drift motion). Any minority carrier reaching the depletion region will be accelerated across the junction. This results in a *minority carrier reverse current.*

As shown in Fig. 6.2.3, the contact potential V_{bi} adjusts itself to that precise value that permits just enough high-energy majority carriers (B in the figure) to make it through the depletion region by diffusion to compensate for this minority carrier drift current.

Now that the phenomenon has been described qualitatively, there remains to compute the contact potential. This requires a quantitative description, which, as usual, can easily be achieved by repeating the same logical sequence that led to the qualitative description of the phenomenon. In this instance, however, the argument must be repeated using the language of mathematics.

As stated, the current transient resulting from the formation of the junction consists of a diffusion and a conduction component. Mathematically (cf. Sec. 4.3), the electron current density must therefore be expressed as

$$J_e = q\mu_e n\mathscr{E} + qD_e \frac{dn}{dx} \tag{6.2.2}$$

Equilibrium is reached when the diffusion component balances out the conduction component, so that the total electron current density equals zero. The equilibrium condition, therefore, is

$$q\mu_e n\mathscr{E} + qD_e \frac{dn}{dx} = 0 \tag{6.2.3}$$

where $\mathscr{E}$ must be interpreted as the electric field resulting from the electric potential distribution $\psi(x)$ generated by the charge displacement. Remembering that, in our one-dimensional example, the gradient can be represented by the total derivative because at equilibrium $\psi(x)$ does not depend on time,

$$\mathscr{E} = -\frac{d\psi}{dx} \tag{6.2.4}$$

In order to interpret (6.2.3) correctly, all quantities must be expressed in accordance with the prevailing conditions. For instance, at equilibrium, from (3.4.3),

$$n = n_i\, e^{(E_f - E_i)/(kT)} \tag{6.2.5}$$

Substituting this expression in (6.2.3) and using the Einstein relationship (4.3.2),

$$-q\mu_e n \frac{d\psi}{dx} + q\frac{kT}{q}\mu_e \frac{n}{kT}\left(\frac{dE_f}{dx} - \frac{dE_i}{dx}\right) = 0 \tag{6.2.6}$$

from which, using (6.2.1) and after some simplification,

$$\frac{dE_f}{dx} = 0 \tag{6.2.7}$$

indicating that, at equilibrium, the graph of E_f must be a horizontal straight line over all the device. In other words, when equilibrium is reached *the Fermi levels align*.

This phenomenon is not peculiar to semiconductors: it occurs whenever an intimate contact is generated between solid objects containing current carriers. A transient takes place, resulting in the alignment of the Fermi levels, so that *a contact potential is generated*. This physical principle accounts for the behavior of such devices as batteries, thermocouples, etc.

The contact potential can evidently be computed as

$$V_{bi} = \frac{E_{fN} - E_{fP}}{q} \tag{6.2.8}$$

For the PN junction, remembering (3.4.7) and (3.4.9),

$$\boxed{V_{bi} = \frac{kT}{q} \ln \frac{N_A N_D}{n_i^2}} \tag{6.2.9}$$

justifying Fig. 6.2.3, in which, at equilibrium, the Fermi levels of the two semiconductors are shown aligned (dash-dot line).

The above discussion analyzed the equilibrium of the electron currents. The same argument could be repeated for the hole current. The same result would be reached.

6.3   QUANTITATIVE JUNCTION ANALYSIS AT EQUILIBRIUM

It has already been noticed that, in the depleted region, the charges stored in the impurity ions are no longer compensated by the opposite charges of the majority carriers. This results in a net charge density distribution which, assuming complete depletion, equals the charge per cubic centimeter stored in the impurity ions. In the P semiconductor there are N_A ions per cubic centimeter and each carries a charge $-q$, yielding a net charge density $\rho_v = -qN_A$. Analogously, in the N semiconductor, $\rho_v = qN_D$.

Notice that, in the process of depletion, whenever the N side acquired a positive charge, the P side acquired an equal and opposite negative charge: the *total net charge in the entire device remains zero* (condition of charge neutrality).

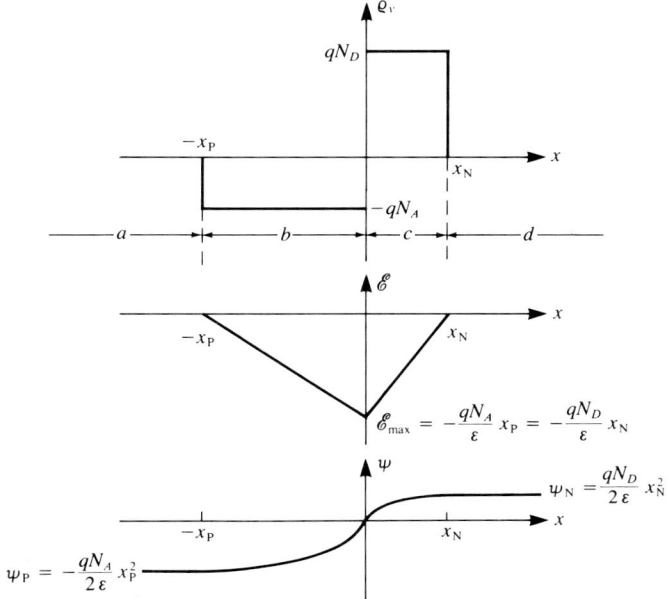

FIGURE 6.3.1
Charge density, electric field, and potential distribution in an abrupt junction with $N_D = 2N_A \, \text{cm}^{-3}$.

In mathematical terms,

$$\int_{v_P} -qN_A \, dv + \int_{v_N} qN_D \, dv = 0 \tag{6.3.1}$$

where v_P and v_N are the volumes of the P, respectively N, depletion regions.

The charge density distribution generated by depletion is illustrated in Fig. 6.3.1 (top graph), where the doping is assumed to be uniform throughout each semiconductor up to the junction (*abrupt junction*) with $N_D = 2N_A$ in the example of the figure. In Fig. 6.3.2, instead, the doping is assumed to vary linearly with distance from the junction (*linearly graded junction*). Several fabrication techniques (cf. Sec. 5.5 and Fig. 5.5.4) often result in charge density distributions approximating the linearly graded junction.

Abrupt Junction

For the abrupt junction, the condition of charge neutrality becomes

$$N_A x_P = N_D x_N \tag{6.3.2}$$

A special case is of some interest because it is frequently encountered in practice. Suppose $N_D \gg N_A$; from (6.3.2), then $x_N \ll x_P$. In general, if one semiconductor is much more heavily doped than the other, its depletion region is much narrower. When the difference in doping is very large, practically the depletion region exists

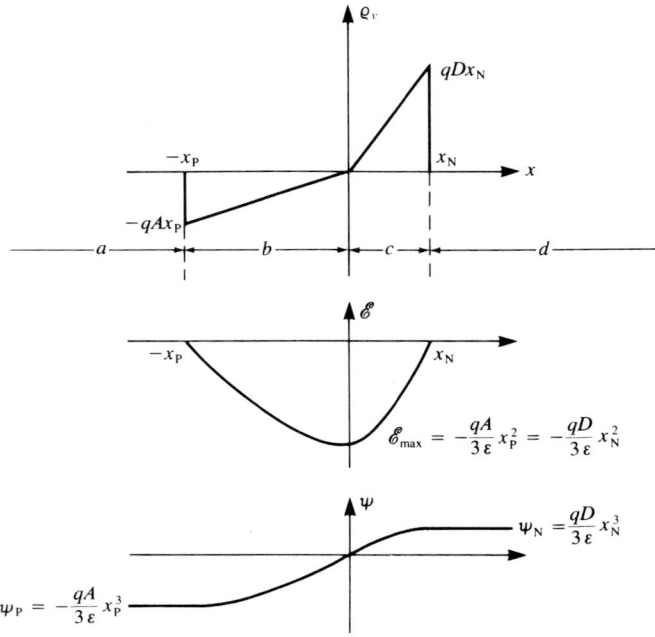

FIGURE 6.3.2
Charge density, electric field, and potential distribution in a linearly graded junction with $D = 4A$ cm^{-4}.

only in the low-doped semiconductor. Junctions of this type are said to be *one-sided*. This characteristic is a consequence of the general condition of charge neutrality (6.3.1), so all types of junction, not only the abrupt ones, can be one-sided.

With reference to Fig. 6.3.1, the charge density distribution defines four regions within the device. Applying Poisson's theorem to regions a and d,[1]

$$\frac{d^2\psi}{dx^2} = 0 \tag{6.3.3}$$

At equilibrium no current must flow, so the gradient of ψ must be zero in these nondepleted (and therefore conducting) regions; therefore $d\psi/dx = 0$.

---

[1] Poisson's equation:

$$\nabla^2\psi = -\frac{\rho_v}{\varepsilon}$$

for the case of one-dimensional systems becomes

$$\frac{d^2\psi}{dx^2} = -\frac{\rho_v}{\varepsilon}$$

Under this boundary condition, the solution to (6.3.1) is $\psi = $ constant or

$$\psi = \psi_N \qquad \text{in the N semiconductor} \qquad (6.3.4)$$

$$\psi = \psi_P \qquad \text{in the P semiconductor} \qquad (6.3.5)$$

Notice that, by our definition of contact potential and by (6.2.9),

$$\psi_N - \psi_P = V_{bi} = V_T \ln \frac{N_A N_D}{n_i^2} \qquad (6.3.6)$$

In region b, where the charge density is $-qN_A$,

$$\frac{d^2\psi}{dx^2} = \frac{qN_A}{\varepsilon} \qquad (6.3.7)$$

with boundary conditions:

$$\text{at } x = -x_P \to \frac{d\psi}{dx} = 0; \qquad \text{at } x = 0 \to \psi = 0 \qquad (6.3.8)$$

Then, by successive integrations:

$$\frac{d\psi}{dx} = \frac{qN_A}{\varepsilon} (x + x_P) \qquad (6.3.9)$$

$$\psi = \frac{qN_A}{2\varepsilon} (x + x_P)^2 - \frac{qN_A}{2\varepsilon} x_P^2 \qquad (6.3.10)$$

In region c, with charge density qN_D,

$$\frac{d^2\psi}{dx^2} = -\frac{qN_D}{\varepsilon} \qquad (6.3.11)$$

with boundary conditions:

$$\text{at } x = x_N \to \frac{d\psi}{dx} = 0; \qquad \text{at } x = 0 \to \psi = 0 \qquad (6.3.12)$$

Then

$$\frac{d\psi}{dx} = -\frac{qN_D}{\varepsilon} (x - x_N) \qquad (6.3.13)$$

$$\psi = -\frac{qN_D}{2\varepsilon} (x - x_N)^2 + \frac{qN_D}{2\varepsilon} x_N^2 \qquad (6.3.14)$$

The electric field $\mathscr{E}$ and potential ψ are plotted in Fig. 6.3.1, remembering that

$\mathscr{E} = -d\psi/dx$. The above formulas also indicate that the maximum electric field intensity occurs at $x = 0$ and is

$$\mathscr{E}_{max} = -\frac{qN_D}{\varepsilon} x_N = -\frac{qN_A}{\varepsilon} x_P \tag{6.3.15}$$

satisfying (6.3.1). The negative sign simply indicates that for the choice of boundaries made in this one-dimensional case, the electric field points in the negative x direction, as shown in Fig. 6.3.1.

Similarly, as also indicated in Fig. 6.3.1,

$$\psi_P = -\frac{qN_A}{2\varepsilon} x_P^2 \tag{6.3.16}$$

$$\psi_N = \frac{qN_D}{2\varepsilon} x_N^2 = \frac{qN_A}{2\varepsilon} x_P^2 \frac{N_A}{N_D} \tag{6.3.17}$$

where use was made of (6.3.2). From this, after some algebra,

$$V_{bi} = \psi_N - \psi_P = \frac{qN_A}{2\varepsilon} x_P^2 \frac{N_A + N_D}{N_D} \tag{6.3.18}$$

$$x_P = \sqrt{\frac{2\varepsilon}{qN_A} \frac{N_D}{N_A + N_D} V_{bi}} \tag{6.3.19}$$

$$x_N = \sqrt{\frac{2\varepsilon}{qN_D} \frac{N_A}{N_A + N_D} V_{bi}} \tag{6.3.20}$$

and finally the total width of the depletion region is

$$w = x_N + x_P = x_P\left(1 + \frac{N_A}{N_D}\right) = \sqrt{\frac{2\varepsilon}{q} \frac{N_A + N_D}{N_A N_D} V_{bi}} \tag{6.3.21}$$

Example 6.3.1. An abrupt PN junction in Si has $N_A = 10^{15}$ cm^{-3} and $N_D = 10^{16}$ cm^{-3}. The temperature is 300 K. Compute contact potential, N- and P-type depletion region widths, maximum electric field, maximum potential differences between the junction, and the N-type, respectively P-type, semiconductor, charge stored in each depletion region per square centimeter of junction interface.

Solution. From (6.2.9),

$$V_{bi} = 0.0259 \ln \left(\frac{10^{15} \times 10^{16}}{2.25 \times 10^{20}}\right) = 0.6325 \text{ V}$$

From (6.3.20),

$$x_N = \sqrt{\frac{2 \times 8.854 \times 10^{-14} \times 11.8}{1.6 \times 10^{-19} \times 10^{16}} \frac{10^{15}}{10^{15} + 10^{16}} 0.6325} = 0.0866 \ \mu m$$

From (6.3.1),

$$x_P = 0.0866 \times \frac{10^{16}}{10^{15}} = 0.866 \ \mu m$$

From (6.3.15),

$$\mathscr{E}_{max} = \frac{-1.6 \times 10^{-19} \times 10^{16} \times 8.666 \times 10^{-6}}{11.8 \times 8.854 \times 10^{-14}}$$

$$= -13,271 \ V/cm$$

From (6.3.16),

$$\psi_P = \frac{-0.00016 \times (8.666 \times 10^{-5})^2}{2 \times 1.045 \times 10^{-12}} = -0.575 \ V$$

From (6.3.17) analogously,

$$\psi_N = 0.0575 \ V$$

The charge stored equals the charge density times the volume. For instance, in the N region the charge density is $qN_D = 1.6 \times 10^{-19} \times 10^{16} = 0.0016 \ C/cm^3$ and the volume of a 1-cm^2 cross section of the N depletion region is $1 \times 8.666 \times 10^{-6}$; therefore the charge requested is $0.0016 \times 8.666 \times 10^{-6} = 1.39 \times 10^{-8} \ C/cm^2$.

Linearly Graded Junction

For the linearly graded junction of Fig. 6.3.2 the depletion region charge neutrality condition becomes

$$Ax_P^2 = Dx_N^2 \tag{6.3.22}$$

where A and D are the rates of variation of the acceptor, respectively donor, doping concentrations in cm^{-4}. In the example of the figure $D = 4A$.

As in the previous case, applying Poisson's equation to regions a and d, it can be concluded that

$$\psi = \psi_P \qquad \text{in the P semiconductor} \tag{6.3.23}$$

$$\psi = \psi_N \qquad \text{in the N semiconductor} \tag{6.3.24}$$

In region b the charge density is $\rho_v = qAx$ (note that $x < 0$), so that

$$\frac{d^2\psi}{dx^2} = -\frac{qAx}{\varepsilon} \tag{6.3.25}$$

with boundary conditions:

$$\text{at } x = -x_P \to \frac{d\psi}{dx} = 0; \qquad \text{at } x = 0 \to \psi = 0 \tag{6.3.26}$$

then, again by successive integrations:

$$\frac{d\psi}{dx} = -\frac{qA}{2\varepsilon}(x^2 - x_P^2) \tag{6.3.27}$$

$$\psi = -\frac{qA}{2\varepsilon}\left(\frac{x^3}{3} - x_P^2 x\right) \tag{6.3.28}$$

In region c, with charge density $\rho_v = qDx$,

$$\frac{d^2\psi}{dx^2} = -\frac{qDx}{\varepsilon} \tag{6.3.29}$$

with boundary conditions:

$$\text{at } x = x_N \to \frac{d\psi}{dx} = 0; \quad \text{at } x = 0 \to \psi = 0 \tag{6.3.30}$$

Then

$$\frac{d\psi}{dx} = -\frac{qD}{2\varepsilon}(x^2 - x_N^2) \tag{6.3.31}$$

$$\psi = -\frac{qD}{2\varepsilon}\left(\frac{x^3}{3} - x_N^2 x\right) \tag{6.3.32}$$

which is plotted in Fig. 6.3.2. Maximum field intensity occurs at $x = 0$:

$$\boxed{\mathscr{E}_{\max} = -\frac{qA}{2\varepsilon}x_P^2 = -\frac{qD}{2\varepsilon}x_N^2} \tag{6.3.33}$$

in accordance with (6.3.22). For the physical meaning of the negative sign see the analogous considerations valid for the abrupt junction. Then

$$\boxed{\psi_P = -\frac{qA}{3\varepsilon}x_P^3} \tag{6.3.34}$$

$$\boxed{\psi_N = \frac{qD}{3\varepsilon}x_N^3 = \frac{qA}{3\varepsilon}x_P^3\sqrt{\frac{A}{D}}} \tag{6.3.35}$$

finally yielding, after some algebra,

$$\boxed{x_P = \sqrt[3]{\frac{3\varepsilon}{qA}\frac{\sqrt{D}}{\sqrt{A}+\sqrt{D}}V_{bi}}} \tag{6.3.36}$$

$$\boxed{x_N = \sqrt[3]{\frac{3\varepsilon}{qD}\frac{\sqrt{A}}{\sqrt{A}+\sqrt{D}}V_{bi}}} \tag{6.3.37}$$

and

$$w = \sqrt[3]{\frac{3\varepsilon}{q} \frac{(\sqrt{A} + \sqrt{D})^2}{AD} V_{bi}} \qquad (6.3.38)$$

Electric field and potential distributions for the linearly graded junction are shown in Fig. 6.3.2.

The equilibrium conditions depicted in Figs. 6.2.3, 6.3.1, and 6.3.2 prevail when the device is standing alone, free from external actions and influences. If, however, during operation, an external voltage source V is applied between the P and N semiconductors, this equilibrium configuration is distorted: an additional relative displacement $-qV$ is introduced between the P and the N side of the energy-position diagram (cf. Sec. 1.4).

Device behavior under external bias depends strongly on the polarity of the applied voltage. In the following, when the P-type semiconductor is made positive with respect to the N-type (*forward bias*), V will be conventionally assigned the positive sign. For opposite polarity (*reverse bias*), V will be negative by convention.

6.4 THE PN JUNCTION UNDER REVERSE BIAS

The reverse bias circuit configuration is depicted in Fig. 6.4.1 both in terms of the usual circuit schematic notation and in a symbolic representation of the depletion region and current carriers within the physical device.

The bias generator V establishes a field $\mathscr{E}$ in the system, so that an electrostatic force $\mathbf{F}$ is exerted on the current carriers. As shown in the figure, the direction of this force is such that the majority carriers, both in the P and the N semiconductor, are pulled away from the junction. Their consequent *drift motion* constitutes a current transient I, which must be provided to the circuit by the bias battery V. The student should verify that the opposite motions of electrons and holes both contribute to the current transient in the direction indicated in the figure.

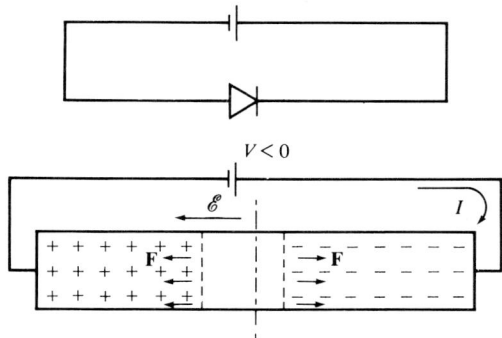

FIGURE 6.4.1
Reverse biased junction. Circuit diagram and a representation of the structure in principle, showing the direction of the electric field and of the forces acting on the majority carriers.

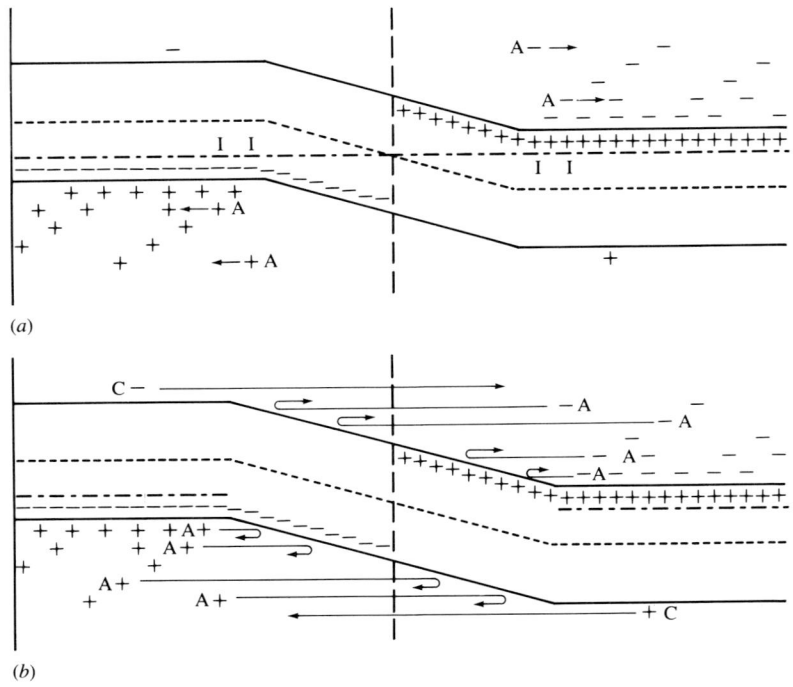

(a)

(b)

FIGURE 6.4.2
Energy-position diagram of reverse biased junction. (a) At the instant of application of the reverse bias. The majority carriers (A) under the applied electrostatic force tend to move in the direction of the arrows. The ionic charges (I) are left uncompensated, extending the depletion region. (b) Steady state. The barrier is so high that none of the majority carriers (A) can overcome it. The minority carrier (C) current is now uncompensated and constitutes the inverse saturation current. Compare with Fig. 6.2.3.

Figure 6.4.2a represents this condition in the energy-position diagram of the system, showing that the majority carriers A, in the drift motion (indicated by the arrows in the figure), leave behind the uncompensated ionic charges of their respective impurity ions (I in the figure), extending the depletion region and increasing the voltage across it, thereby raising the barrier.

Voltage equilibrium around the loop is established when the junction voltage reaches $V_{bi} - V$.[2] The majority carrier transient current then dies out, the depletion region stops expanding, and a steady state is maintained. The equilibrium around the circuit loop is considered in some detail in Probs. 6.14 through 6.16. Figure 6.4.2b shows that, at steady state, even under moderate reverse bias, the majority carriers (A) entering the depletion region under the

---

[2] It should be borne in mind that the reverse bias voltage V is negative in accordance with the convention adopted, so that the junction voltage $V_{bi} - V > V_{bi}$, the contact potential.

push of diffusion are *all* turned back by the opposing junction field. Not even the most energetic among them can overcome the higher potential barrier and the majority carrier diffusion current is fully neutralized by their own drift.

The junction field, however, still tends to help the minority carriers (C) to cross the junction and, for sufficiently high reverse bias, pushes them into the adjacent semiconductor as fast as they reach the edge of the depletion region. The resulting *minority carrier reverse current* involves all of the available minority carriers; it is therefore *saturated* and does not increase with further increments of the reverse bias voltage. If the reverse bias is greater than a few tenths of a volt, this minority carrier current is no longer balanced by a trickle of high-energy majority carrier diffusion, as it was when the barrier was lower (cf. Fig. 6.2.3), and so constitutes a net reverse saturation current (reverse, because it is in the opposite direction to the forward bias current, cf. Sec. 6.5).

In conclusion:

1. To establish reverse bias conditions, the battery must provide a *current transient*, which stores a charge in the depletion region. This charge is large enough to raise the junction barrier voltage to the steady state level $V_{bi} - V > V_{bi}$.
2. To maintain reverse bias, the battery must also furnish a small steady state *reverse saturation current*.

The above evidently means that the reverse biased PN junction is equivalent to a load consisting of a constant current generator (the reverse saturation current I_0) shunted across a capacitor (the depletion capacitance C_d), as shown in Fig. 6.4.3.

If very large reverse bias voltages are applied, then the resulting electric field in the depletion region can become very strong and imparts a large amount of electrostatic energy to any charge within it.

Current carriers may then acquire enough kinetic energy to ionize depletion region atoms by collision. Conversely, bound charges may escape from the barrier keeping them tied to their atoms. In both cases new current carriers are

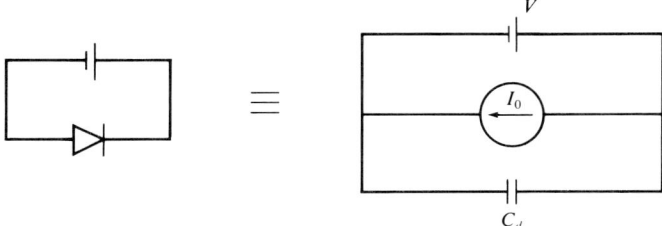

FIGURE 6.4.3
Reverse biased junction. Circuit schematic and equivalent circuit. The bias battery must provide the inverse saturation current, so a current generator I_0 is part of its load. To reach steady state reverse biased conditions, the battery must also provide the transient current required to charge the depletion region. The depletion capacitor C_d represents this part of the battery load.

generated and the current in the depletion region (and therefore through the whole device) correspondingly increases.

Under the proper conditions this current can increase to the point where it is limited only by the circuit resistances. This condition of *junction breakdown* is characterized by a very large flow of current (cf. Sec. 6.5).

6.5   QUANTITATIVE JUNCTION ANALYSIS AT REVERSE BIAS

To characterize the performance of a reverse biased PN junction quantitatively, all of its pertinent characteristics must be computed. These include not only the depletion region parameters (as in the case of the equilibrium condition of Sec. 6.3) but also the parameter of the equivalent circuit of Fig. 6.4.3.

Depletion Region Parameter Computation

Poisson's equation and the space charge neutrality condition, together with the considerations used in Sec. 6.3 for the discussion of the equilibrium case, still apply under reverse bias conditions, so that the relationships between depletion region parameters and junction voltage are still represented by expressions similar to (6.3.15) to (6.3.21) [or (6.3.33) to (6.3.38)]. Depletion region widths, charge density configurations, electric field, and potential distributions can therefore be computed by substituting $V_{bi} - V$ for V_{bi} in those equations.

> **Example 6.5.1.** An abrupt Si PN junction at 300 K is characterized by $N_D = 10^{16}$ cm^{-3} and $N_A = 2.5 \times 10^{15}$ cm^{-3}. Compute the unbiased junction voltage, maximum electric field, and depletion region width. Repeat the computation with an applied reverse bias voltage of 1 V.
>
> **Solution.** From (6.2.9),
>
> $$V_{bi} = 0.0259 \ln \left(\frac{10^{16} \times 2.5 \times 10^{15}}{2.25 \times 10^{20}} \right) = 0.658 \text{ V}$$
>
> From (6.3.21),
>
> $$w = \sqrt{\frac{2 \times 1.04 \times 10^{-12}}{1.6 \times 10^{-19}} \frac{10^{16} + 2.5 \times 10^{15}}{10^{16} \times 2.5 \times 10^{15}} 0.658} = 0.65 \ \mu m$$
>
> From (6.3.2),
>
> $$x_P = \frac{w}{1.25} = 0.52 \ \mu m$$
>
> From (6.3.15),
>
> $$\mathscr{E}_{max} = 5.2 \times 10^{-5} \frac{1.6 \times 10^{-19} \times 2.5 \times 10^{15}}{1.04 \times 10^{-12}} = 20 \text{ kV/cm}$$
>
> For the biased condition, a reverse bias voltage of 1 V defines $V = -1$ so that the junction bias voltage is
>
> $$V_{bi} - V = 0.658 - (-1) = 1.658 \text{ V}$$

Expression (6.3.21) and the considerations of Sec. 6.5 imply that the reverse biased depletion region width w_{rb} is related to the unbiased width w by

$$w_{rb} = w\sqrt{\frac{V_{bi} - V}{V_{bi}}} = 0.65\sqrt{\frac{0.658 - (-1)}{0.658}} = 1.04 \ \mu m$$

To compute the maximum electric field, as before, from (6.3.2),

$$x_P = \frac{w}{1.25} = 8.3 \times 10^{-5} \text{ cm}$$

and from (6.3.15),

$$\mathscr{E}_{max} = \frac{1.6 \times 10^{-19} \times 2.5 \times 10^{15}}{1.04 \times 10^{-12}} \ 8.3 \times 10^{-5} = 32 \text{ kV/cm}$$

The above example is illustrated graphically in Fig. 6.5.1.

Small-Signal Parameter Computation

Expressions for the theoretical computation of the minority carrier reverse saturation current will be obtained in Sec. 6.7 [cf. Eq. (6.7.20)].

Before proceeding to the computation of the capacitance C_d in Fig. 6.4.3, its physical significance should be visualized. This can best be done with the help of

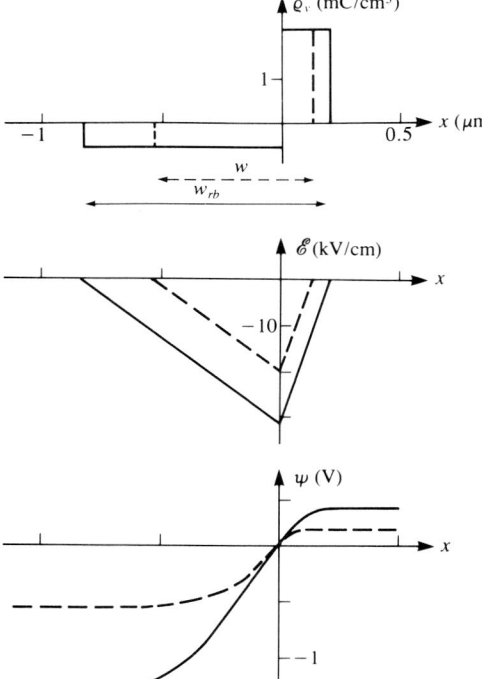

FIGURE 6.5.1
Charge density, electric field, and potential distribution in the junction of Example 6.5.1. Unbiased (dashed line) and under 1 V negative bias (solid line). Computed values are given in the example; additional values are: $\rho_{vN} = 1.6 \times 10^{-3}$ C/cm^3, $\psi_P = -0.525$ V, $\psi_N = 0.131$ V, $\psi_{Prb} = -1.34$ V, $\psi_{Nrb} = 0.31$ V.

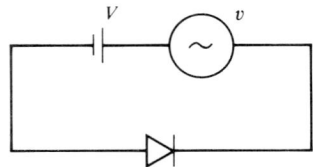

FIGURE 6.5.2
Circuit schematic of reverse biased junction with small ac generator v for small-signal equivalent parameter computation.

the circuit of Fig. 6.5.2, where the bias generator is split into a bias battery V, determining a dc work point, and an ac small-signal generator v, providing a small, time-dependent bias variation (or modulation). The student will recognize the familiar artifice used in equivalent small-signal linear circuit analysis, as applied in his or her circuit courses.

The ac load on the small-signal generator is only the capacitance C_d, which therefore has the significance of a small-signal (or incremental) capacitance.

Assuming, as usual, the small-signal voltage to be infinitesimal, the incremental capacitance is, by definition,

$$C_d = \frac{dQ_s}{dV} \tag{6.5.1}$$

where dQ_s is the infinitesimal charge that must be stored in the depletion region in order to produce the infinitesimal voltage variation $v = dV$. For the abrupt junction, the total charge stored in the depletion region is $Q_s = qN_D A x_N$ or, remembering (6.3.20),

$$Q_s = qN_D A \sqrt{\frac{2\varepsilon}{qN_D}\frac{N_A}{N_A + N_D}(V_{bi} - V)} \tag{6.5.2}$$

where A is the area of the interface between the N and the P semiconductors (the junction cross section). Then

$$\boxed{C_d = \frac{dQ_s}{dV} = A\sqrt{\frac{\varepsilon qN_A N_D}{2(N_A + N_D)}}\frac{1}{\sqrt{V_{bi} - V}}} \tag{6.5.3}$$

This capacitor, it should be noticed, is not a linear circuit element; indeed, its value depends on the bias voltage V and can therefore be changed by varying the bias voltage applied. It constitutes an unusual and useful circuit element: *a voltage-controlled variable capacitor.*

Remembering (6.3.21) and the previous remarks about the computation of the biased depletion region width, expression (6.5.3) can be rewritten in the form:

$$C_d = \frac{\varepsilon A}{w_{rb}} \tag{6.5.4}$$

showing that C_d equals the capacitance of a parallel plate capacitor having a dielectric of thickness w_{rb} (the width of the reverse biased depletion region). Intu-

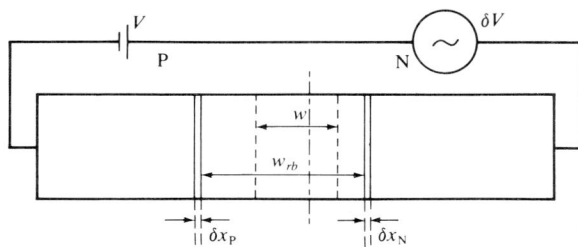

FIGURE 6.5.3
Reverse biased junction. The unbiased depletion region width w expands to w_{rb} after application of the reverse bias V. A small variation δV of bias voltage changes the depletion region width by the small amount $\delta x_P + \delta x_N$, adding small ionic charges *at the edges* of the depletion region.

itively this makes a lot of sense, in the light of Fig. 6.5.3, which shows that the extra charge deposited by the small-signal generator is stored in two thin layers δx_P and δx_N at the edge of the depletion region and therefore at a distance w_{rb} from each other.

Example 6.5.2. An abrupt Si PN junction characterized by $N_D = 10^{16}$ cm^{-3} and $N_A = 2.5 \times 10^{15}$ cm^{-3} with a junction cross section of 1 mm^2 is reverse-biased by 1 V. Compute the depletion capacitance at 300 K presented to a small-signal generator of 0.1 V step voltage making the P semiconductor more negative.

Solution. As computed in Example 6.5.1, under the given bias conditions, the PN junction has $V_{bi} = 0.658$ V, $w_{rb} = 1.04$ μm, so that the theoretical depletion capacitance is, from (6.5.4),

$$C_d = \frac{1.04 \times 10^{-12} \times 10^{-2}}{1.04 \times 10^{-4}} = 100 \text{ pF}$$

This capacitance, however, is the one that would be presented to a generator of *infinitesimally small voltage*. For finite voltage, a better approximation can be obtained by computing the variation of charge δQ_s stored when the additional bias is applied and by dividing it by the voltage swing $v = 0.1$ V. Using the criteria developed in Example 6.5.1, the variation of depletion region width, after the application of the extra 0.1 V negative bias, is

$$dw_{rb} = w_{rb}\left(\sqrt{\frac{V_{bi} - V + v}{V_{bi} - V}} - 1\right) = 1.04 \times 10^{-4}\left(\sqrt{\frac{1.758}{1.658}} - 1\right) = 3.1 \times 10^{-6} \text{ cm}$$

The variation of x_N is, from (6.3.1), $\delta x_N = \delta w_{rb}[N_A/(n_A + N_d)] = 6.2 \times 10^{-7}$ cm. The variation of stored charge in C_d is therefore $\delta Q_s = \delta x_N A N_D q$ or $\delta Q_s = 9.9 \times 10^{-12}$ C. Finally, the capacitance C_d is

$$C_d = \frac{\delta Q_s}{v} = \frac{9.9 \times 10^{-12}}{0.1} = 99 \text{ pF}$$

which compares quite well with, but is slightly smaller than, the value computed for infinitesimal signal variation.

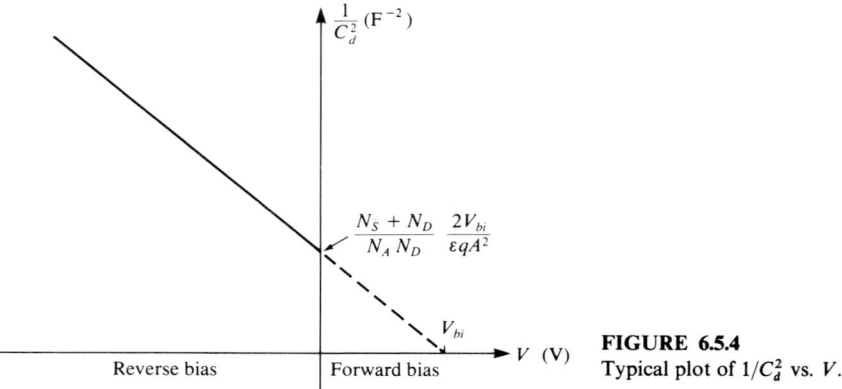

FIGURE 6.5.4
Typical plot of $1/C_d^2$ vs. V.

Measurements of the depletion capacitance can provide a useful analytical tool for the analysis of PN junctions. From (6.5.3),

$$\left(\frac{1}{C_d}\right)^2 = \left[\frac{2(N_A + N_D)}{\varepsilon q N_A N_D A^2}\right](V_{bi} - V) \tag{6.5.5}$$

showing that, for abrupt junctions, a plot of the inverse square depletion capacitance vs. the bias voltage is a straight line, as shown in Fig. 6.5.4.

In engineering practice, this linear graph lends itself to several useful applications.

From (6.5.5), $(1/C_d)^2$ becomes zero when $V = V_{bi}$. In the graph this is represented by the intersection of the plotted line with the V axis.[3] The construction shown in Fig. 6.5.4 is an often-used technique for the experimental determination of the contact potential.

The slope of the graph equals the expression in square brackets in Eq. (6.5.5) and can be used to measure one of the parameters appearing in the expression when the others are known.

In the literature it is customary to plot the *capacitance per square centimeter* c_d (rather than the actual capacitance of the junction) vs. the reverse bias voltage $V_R = -V$ (rather than the forward bias voltage V). For one-sided abrupt junctions, (6.5.5) becomes

$$\left(\frac{1}{c_d}\right)^2 = \left(\frac{2}{\varepsilon q N}\right)(V_{bi} + V_R) \tag{6.5.6}$$

---

[3] Expression (6.5.5) (and so its graph, Fig. 6.5.4) was obtained for reverse bias conditions. As previously noted, this implies that $V < 0$, so that, rigorously, the plot does not extend to positive values of V. In the construction of the figure, the plotted straight line has been *geometrically extended* into the $V > 0$ region. This process of extrapolation (indicated in the figure by the dotted line) determines the point of intersection with the V axis, the abscissa of which, as shown, measures the contact potential V_{bi}.

where N represents the concentration of minority carriers *in the lightly doped side* of the one-sided junction.

Expression (6.5.6) suggests a diagnostic technique of particular interest when N is not uniform, but varies with the depth from the junction. In this case, the coefficient in brackets is not constant and the graph is not linear any more. The plot can be obtained experimentally by measuring c_d at different values of dc bias. The capacitance meter used should introduce only a small ac signal. From the measured values, a graph of (6.5.6) can easily be drawn and its slope at any value of V_R can be evaluated graphically. Let this slope be S; then, from (6.5.6),

$$N = \frac{2}{\varepsilon q S}$$

(6.5.7)

where N is the impurity concentration in the lightly doped side of the junction at a distance from the junction corresponding to the chosen value of V_R. This technique yields the *doping profile* of the semiconductor near the junction.

Example 6.5.3. A PN junction in Si is assumed to be one-sided. The temperature is 300 K. Figure 6.5.5 is a plot of c_d (the junction's depletion capacitance per unit interface cross section) vs. V_R (the reverse bias voltage).
(a) Is the junction abrupt?
(b) What is the contact potential?
(c) What is N_L (impurity concentration in the low-doped side)?
(d) What is N_H (impurity concentration in the high-doped side)?
(e) What is the ratio x_L/x_H of low-doped to high-doped depletion region width?
(f) Is the one-sided assumption justified?

Solution
(a) The plot of Fig. 6.5.5 is linear; therefore N_L is constant, so that finally: *yes*, the junction is abrupt.
(b) The dotted extension of the plot into the forward bias zone intercepts the V_R axis at $V_R = 0.695$ V; therefore $V_{bi} = 0.695$ V.
(c) The slope of the plot of Fig. 6.5.5 is $S = 3.2 \times 10^{18}/2.695 = 1.19 \times 10^{18}$, so that, from (6.5.7), $N_L = 2/(1.04 \times 10^{-12} \times 1.6 \times 10^{-19} \times 1.19 \times 10^{18}) = 10^{13}$ cm^{-3}.
(d) From (6.2.9), after some algebra,

$$N_H = \frac{n_i^2}{N_L} e^{V_{bi}/V_T} = \frac{2.25 \times 10^{20}}{10^{13}} e^{0.695/0.0259} = 10^{19} \text{ cm}^{-3}$$

In the above V_T represents the quantity kT/q, i.e., the available thermal energy in electronvolts. This notation is quite common and we shall use it frequently.
(e) From (6.3.1),

$$\frac{x_H}{x_L} = \frac{N_L}{N_H} = \frac{10^{13}}{10^{19}} = 10^{-6}$$

(f) The high-doped semiconductor's depletion region is six orders of magnitude smaller than the other and so is negligible: *yes*, the one-sided assumption is reasonable.

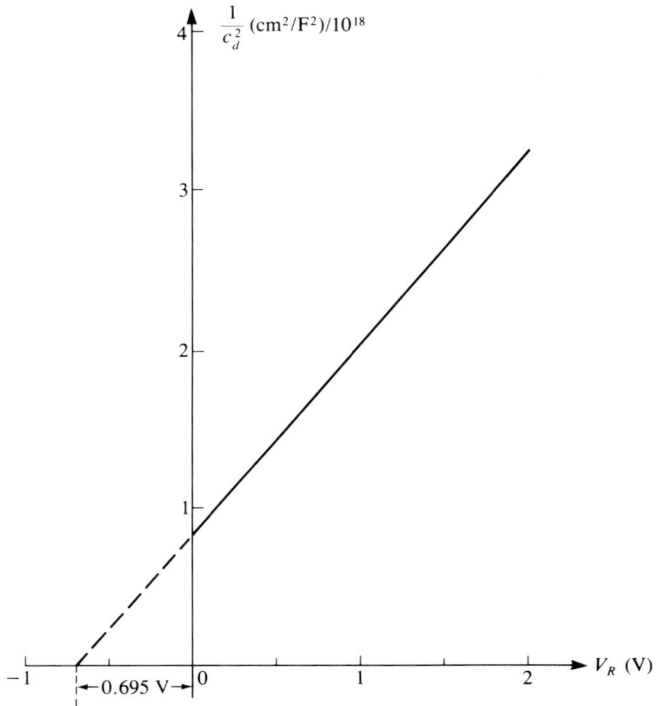

FIGURE 6.5.5
Inverse square junction specific capacitance c_d vs. reverse bias voltage V_R for Example 6.5.3.

Breakdown Voltage Computation

The breakdown phenomenon described in Sec. 6.4 can be mediated by two different phenomena: Zener and avalanche breakdown.

The quantum conditions keeping a bound electron within the confines of its atom can be affected by the presence of a strong electric field, such as may exist within a reverse biased depletion region. If the field is large enough, there may arise a strong probability that the valence electron of an atom in the depletion region may reach some point beyond the atom's potential barrier, to occupy there a permissible state at its own energy level.

This is a tunneling phenomenon (cf. Chap. 1, App. 1A.2, and Sec. 2.5), because the electron does not have enough energy to "jump above the barrier" to reach its new state, but rather "tunnels under it." The strong electric field simply increases the probability of tunneling.

When such a transition occurs, an electron-hole pair is generated within the depletion region and is swept across it by the electric field, so that the reverse saturation current increases. If the rate of carrier generation is large enough, the current can increase without bounds and *Zener breakdown* occurs. In semiconductors this may happen when the electric field becomes greater than about 10^6 V/cm (or 10^{-2} V/Å) [1].

Conversely, a current carrier being swept across the reverse biased depletion region by the strong electric field existing there may acquire enough energy between two consecutive collisions with lattice atoms to ionize an atom on impact. This, in turn, generates a carrier pair, which now contributes both to the reverse current and to the impact ionization phenomenon.

The stronger the field, the greater the probability of impact ionization and the greater the rate of increment of the current, so that the flow of carriers builds up like an avalanche as the carriers move across the depletion region. When the electric field is large enough for the current to tend to infinity then *avalanche breakdown* occurs.

A detailed analysis of the mechanism described (cf. App. 6A) shows that breakdown occurs whenever the reverse bias voltage across a junction exceeds a critical *breakdown voltage* V_{bd}. Figure 6.5.6 plots breakdown voltages for Si and GaAs abrupt and linearly graded junctions.

Example 6.5.4. A one-sided abrupt PN junction in Si at 300 K is characterized by $N_A = 10^{16}$ cm^{-3}, $N_D = 10^{19}$ cm^{-3}. Compute the breakdown voltage.

Solution. The lowest doped semiconductor is the P-type and the junction is one-sided, so from Fig. 6.5.6, using the value of N_A, $V_{bd} = 60$ V.

6.6 THE PN JUNCTION UNDER FORWARD BIAS

Under forward bias conditions V is positive, so that the steady state junction voltage $V_{bi} - V$ is less than the contact potential V_{bi}. Comparing the forward biased energy band diagram of Fig. 6.6.1 with Fig. 6.2.3 (for the unbiased

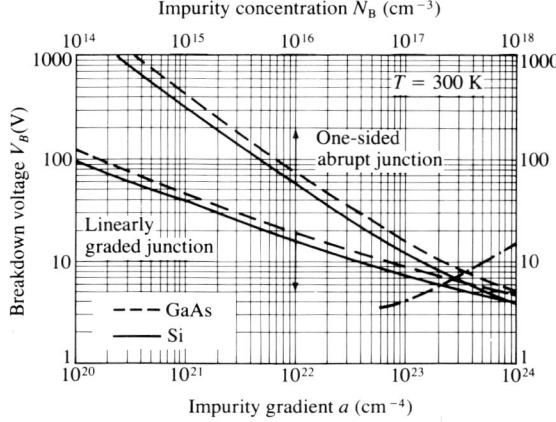

FIGURE 6.5.6
Breakdown voltage vs. impurity concentration (abrupt junction) or impurity gradient (linearly graded junction) in Si and GaAs at 300 K. Zener breakdown occurs along the dash-dot line [1].

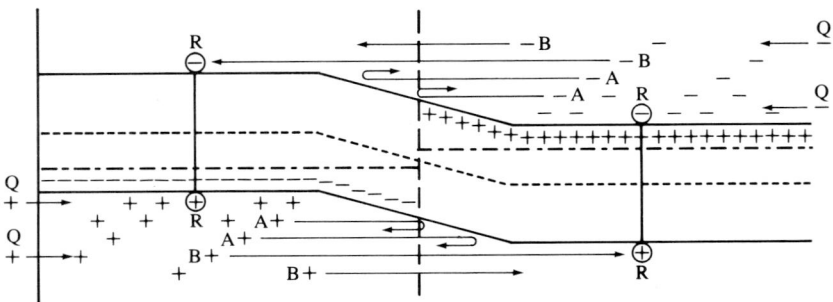

FIGURE 6.6.1
Band diagram for a forward biased junction. In comparison with Fig. 6.2.3 the lower barrier permits a larger portion of the majority carriers (B) to cross the junction, diffuse, and recombine, while fewer carriers (A) rebound back. To maintain the narrow depletion region characteristic of forward bias, the battery must provide charges (Q) to fill in for the loss of the migrating majority carriers (B). This flow of charge constitutes the forward diode current.

junction), it is observed that the Fermi levels are not aligned, the depletion region is narrower than in the unbiased case, the barrier height is lower, and so a larger percentage of the majority carriers (B in the figure) has enough energy to diffuse across the junction into the adjacent semiconductor, and the diffusion current becomes much larger than the minority carrier reverse current. This results in a net forward current.

The main features of the current carrier motion can be observed in Fig. 6.6.1, where the arrows symbolize carrier displacement:

1. *Diffusion* pushes majority carriers of sufficient energy (such as B in the figure) over the barrier into the adjacent semiconductor, where they now become minority carriers and so increase the minority carrier concentration in the immediate vicinity of the depletion region.
2. This localized pileup of minority carriers results in a concentration gradient, which causes further diffusion of the carriers away from the junction toward the body of the N semiconductor.
3. In the course of this migration they meet and recombine with the very numerous majority carriers. This is symbolized in the figure by encircling the recombining charges within circles joined by a line labeled R–R (for recombination).
4. If the bias battery were not connected, this would simply further deplete the semiconductors near the junction, extending the depletion region and finally raising the barrier height back to the unbiased level V_{bi}, but the battery forces the junction voltage to remain at the low level $V_{bi} - V$. To achieve this result, *the battery injects carriers* (Q) from its terminals into the semiconductors, and so replenishes the majority carrier population compensating for the effects of recombination.

In the body of the N semiconductor, the majority carrier electrons, under the push of the battery's electric field, *drift* toward the vacancies due to the holes diffusing in the opposite direction, finally replenishing these vacancies. Because electrons and holes carry opposite charges, both of these two migrations in opposite directions constitute a conventional electric current from the junction into the negative battery terminal. The current, of course, flows continuously through all the device, but it is mediated first by electrons and then by holes. A similar phenomenon occurs in the P-type semiconductor.

To visualize how the current in a forward biased PN junction is mediated by different carriers at different points along the length of the device, Fig. 6.6.2 shows a cross section taken along the length of the device, indicating the paths of electron and hole motion by means of *carrier flow lines*.

This symbolic representation of current flow is similar to the magnetic flux lines representation of the magnetic field. Each carrier flow line represents a current flow carrying a "unit" amount of current. Consequently, the total current through a cross section at any point along the device equals the number of flow lines through that cross section times the arbitrary "unit" of current associated with each flow line. In the figure dotted lines represent hole flow and solid lines electron flow. Notice that electron velocity is opposite in direction to hole velocity.

In reality, of course, electrons and holes within the device are distributed and move in a very disorderly way; however, in the figure the flow lines are depicted in an orderly distribution to help visualize the contribution of different carriers to the total current at different points along the device.

The total current (total number of carrier flow lines through a given cross section) is, of course, constant, but, progressing along the device, the relative intensity of electron-mediated current (solid flow lines) to hole-mediated current (dotted flow lines) varies. Proceeding from left to right in Fig. 6.6.2:[4]

Region A. At the extreme left, deep in the body of the P semiconductor, all the current is mediated by a *drift current of holes* from left to right.

Region B. As we approach the junction, the drifting holes reach a region where electrons, diffusing from the right, recombine with majority carriers, neutralizing them. Here the holes drifting in substitute the neutralized majority carriers taking on their role, thus restoring charge neutrality. Their drift motion stops, so they do not mediate the current any more.

For each flow line, to the right of the point of recombination, the current is instead mediated by electrons moving by diffusion from the junction into the body of the P semiconductor, so that, in region B, the total current is partly due

---

[4] Notice that, to minimize confusion, only a few of the majority carriers are shown. In some regions, however, their concentration is actually so high that recombinations are very probable. This fact must be borne in mind, to understand where and how minority carrier recombination occurs.

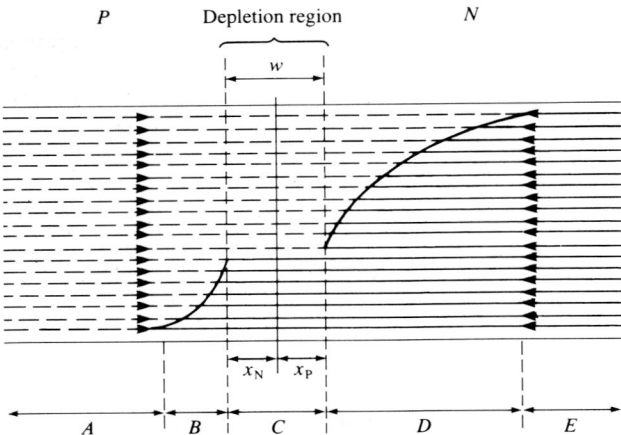

FIGURE 6.6.2
Forward biased junction. Carrier flow line representation of the current flow. Each line represents the flow of a "unit" of current. Dotted lines indicate hole-mediated and solid lines electron-mediated current. The orderly appearance of the diagram does not correspond to the actual random distribution of the carrier flow. If properly interpreted, however, the graph has statistical validity.

to holes drifting to the right and partly to electrons diffusing to the left. Progressing toward the junction, the hole-mediated current decreases and the electron-mediated current increases (the sum remains constant to satisfy Kirchoff's law at steady state).

Region C. Within the depletion region both hole and electron currents are due to diffusion and their ratio remains constant, because, in this depleted region, the concentration of carriers is so low that the probability of recombination is extremely small.

Region D. Entering the N semiconductor, the hole-mediated diffusion current component decreases by recombination with the plentiful electron majority carriers, while the current of electrons, drifting in to replace the neutralized majority carriers, correspondingly increases.

Region E. Finally, deep within the body of the N semiconductor, all diffusing holes have recombined and the current is mediated exclusively by *electron drift.*

The perceptive reader will have observed that, in the above description of carrier motion, it has been implied that minority carriers move by diffusion and majority carriers by drift (conduction). This assumption is easily justified by considering the carrier concentrations in the various regions.

When holes diffuse into the N semiconductor, the local hole concentration is increased from the equilibrium value p_{oN} to a new value p_N (using the notation convention defined in Sec. 4.4). The equilibrium minority carrier concentration is generally very low, so even a moderate hole injection can increase it appreciably, generating a sizable *minority carrier concentration gradient*, so the diffusion component is large.

Under *low-level injection conditions*, the number of injected minority carriers is negligible compared to the high concentration of the majority carriers, so it is legitimate to assume that this concentration remains essentially equal to the equilibrium value, $n_N \approx n_{oN}$, and no appreciable gradient of majority carrier concentration is generated. Diffusion therefore constitutes a negligible component of the current mediated by the majority carriers. Their motion is instead due to the electric field set up by the accumulation of excess minority carriers near the junction and maintained by the action of the battery. In other words, the majority carriers drift in to neutralize the opposite charge built up by the minority carrier injection.

The same line of reasoning can be repeated for electron injection into the P semiconductor. Remembering the considerations of Chap. 4, it can then be safely assumed that, in the body of each semiconductor, near the depletion region, *minority carriers move mainly by diffusion and majority carriers by drift.*

6.7 QUANTITATIVE JUNCTION ANALYSIS AT FORWARD BIAS

Diffusion currents are produced by the gradient of the carrier concentration. To compute their intensities, one must determine quantitatively the concentration of the minority carriers at the edges of the depletion region resulting, as previously mentioned, from the injection of carriers across the junction.

At equilibrium, from (6.2.5) using (6.2.1),

$$n_{oN} = n_i \, e^{(E_{fN} - E_{io} + q\psi_N)/(kT)} \tag{6.7.1}$$

$$p_{oP} = n_i \, e^{(E_{io} - q\psi_P - E_{fP})/(kT)} \tag{6.7.2}$$

by multiplication, remembering that, at equilibrium, $E_{fN} = E_{fP}$ and $\psi_N - \psi_P = V_{bi}$:

$$p_{oP} \, n_{oN} = n_i^2 \, e^{V_{bi}/V_T} \tag{6.7.3}$$

where, as usual, $V_T = kT/q$ is the thermal energy voltage, measuring the thermal electron energy in electronvolts. Remembering (3.3.6),

$$p_{oP} \, n_{oN} = p_{oP} \, n_{oP} \, e^{V_{bi}/V_T} = n_{oN} \, p_{oN} \, e^{V_{bi}/V_T} \tag{6.7.4}$$

from which, after simplification and some algebra,

$$p_{oN} = p_{oP} \, e^{-V_{bi}/V_T} \tag{6.7.5}$$

$$n_{oP} = n_{oN} \, e^{-V_{bi}/V_T} \tag{6.7.6}$$

These equations show that, at equilibrium, *the minority carrier concentration at the edge of a depletion region depends exponentially on the voltage across it.*

Similar relationships can be assumed if the junction is biased. Under biased conditions, the junction voltage is $V_{bi} - V$ and using the appropriate notation for this nonequilibrium condition, (6.7.5) and (6.7.6) become

$$p_N = p_P \, e^{-(V_{bi}-V)/V_T} = p_{oP} \, e^{-V_{bi}/V_T} \, e^{V/V_T} = p_{oN} \, e^{V/V_T} \qquad (6.7.7)$$

$$n_P = n_{oP} \, e^{V/V_T} \qquad (6.7.8)$$

where use has been made of the low injection condition $p_P \approx p_{oP}$.

The important relationships (6.7.7) and (6.7.8) are general and permit computation of the minority carrier concentrations at the edge of a biased depletion region.[5]

To determine the complete concentration distribution, apply the continuity equation (4.5.13) to the minority carriers in the body of the semiconductors near the depletion region. Under the stated conditions (steady state, current due to diffusion, one-dimensional system, and low-level injection), (4.5.13) becomes

$$D_{hN} \frac{d^2}{dx^2} (p_N - p_{oN}) - \frac{p_N - p_{oN}}{\tau_{hN}} = 0 \qquad \text{for } x \geq x_N \qquad (6.7.9)$$

for holes in the N semiconductor and

$$D_{eP} \frac{d^2}{dx^2} (n_P - n_{oP}) - \frac{n_P - n_{oP}}{\tau_{eP}} = 0 \qquad \text{for } x \leq -x_P \qquad (6.7.10)$$

for electrons in the P semiconductor. In these equations the continuity condition has been expressed exclusively in terms of the *excess minority carrier concentration* $p_N - p_{oN}$ and $n_P - n_{oP}$ by remembering that, as the equilibrium concentrations are constant, $d^2 p_{oN}/dx^2 = d^2 n_{oP}/dx^2 = 0$.

Solving (6.7.9) and (6.7.10) for the excess minority carrier concentrations

$$p_N - p_{oN} = A \exp\left(\frac{x}{\sqrt{D_{hN}\,\tau_{hN}}}\right) + B \exp\left(\frac{-x}{\sqrt{D_{hN}\,\tau_{hN}}}\right) \qquad (6.7.11)$$

for holes in the N semiconductor and

$$n_P - n_{oP} = C \exp\left(\frac{x}{\sqrt{D_{eP}\,\tau_{eP}}}\right) + D \exp\left(\frac{-x}{\sqrt{D_{eP}\,\tau_{eP}}}\right) \qquad (6.7.12)$$

for electrons in the P semiconductor. The arbitrary constants should be com-

---

[5] In the case of reverse bias, V is a negative voltage, so (6.7.7) and (6.7.8) predict zero minority carrier concentrations at the edge of a sufficiently strongly reverse biased depletion region. This confirms the previous statement (Sec. 6.4) that, under sufficiently strong reverse bias, all minority carriers are swept into the adjacent semiconductor as fast as they reach the edge of the depletion region.

puted from the boundary conditions, which, in this case, thanks to (6.7.7) and (6.7.8), are

$$x = x_N \rightarrow p_N(x_N) - p_{oN} = p_{oN} (e^{V/V_T} - 1); \qquad x = \infty \rightarrow p_N = p_{oN}$$

$$x = -x_P \rightarrow n_P(-x_P) - n_{oP} = n_{oP} (e^{V/V_T} - 1); \qquad x = -\infty \rightarrow n_P = n_{oP}$$

finally obtaining for the minority carrier concentration distributions:

$$p_N = p_{oN} + p_{oN} e^{-(x - x_N)/L_{hN}} (e^{V/V_T} - 1) \qquad \text{for } x \geq x_N \qquad (6.7.13)$$

and

$$n_P = n_{oP} + n_{oP} e^{(x + x_P)/L_{eP}} (e^{V/V_T} - 1) \qquad \text{for } x \leq -x_P \qquad (6.7.14)$$

where we have introduced the constants

$$L_{hN} = \sqrt{D_{hN} \tau_{hN}} \qquad \text{and} \qquad L_{eP} = \sqrt{D_{eP} \tau_{eP}} \qquad (6.7.15)$$

These parameters have the dimensions of length and are designated as the *diffusion lengths* of holes in the N, respectively electrons in the P, semiconductor.

The minority carrier distributions of Eqs. (6.7.13) and (6.7.14) are plotted in Fig. 6.7.1, for forward bias (*a*) and reverse bias (*b*) (cf. Prob. 6.26).

As the minority carrier current is due only to diffusion, then, by (4.3.1) and because, by definition, the diffusion current density equals the particle flow-rate density times the particle charge,

$$J_e = q n_{oP} \frac{D_{eP}}{L_{eP}} e^{(x + x_P)/L_{eP}} (e^{V/V_T} - 1) \qquad \text{for } x \leq -x_P \qquad (6.7.16)$$

for the electron current density and

$$J_h = q p_{oN} \frac{D_{hN}}{L_{hN}} e^{-(x - x_N)/L_{hN}} (e^{V/V_T} - 1) \qquad \text{for } x \geq x_N \qquad (6.7.17)$$

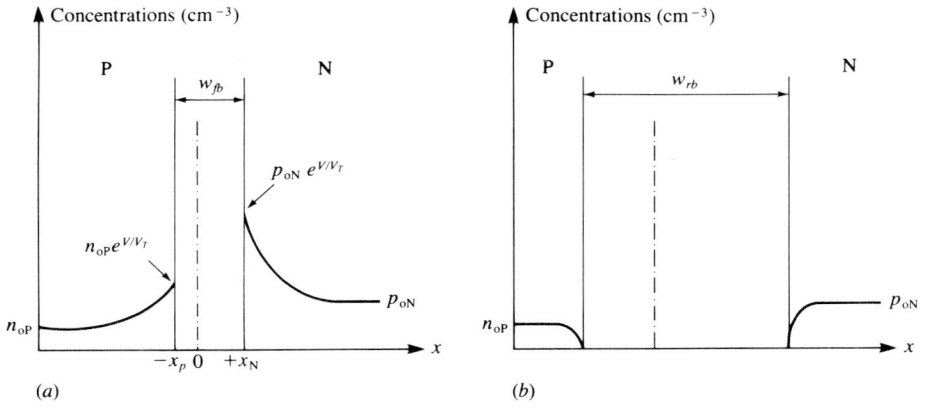

FIGURE 6.7.1
Minority carrier distribution near a biased depletion region ($N_D > N_A$): (*a*) forward biased, (*b*) reverse biased.

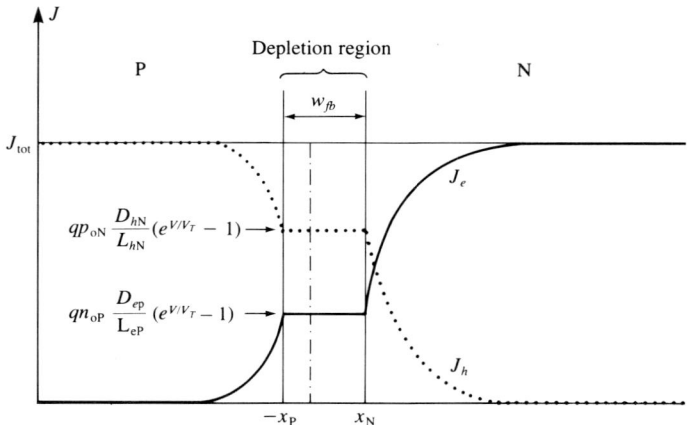

FIGURE 6.7.2
Electron (solid line) and hole (dotted line) current density component distributions in forward biased PN junction for $N_A = 2N_D$. In the depletion region the magnitude of each component is proportional to the slope of the corresponding minority carrier concentration at the edge of the depletion region. It is assumed that no recombination occurs within the depletion region, the total current density J_{tot} is the sum of the two components, and, for uniform cross section, is constant over all the device (solid horizontal plot).

for the hole current density. The current density distributions are shown in Fig. 6.7.2.

It is well worth noticing that the electron and hole current densities in the depletion region are determined by the *slope of the corresponding minority carrier concentration plot at the edge of the depletion region* (cf. Fig. 6.7.1).

Electron and hole current densities are shown along the whole length of the device, including the regions in which (6.7.16) and (6.7.17) are not valid, namely:

1. *In the depletion region*, where recombination is assumed to be negligible, because of the low carrier concentrations.
2. *In the P region for J_h, N region for J_e*, where the carriers mediating the current component are mainly majority carriers.

The total current is, of course, constant over all the length of the device, so that, assuming constant cross section,

$$J_t = J_h + J_e = \text{constant} \tag{6.7.18}$$

as implied in Fig. 6.7.2. J_t can then be easily computed as the sum of the two components within the depletion region, where the components remain constant:

$$J_t = J_h(x_N) + J_e(-x_P) = q\left(\frac{D_{hN}}{L_{hN}} p_{oN} + \frac{D_{eP}}{L_{eP}} n_{oP}\right)(e^{V/V_T} - 1) \tag{6.7.19}$$

For a junction cross section A, the diode current is

$$I_D = qA\left(\frac{D_{hN}}{L_{hN}} p_{oN} + \frac{D_{eP}}{L_{eP}} n_{oP}\right) (e^{V/V_T} - 1) = I_0\, (e^{V/V_T} - 1) \qquad (6.7.20)$$

where $I_0 = J_0 A$ is the magnitude of the *inverse saturation current* (cf. Sec. 6.4). This is easily verifed by letting V approach $-\infty$ in (6.7.20).

Analyzing the role of each parameter of Eq. (6.7.20), the student should notice that the most significant of these in determining the inverse saturation current density in a PN junction is the impurity concentration *in the low-doped semiconductor*.

With the aid of (6.7.20), it is possible to plot the *I/V diode characteristic* curve.

Example 6.7.1. A PN junction in Si at 300 K is characterized by $N_A = 10^{16}$ cm^{-3} and $N_D = 10^{13}$ cm^{-3}. For each semiconductor, net and total doping are the same. The minority carrier lifetimes are $\tau_{hN} = \tau_{eP} = 1$ μs. The junction cross section is 1 mm^2. Compute the current at an applied voltage of (a) $V = -1.5$ V and (b) $V = 0.5$ V.

Solution. From (4.2.9), under the given impurity concentrations, $\mu_{eP} = 1329$ cm^2/ (V·s) and $\mu_{hN} = 401$ cm^2/(V·s); therefore, from (4.3.2), $D_{eP} = 34.3$ cm^2/s and $D_{hN} = 10.35$ cm^2/s. From (3.3.13) and (1.3.14), $n_{oP} = 2.25 \times 10^4$ cm^{-3} and $p_{oN} = 2.25 \times 10^7$ cm^{-3}.

(a) From (6.7.20), noticing that, for $V = -1.5$, the exponential is essentially equal to zero,

$$I_D(-1.5) \approx -I_0 = -1.6 \times 10^{-19} \times 10^{-2}$$

$$\times \left(\sqrt{\frac{10.35}{10^{-6}}}\, 2.25 \times 10^7 + \sqrt{\frac{34.3}{10^{-6}}}\, 2.25 \times 10^4\right)$$

$$= -0.116 \text{ nA}$$

where advantage was taken of the identity

$$\frac{D_{eP}}{L_{eP}} = \sqrt{\frac{D_{eP}}{\tau_{eP}}}$$

obtained from (6.7.15) and also valid for holes. It should be noticed that, of the two addends under parentheses in the computation of I_0, the second is three orders of magnitude smaller than the first and could therefore be omitted in the computation. This happens whenever the junction is essentially unilateral.

(b) From (6.7.20) for $V = 0.5$,

$$I_D(0.5) = 0.116 \times 10^{-9}\, (e^{0.5/0.0259} - 1) = 28 \text{ mA}$$

Figure 6.7.3 illustrates the previous example, showing the *I/V* curve of the PN junction under analysis. Figure 6.7.4 shows the published characteristics and data for a commercially available diode.

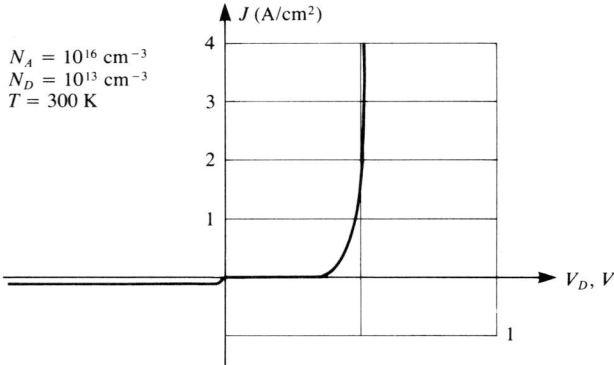

$N_A = 10^{16}$ cm^{-3}
$N_D = 10^{13}$ cm^{-3}
$T = 300$ K

FIGURE 6.7.3
Diode I/V characteristic plotted for the data of Example 6.7.1.

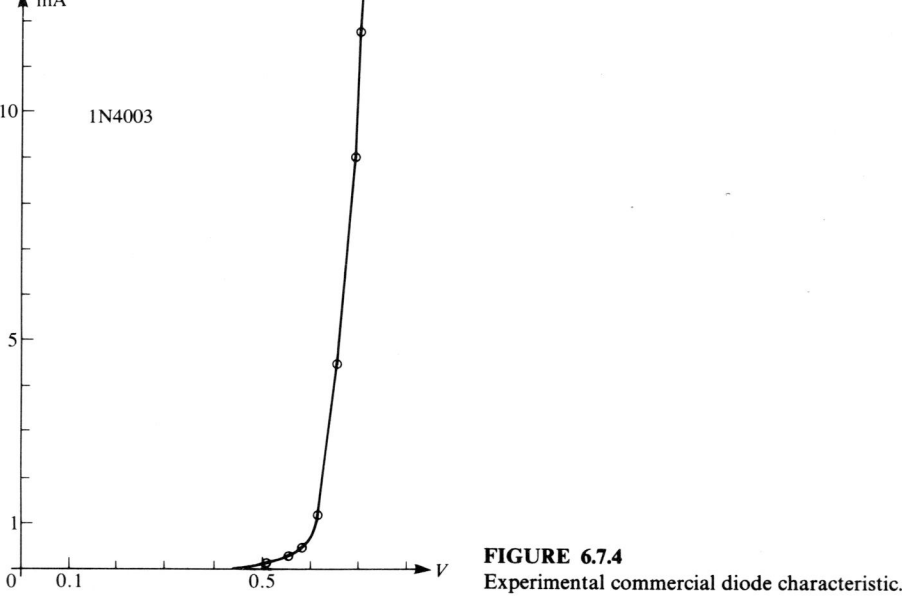

FIGURE 6.7.4
Experimental commercial diode characteristic.

6.8 SMALL-SIGNAL PARAMETERS AT FORWARD BIAS

Figure 6.8.1 shows a small-signal ac generator v applied to a diode forward-biased by a dc voltage V. This is equivalent to the schematic of Fig. 6.5.2 except for the polarity of the bias voltage. The figure also uses the familiar circuit symbol for the diode, in which the arrow points in the direction from the P to the

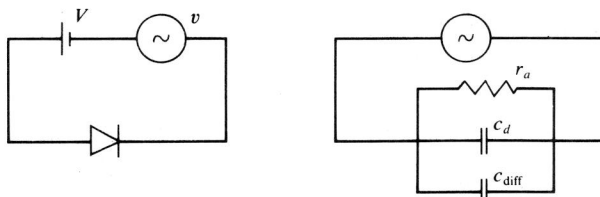

FIGURE 6.8.1
Circuit schematic of a forward biased diode with a small time-varying signal and small-signal equivalent circuit.

N semiconductor. The equivalent small-signal linear circuit contains three parameters: a dynamic forward bias resistance, the depletion capacitance, and a diffusion capacitance.

Dynamic Forward Bias Resistance Computation

As usual (cf. Sec. 6.5), in computing small-signal (dynamic) circuit parameters, voltage and current variations are taken as infinitesimal. From (6.7.20), the dynamic forward bias conductance can be computed as

$$g_d = \frac{dI_D}{dV} = \frac{I_0}{V_T} e^{V/V_T} \approx \frac{I_D}{V_T} \tag{6.8.1}$$

The dynamic resistance is, of course, the inverse of the conductance, yielding $r_d = V_T/I_D$, i.e., a nonlinear resistance inversely proportional to the current, showing that a forward biased diode can be used as a *current controlled resistor*. At room temperature, where $V_T \approx 26$ mV, a commonly used expression for the dynamic resistance is $r_d \approx 1/(40I_D)$.

Figure 6.8.2 plots the small-signal resistance r_d of a forward biased diode vs. the dc diode current I_D computed in accordance with (6.8.1). Experimental values at 300 K are also shown.

Diffusion Capacitance Computation

In the forward biased condition, in accordance with the previous analysis, charge is not only stored in the depletion region but also next to it, in each of the adjacent semiconductor regions, in the form of excess minority carriers. To account for this charge accumulation, in the small-signal mode, another capacitance must be added to the depletion capacitance. Because this stored charge is the cause of the minority carrier diffusion, this extra capacitance is known as the *diffusion capacitance*.

The excess hole concentration injected into the N semiconductor is $p_N - p_{oN}$; therefore, the corresponding charge density is $q(p_N - p_{oN})$. Using

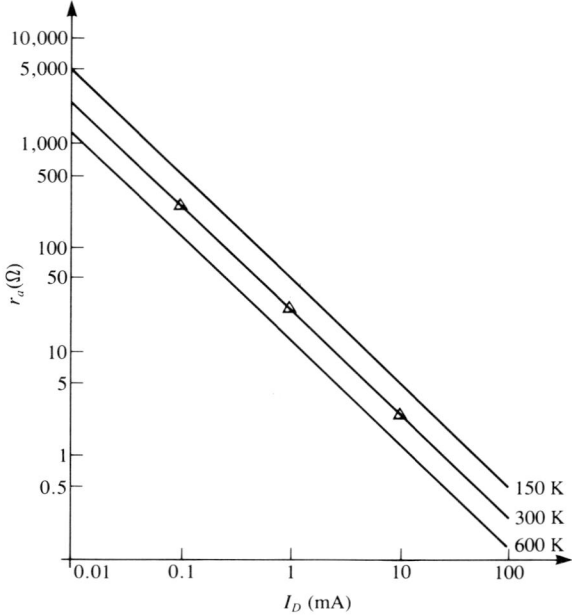

FIGURE 6.8.2
Small-signal dynamic resistance r_d of forward biased PN diode vs. dc component of diode current.

(6.7.13), the total charge accumulated as excess minority carriers in the N semiconductor is then

$$Q_{hN} = \int_{x_N}^{\infty} q p_{oN}\, e^{-(x-x_N)/L_{hN}}\, (e^{V/V_T} - 1)A\ dx = q p_{oN} L_{hN}\, (e^{V/V_T} - 1)A \quad (6.8.2)$$

so that the corresponding dynamic capacitance is

$$C_{\text{dif}} = \frac{dQ_{hN}}{dV} = \frac{q p_{oN}\, A L_{hN}}{V_T}\, e^{V/V_T} \quad (6.8.3)$$

showing that the diffusion capacitance varies exponentially with the forward bias; therefore, at strong reverse bias ($V \ll 0$) it vanishes.

Expression (6.8.3) assumes the small signal applied to be infinitesimally small. If the signal is finite, some adjustment is necessary. It turns out that this adjustment depends on the waveshape of the signal. For sinusoidal signals C_{dif} is approximately computed as half of the value given by (6.8.3).

In many practical cases, however, Fig. 6.8.1 adequately represents the small-signal equivalent circuit of a forward biased diode, with C_{dif} given with sufficient approximation by (6.8.3).

While the circuits of Figs. 6.4.3 and 6.8.1 can be validly used to compute the small-signal frequency dependence of the behavior of the biased diode, they do not solve the problem of predicting the transient behavior of the device under large signals (e.g., switching from on to off conditions and vice versa).

Each biasing condition implies that minority carrier charges have accumulated near the junction, as shown in Fig. 6.7.1. If the bias is changed from one value to another, the charge configuration must be accordingly changed.

While the required charge is being either accumulated or removed the system undergoes a transient, the duration of which can be shown to depend on the minority carrier's recombination or generation time constants, τ_r and τ_g. Usually the most important parameter is the turnoff time, which can be shown to increase with τ_r. Minimizing τ_r is therefore a design goal for fast devices (cf. Sec. 4.4).

The switching response of bipolar semiconductor devices is discussed in greater detail in Sec. 10.9 for the more interesting case of the BJT.

6.9 RECOMBINATION CURRENT

The analysis of Sec. 6.7 is based on the assumption that no recombination or generation phenomena occur within the depletion region, so the carriers simply cross this region uneventfully. However, as discussed in Sec. 4.4, such phenomena do occur under nonequilibrium conditions. In the forward biased junction, recombination phenomena are induced by carrier injection; in the reverse biased junction, carrier depletion gives rise to generation phenomena.

For the forward biased junction (cf. Sec. 4.4), a quantitative analysis [2] shows that recombination within the depletion region results in a *recombination current* component expressed by

$$I_{\text{rec}} = I_{rec_0} \left(e^{V/(2V_T)} - 1 \right) \tag{6.9.1}$$

with
$$I_{\text{reco}} = \frac{qW A n_i}{2\tau_r} \tag{6.9.2}$$

where W is the depletion region width and τ_r is the recombination time constant (cf. Sec. 4.4). The total current is the sum of this recombination component plus the diffusion component discussed in Sec. 6.7:

$$I = I_{\text{diff}_0} \left(e^{V/V_T} - 1 \right) + I_{\text{reco}} \left(e^{V/(2V_T)} - 1 \right) \tag{6.9.3}$$

or, remembering (6.7.20),

$$I = qA \left[\left(\frac{D_{hN}}{L_{hN}} p_{oN} + \frac{D_{eP}}{L_{eP}} n_{oP} \right) \left(e^{V/V_T} - 1 \right) + \frac{W n_i}{2\tau_r} \left(e^{V/(2V_T)} - 1 \right) \right] \tag{6.9.4}$$

The role played by the different recombination time constants in determining each of these two components of the diode current can be estimated by replacing the diffusion lengths in (6.9.4) with their expressions—Eqs. (6.7.15).

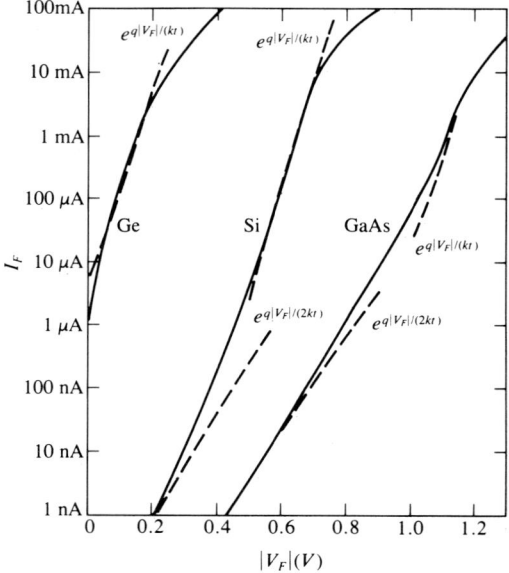

FIGURE 6.9.1
Semilogarithmic plot of diode current vs. applied voltage for Si, Ge, and GaAs diodes, showing voltage dependence of ideality factor η [2].

The recombination component is usually negligible in Ge junctions, but in Si and GaAs transistors at low current densities it becomes prevalent. At higher currents, instead, the diffusion component prevails, while at intermediate values the contributions of the two components are comparable. Consequently, in these diodes, the I/V curve follows the expression:

$$e^{V/(\eta V_T)} \tag{6.9.5}$$

where η is an *ideality factor* ranging from $\eta = 2$ at low currents to $\eta = 1$ at high current densities, as shown in Fig. 6.9.1. In the semilogarithmic graph of the figure, the exponential expression (6.9.5), to which the diode current is proportional, translates into a linear plot of current vs. applied voltage. The slope of this plot is the exponent of (6.9.5). The plot of Fig. 6.9.1 clearly displays a change of slope corresponding to the predicted gradual variation of η from 1 to 2 as the condition of operation varies from high to low applied voltages.

In reverse biased junctions, carrier depletion induces generation phenomena, both in the bulk of the depletion region and on the surface. The resulting *generation currents* are inversely proportional to the bulk and surface generation time constants τ_g (cf. Sec. 4.4). They are added to the reverse saturation current I_0 computed in Sec. 6.7, Eq. (6.7.20), and often represent the dominant factor in determining the total reverse saturation current, sometimes becoming so large that the device is made inoperative.

Minimization of the generation component of the leakage current requires maximization of τ_g. Some elementary considerations regarding the approach to the design problem of maximizing τ_g while keeping τ_r reasonably small (a requirement for fast transient response) have been offered in Sec. 4.4.

Still another possible component of the forward biased current is provided by *tunneling through the depletion region potential barrier*. In accordance with the description of this quantum phenomenon introduced in Chap. 1 (also cf. App. 1A.2), for tunneling to occur two conditions must be met:

1. The potential barrier must be very narrow and
2. The initial and final state of the transition must be essentially at the same energy level.

The tunneling we are considering occurs (1) across the depletion region and (2) between the conduction electrons and the valence band holes. For tunneling to be at all significant, therefore:

1. The depletion region must be very narrow (a characteristic of forward biasing) and
2. The electrons near the bottom of the conduction band in the N-type semiconductor must be essentially at the same energy level as the holes near the top of the valence band of the P-type semiconductor.

Figure 6.9.2 shows the band diagram of a PN junction in which the Fermi level falls within the gap in both the P and the N semiconductors. The electrons

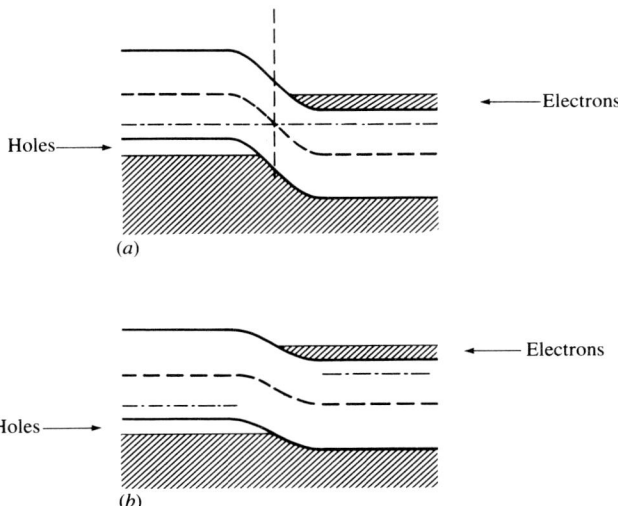

FIGURE 6.9.2
PN junction band diagram with Fermi level within gap. Clear and dark regions indicate vacant, respectively occupied, states. (*a*) No bias. Notice comparatively wide depletion region and sizable energy difference between holes on one side of the junction and electrons on the other side. (*b*) Forward bias. The depletion region is much narrower, but the energy gap has increased: no tunneling.

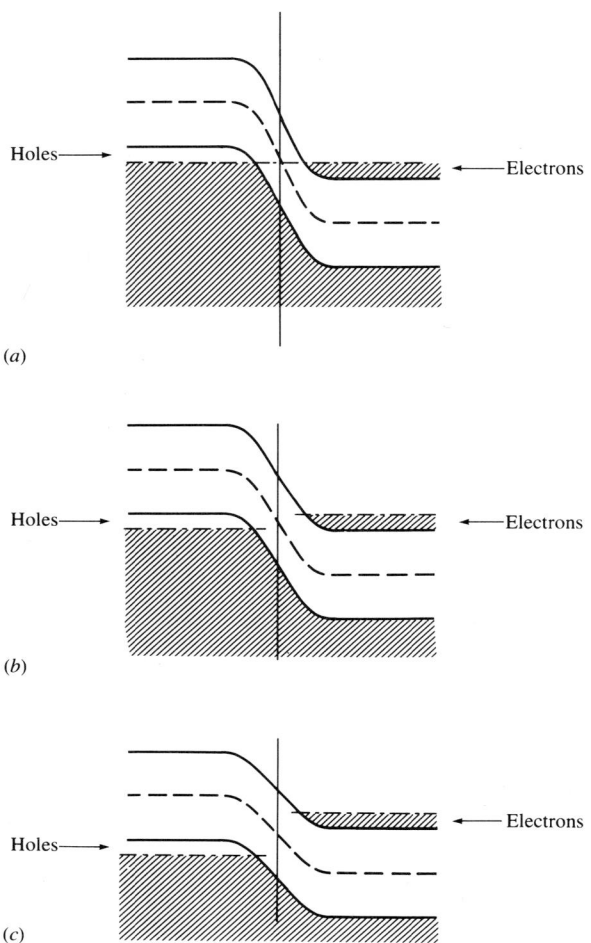

Holes──→   ←──Electrons

(a)

Holes──→   ←──Electrons

(b)

Holes──→   ←──Electrons

(c)

FIGURE 6.9.3
Band diagram of doubly degenerate junction (Fermi levels outside of gap). (a) No bias. Notice small energy difference between holes and electrons but rather wide depletion region: marginal conditions for tunneling. (b) Moderate forward bias. No energy difference between electrons and holes, narrower depletion region: intense tunneling. (c) Larger forward bias. The depletion region is still narrower, but the energy difference between electrons and holes is greater: tunneling stops.

occupy the narrow dark strip near the bottom of the conduction band, the holes the narrow light strip near the top of the valence band. In Fig. 6.9.2a the junction is not biased, in Fig. 6.9.2b it is forward biased.

The figure clearly shows that condition 2 above is not met in either case and would require reverse biasing, which would not meet condition 1.

In the band diagram of Fig. 6.9.3, instead, the Fermi level falls within the conduction band of the N semiconductor and within the valence band of the P semiconductor (the semiconductors are degenerately doped, cf. Sec. 3.4).

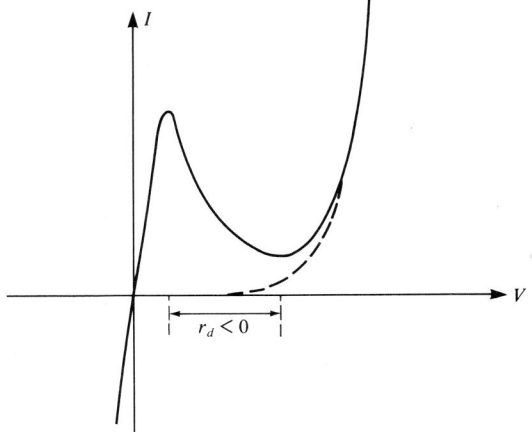

FIGURE 6.9.4
Tunnel diode characteristic (solid line) and diffusion current component (dashed line), indicating the range of negative resistance operating conditions. The large negative currents at reverse bias are due to tunneling of electrons from the valence band to the conduction band.

It is easy to see that[6] in Fig. 6.9.3a (nonbiased junction) condition 1 is marginally met and that a small forward bias, as in Fig. 6.9.3b aligns the two strips while at the same time decreasing the depletion region. Under these conditions tunneling can be very significant.

If the forward bias voltage is further increased, as in Fig. 6.9.3c, the depletion region becomes still narrower, but, although condition 1 is met, the energy difference between electrons and holes becomes considerable, so tunneling must stop.

The I/V characteristic of this diode is shown in Fig. 6.9.4, which also plots (dashed line) the diffusion current component. This device is called a *tunnel diode*. It will be noticed that its I/V characteristic displays a negative slope over a range of operating conditions. Equation (6.8.1) then shows that, in this region of operation, the tunnel diode is characterized by a *negative dynamic resistance r_d*. This property leads to very useful applications of the device, for instance, in oscillators and high-speed switching circuits.

The large reverse current at reverse bias is due to electron tunneling from the valence to the conduction band. Remembering that the total current is the sum of the diffusion and tunneling component and considering Fig. 6.9.3, the student should easily reach an intuitive justification of the shape of the characteristics of Fig. 6.9.4.

6.10 HETEROJUNCTIONS

In discussing the exchange of carriers across a PN junction, it has been assumed that the two semiconductors in contact differ only by their doping, but that the host materials are the same (homojunctions). Sometimes it is expedient to form

---

[6] As a first approximation, states below the Fermi level are assumed to be fully occupied while above it, empty, as if the temperature were very low.

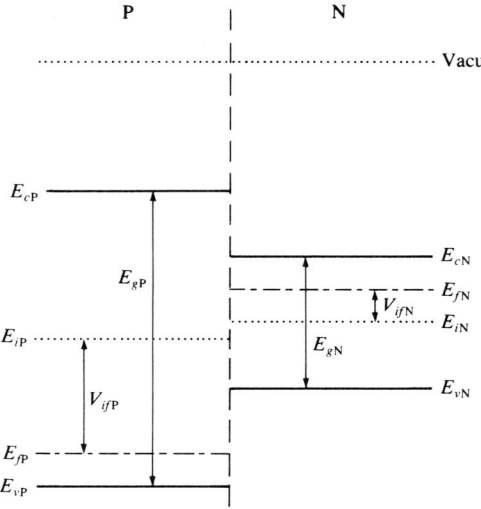

FIGURE 6.10.1
Band diagrams of P and N semiconductors with different host materials before a heterojunction is made. Notice the difference in E_g.

heterojunctions between extrinsic semiconductors fabricated from different host materials.

Figure 6.10.1 shows the band diagrams of the two materials next to each other before the heterojunction is made. All energies are expressed in electronvolts. The vacuum level (minimum energy of electrons outside the semiconductor) is shown as a horizontal straight line (dotted), indicating that the two materials are at the same potential. In the figure the characteristic energy levels of the N and P semiconductors are identified by the subscripts N and P respectively. Contrary to the homojunction case depicted in Fig. 6.2.1a, in the heterojunction these levels do not necessarily coincide and $E_{g\mathrm{N}} \neq E_{g\mathrm{P}}$.

Figure 6.10.2 shows the distortion of the band diagrams after the junction has been made and carrier exchange has taken place. The Fermi levels are aligned, the contact potential is $V_{bi} = \psi_\mathrm{N} - \psi_\mathrm{P} = E_{f\mathrm{N}} - E_{f\mathrm{P}}$, and the vacuum level follows the potential distribution around the junction; however, the graphs of E_c and E_v *maintain the discontinuities at the junction* arising from the differences between the limits of the energy bands of the two materials.

Quantitatively, with the notation of the figures and assuming dopings N_A and N_D, electric permittivities ε_P and ε_N, and intrinsic concentrations $n_{i\mathrm{P}}$ and $n_{i\mathrm{N}}$, an analysis entirely analogous to that of Sec. 6.3 yields

$$V_{bi} = E_{i\mathrm{N}} - E_{i\mathrm{P}} + kT \ln \frac{N_A N_D}{n_{i\mathrm{N}} n_{i\mathrm{P}}} \tag{6.10.1}$$

For the depletion region widths,

$$x_\mathrm{N} = \sqrt{\frac{2\varepsilon_\mathrm{N}}{qN_D} \frac{\varepsilon_\mathrm{P} N_A}{\varepsilon_\mathrm{N} N_D + \varepsilon_\mathrm{P} N_A} (V_{bi} - V)} \tag{6.10.2}$$

and

$$x_\mathrm{P} = \sqrt{\frac{2\varepsilon_\mathrm{P}}{qN_A} \frac{\varepsilon_\mathrm{N} N_D}{\varepsilon_\mathrm{N} N_D + \varepsilon_\mathrm{P} N_A} (V_{bi} - V)} \tag{6.10.3}$$

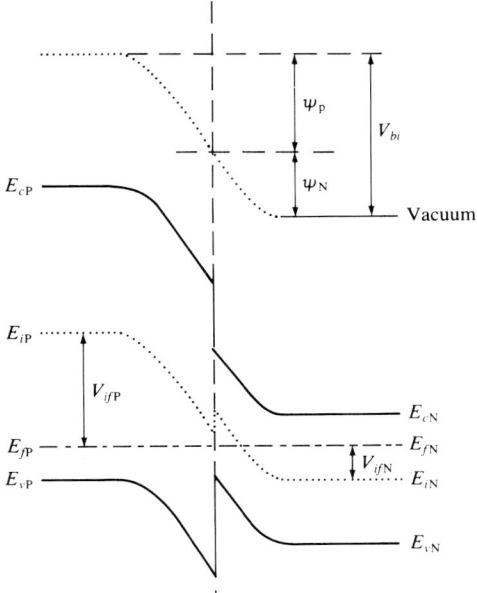

FIGURE 6.10.2
Band diagram of the heterojunction between the materials of Fig. 6.10.1. The potential barrier is higher for the conduction band than for the valence band. Inversion occurs in the N semiconductor near the junction.

where V is the voltage externally applied to the junction (positive for P positive, negative for P negative). The depletion capacitance is

$$C = A \sqrt{\frac{q \varepsilon_N N_D \varepsilon_P N_A}{2(\varepsilon_N N_D + \varepsilon_P N_A)} \frac{1}{V_{bi} - V}} \tag{6.10.4}$$

Figure 6.10.2 assumes that, when the junction is made, the lattice characteristics of the two materials are not disturbed, so no spurious states are generated. This is true only if the lattice constants are very close to each other so that a good match exists between the two crystals. This requirement limits the types of host materials that can be paired in practice to form heterojunctions.

Observation of Fig. 6.10.2 shows that the heights of the junction barriers are different from V_{bi}, being modified by the presence of the discontinuities in the graphs of the potential barriers. This phenomenon can be used to fabricate more efficient MESFETs (cf. Sec. 8.7).

For the heterojunction shown in Fig. 6.10.2 the barrier at the bottom of the conduction band is higher than that at the top of the valence band. This results in changing the proportion of hole-to-electron currents that cross the junction at forward bias. In the case of the figure, the ratio of electron-to-hole current is decreased. It is possible to take advantage of this fact in the fabrication of junction transistors (cf. Sec. 10.6).

Similarly, the gap energies of the two semiconductors are different. In the figure, $E_{gP} > E_{gN}$. This circumstance permits the fabrication of more efficient semiconductor optical detectors by decreasing the attenuation of light rays entering the detecting devices.

Observation of Fig. 6.10.2 also brings to light an interesting possibility. As shown, the choice of an appropriate combination of bandgap locations and doping levels can result in a bending of the bands that causes the intrinsic Fermi level plot to cross the Fermi level line. This occurs in the N-type semiconductor in Fig. 6.10.2, where, in a small region next to the junction, the Fermi level is below the intrinsic Fermi level line. In this region an inversion of the semiconductor type has occurred: the N-type semiconductor has become P-type! Such *inversion layers* and their properties will be discussed at length in Chap. 9. The possibility of generating inversion layers in heterojunctions is sometimes used in the fabrication of MESFET devices (cf. Sec. 8.7).

6.11   SUMMARY

When a PN junction is formed, the crystal surface barrier at the interface is lowered and majority carriers diffuse through it into the next semiconductor, where they become minority carriers and recombine, forming a *depletion region* on both sides of the junction. As the depletion region is charged, a *contact potential* is established across the junction. This generates a displacement of the P and N state diagrams along the energy axis.

The diffusion transient dies out when the *Fermi levels align*. In the condition of *dynamic equilibrium* the diffusion current is balanced by the conduction current generated by the contact potential.

At equilibrium the contact potential varies logarithmically with the product of the dopant concentrations, in accordance with Eq. (6.2.9).

The other parameters depend on the junction profile. For *abrupt junctions* these parameters can be computed from Eqs. (6.3.15) through (6.3.21), showing a square root relationship between the depletion region width and the contact potential.

The electric field $\mathscr{E}$ and potential distribution ψ are plotted in Fig. 6.3.1.

For *linearly graded junctions* the appropriate relationships are given by Eqs. (6.3.33) through (6.3.38), indicating a cube root relationship.

Electric field and potential distributions for the linearly graded junction are shown in Fig. 6.3.2. Maximum field intensity occurs at $x = 0$ in both cases.

Application of a *negative bias* (the N-type semiconductor is negative with respect to the P-type semiconductor) widens the depletion region and raises the junction voltage barrier. The minority carrier drift current is not balanced any more by a trickle of majority carrier diffusion and a small *reverse saturation current* flows through the device.

A depletion capacitance appears across the depletion region. Its dynamic value depends on the reverse bias voltage as expressed by Eq. (6.5.3).

A plot of $1/C_d^2$ is a useful tool for the computation of several characteristic properties of the junction, such as V_{bi} and the doping profile.

Application of reverse bias voltages in excess of the *junction breakdown voltage* results in the generation of current carriers within the depletion region and in the onset of a current limited only by the total ohmic resistance of the circuit.

The junction breakdown voltage can be obtained from the graph of Fig. 6.5.6.

Application of *forward bias* decreases the depletion region width, lowers the junction voltage barrier, and increases majority carrier diffusion destroying the current balance.

Moving along the length of the device from the P to the N terminal, the current is first mediated essentially by holes drifting toward the junction. Starting in the vicinity of the junction, because of recombination, electrons carry an increasing portion of the current until deep within the N region the current is mediated essentially by electrons.

As a first approximation, no recombination occurs within the depletion region and the ratio of electron-to-hole current remains constant.

Because current is mediated by both types of current carriers, the PN junction is a *bipolar device*.

For a junction cross section A, Eq. (6.7.20), in which $I_0 = J_0 A$ is the magnitude of the *inverse saturation current*, expresses the diode current for both forward and reverse biased junctions (cf. Sec. 6.4).

Minority carrier accumulation near the edges of the depletion region is responsible for the diffusion components of the device current and gives rise to a *diffusion capacitance*, the magnitude of which can be computed by Eq. (6.8.3). As is the case for the depletion capacitance, this capacitance is a nonlinear circuit element.

A small-signal equivalent circuit for the PN diode is shown in Fig. 6.8.1. The transient response of the PN diode is often dominated by charge accumulation phenomena, which in turn depend on the recombination time constant τ_r.

Generation and recombination components of the diode current are usually negligible in low-current Ge diode operation, but may become predominantly important in Si and GaAs diodes. Depending predominantly on indirect recombination, these components are particularly significant in the presence of lattice irregularities and so may result in large surface leakage.

A tunneling component can also contribute to leakage and may reach significant values, especially in doubly degenerate junctions.

APPENDIX 6A
AVALANCHE BREAKDOWN

In reverse biased PN junctions, carriers that have reached the depletion region are accelerated by the strong electric field existing there. If sufficient kinetic energy is acquired by these electrons between two successive collisions with depletion region atoms, then extra current carriers are generated by impact.

The phenomenon is characterized by the *rate of ionization*, i.e., the number of electron-hole pairs generated by the accelerating carrier as it travels over a

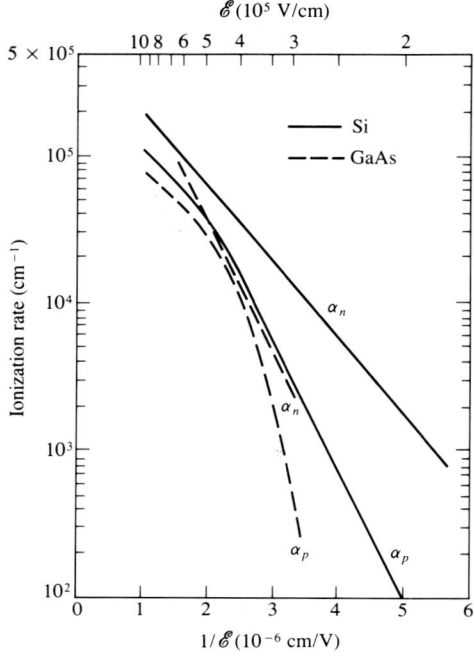

FIGURE 6A.1
Ionization rates vs. inverse electric field magnitude for Ge and GaAs at 300 K [3].

unit distance. It is evident that the rate of ionization increases with the intensity of the accelerating electric field. The relationship of ionization rate to electric field intensity has been determined experimentally and is shown in Fig. 6A.1 [3].

Where newly generated carriers become available for conduction the current increases by an amount proportional to the rate of carrier generation. Suppose that the field is in the $-x$ direction and electrons are accelerated to the right and holes to the left; thus, moving along the x axis, the electron current I_e increases, the hole current I_n decreases, while the total current $I = I_e + I_n$ remains the same, thanks to Kirchoff's law, as shown in Fig. 6A.2.

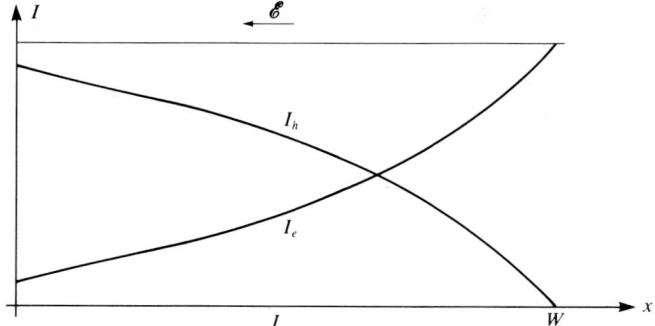

FIGURE 6A.2
Electron and hole currents in depletion region at steady state during impact ionization.

Over a distance dx travelled by the impacting carriers the electron current increment is

$$dI_e = (I_e \alpha_n + I_n \alpha_p) \, dx = I\alpha \, dx \qquad (6A.1)$$

where the rates of ionization due to electron collisions α_n, respectively hole collisions α_p, are assumed equal to minimize complexity. Integrating over the whole depletion region width, i.e., between 0 and W,

$$I_e(W) - I_e(0) = I \int_0^W \alpha \, dx \qquad (6A.2)$$

However, if the collision generation rate is large, then at the right edge of the depletion region in Fig. 6A.2 the current is essentially mediated only by electrons, so that $I_e(W) = I$. Using this equality in (6A.2),

$$I = \frac{I_e(0)}{1 - \int_0^W \alpha \, dx} \qquad (6A.3)$$

Breakdown occurs when $I \to \infty$, so that the breakdown condition is

$$\int_0^W \alpha \, dx = 1 \qquad (6A.4)$$

This integration can be carried out using Fig. 6A.1 and remembering the electric field intensity distribution in the depletion region (6.3.13) for abrupt and (6.3.27) for linearly graded junctions. Using (6.3.15) or (6.3.33), the breakdown condition (6A.4) can then be expressed in terms of the maximum field intensity in the depletion region $\mathscr{E}_{max}$. By this process, breakdown is seen to occur whenever $\mathscr{E}_{max} \geq \mathscr{E}_{crit}$, where $\mathscr{E}_{crit}$ is a critical field intensity depending on the impurity concentration as shown in Fig. 6A.3 [4] for one-sided abrupt junctions [2, 3]. Similar curves can be obtained using the appropriate relationships for other doping profiles.

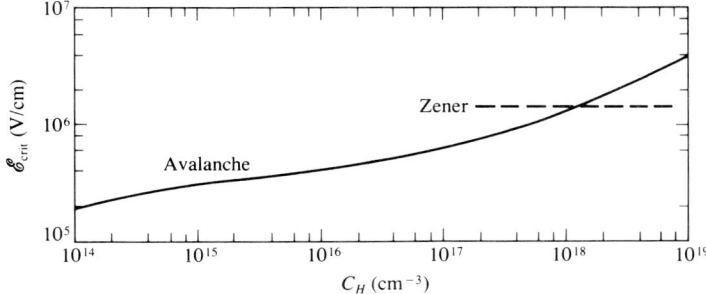

FIGURE 6A.3
Breakdown electric field vs. impurity concentration for one-sided junction in Si [4].

Finally, for the one-sided approximation and assuming the applied voltage to be the breakdown voltage, from (6.3.15) and (6.3.21),

$$\mathscr{E}_{crit} \approx \frac{qN_D}{\varepsilon} \sqrt{\frac{2\varepsilon}{qN_D} V_{bd}} \tag{6A.5}$$

from which

$$V_{bd} = \frac{\varepsilon}{2qN_D} \mathscr{E}_{crit}^2 \tag{6A.6}$$

Conversely, for the linearly graded approximation, from (6.3.33) and (6.3.37) for $A \gg D$,

$$\mathscr{E}_{crit} = \frac{qD}{2\varepsilon} \left(\frac{3\varepsilon}{qD} V_{bd} \right)^{2/3} \tag{6A.7}$$

so that

$$V_{bd} = \frac{4}{3} \sqrt{\frac{2\varepsilon}{qD}} \mathscr{E}_{crit}^{3/2} \tag{6A.8}$$

Expressions (6A.6) and (6A.8) are used in conjunction with the data computed in Fig. 6A.3 (and similar data for the linearly graded junction) to obtain the graphs of Fig. 6.5.6.

PROBLEMS

6.1. Discuss the functional difference between mask and insulation oxide layer in the PN fabrication sequence and comment on the criteria determining their thickness: qualitatively compare them.

6.2. At a certain instant during the process of generation of a PN junction, 10^5 electrons and 10^5 holes have crossed the junction and recombined. At that instant:
  (a) What is the total number of uncompensated ions (both positive and negative) generated in the depletion region?
  (b) What is the total net charge stored in the P side of the depletion region?
  (c) In the N side?

6.3. Identify and analyze the diffusion and drift components of the currents in the PN junction at equilibrium and show qualitatively that they balance each other resulting in zero net current across the junction. In particular, why do the diffusing majority carriers (B in Fig. 6.2.3) not further extend the depletion region by recombination once they cross the junction becoming minority carriers? Why is the minority carrier current said to be saturated?

6.4. Prove Eq. (6.2.7).

6.5. Justify Eq. (6.2.7) following the hole current rather than the electron current.

6.6. A PN junction is made between two Si semiconductors with uniform dopings $N_A = 10^{14}$ cm^{-3} and $N_D = 10^{15}$ cm^{-3} respectively. The temperature is 300 K.

(a) What is the barrier height (in electronvolts) at equilibrium?

(b) Consider an electron entering the depletion region with kinetic ener moving toward the junction and assume no collisions occur. Wil cross the depletion region?

(c) What is the electron's velocity when entering the depletion region? it?

(d) Qualitatively describe the electron's velocity variations occurring during its motion in the depletion region.

6.7. For an abrupt Si PN junction at 300 K the doping levels are $N_A = 2 \times 10^{13}$ cm^{-3} and $N_D = 1.5 \times 10^{15}$ cm^{-3} and the junction cross section is $A = 1$ mm^2.

(a) After equilibrium is reached, how many electrons have crossed the junction and recombined to generate the depletion region?

(b) How many holes?

6.8. An abrupt Si PN junction at 300 K is characterized by $N_A = 10^{13}$ cm^{-3} and $N_D = 2 \times 10^{14}$ cm^{-3}, with a junction cross section $A = 0.1$ mm^2. At unbiased equilibrium compute:

(a) x_N, x_P, w, V_{bi}, ψ_N, ψ_P, E_{max}, net charge in the P-type and the N-type semiconductors.

(b) Repeat for $N_A = 10^{13}$ and $N_D = 2 \times 10^{19}$ cm^{-3}.

(c) Repeat for $N_A = 10^{19}$ and $N_D = 2 \times 10^{14}$ cm^{-3}.

6.9. Compute the contact potential V_{bi} in a Si PN junction with $N_D = 10^{15}$ cm^{-3} and $N_A = 10^{16}$ cm^{-3}:

(a) At 300 K.

(b) At 450 K.

(c) Compare the two values and justify the result in the light of the theory.

(d) Repeat parts (a), (b), and (c) for $N_D = 10^{13}$ cm^{-3} and $N_A = 10^{14}$ cm^{-3}. Critically review and modify the formulas to avoid any physical inconsistency.

6.10. For an abrupt Si PN junction with $N_A = 10^{14}$ cm^{-3} and $N_D = 2 \times 10^{14}$ cm^{-3} draw a plot of the electric field vs. x, as in Fig. 6.3.1 and using the same conventions, assuming:

(a) that the P semiconductor is on the left of the N-type material and

(b) that the P semiconductor is on the right of the junction.

Appropriately draw the axes and draw the plot to scale. Comment.

6.11. A linearly graded PN junction in Si at 300 K has doping concentration rates $A = 10^{17}$ cm^{-4} and $D = 5 \times 10^{16}$ cm^{-4} with bulk dopant concentrations $N_A = 10^{14}$ cm^{-3} and $N_D = 10^{14}$ cm^{-3}. The junction cross section is 1 mm^2. Compute w, x_P, x_N, ψ_P, ψ_N, $\mathscr{E}_{max}$, and the total charge in the P and N semiconductor depletion regions.

6.12. Repeat Prob. 6.11, but assume $N_A = 10^{19}$ cm^{-3} uniform doping. Draw plots of the type of Fig. 6.3.2.

6.13. Prove Eqs. (6.3.36), (6.3.37), and (6.3.38).

6.14. Show that, although across a junction there exists a contact potential $V_{bi} \neq 0$, yet, if the junction is short-circuited, as shown in Fig. P6.14, no current flows through the circuit.

> *Hint:* Consider the contact potentials at the interfaces between the metal conductor and the semiconductors.

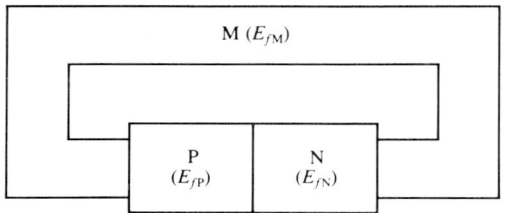

FIGURE P6.14
Short-circuited PN junction.

6.15. Show that, if a battery is inserted in the circuit of Fig. P6.14 as shown in Fig. P6.15, then the junction voltage necessary to maintain voltage equilibrium in the circuit is $V_{bi} - V$ (notice the convention for the battery sign).

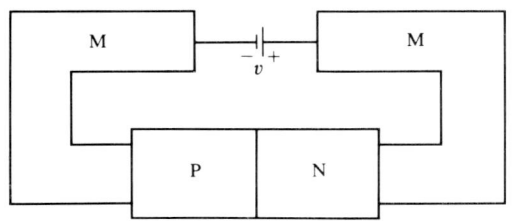

FIGURE P6.15
Reverse biased PN junction.

6.16. Show that, if, in the circuit of Fig. P6.14, the PN junction is kept at a temperature T different from the temperature T_0 of the other contacts in the circuit, a current must flow. Compute this current, assuming that $N_A = 10^{14}$ cm^{-3}, $N_D = 10^{16}$ cm^{-3}, the resistance of the shorting connection is negligible, the semiconductors making up the PN junction are identical prisms of length 1 cm and cross section 1 mm^2, the circuit temperature is $T_0 = 300$ K, and the junction temperature is 330 K and is maintained over a negligibly small region around the junction.

6.17. The three PN junctions described in Prob. 6.8(a), (b), and (c) are subjected to a voltage $V = -2$ V. For each case compute V_{bi}, x_N, x_P, ψ_N, ψ_P, $\mathscr{E}_{max}$, and the charge stored in the P and N side of the depletion region.

6.18. For the one-sided junction of Prob. 6.8(c), compute the reverse bias necessary to store 0.15 nC in each side of the depletion region.

6.19. An abrupt junction in Si with $N_A = 10^{15}$ cm^{-3}, $N_D = 10^{19}$ cm^{-3}, and a junction cross section of 1 mm^2 is used in the circuit of Fig. 6.5.2 with a dc applied voltage $V = -3$ V at 300 K.

(a) Compute the depletion region width and the charge Q_P in the P side of the depletion region.

(b) If a step voltage variation $v = 0.1$ V making the P semiconductor positive is applied, compute the extra charge δQ_P deposited in the P side of the depletion region.

(c) From the above data compute the average capacitance seen by the small-signal generator v.

(d) Compare the capacitance computed in part (c) with the dynamic capacitance of the reverse biased junction defined in the text as the depletion capacitance and comment on any discrepancies.

6.20. Show that, if the small-signal generator of Example 6.5.2 is a symmetrical square wave of 0.1 V amplitude, the average capacitance over a cycle very closely corresponds to the theoretical, infinitesimal signal capacitance.

6.21. Compute the small-signal resonant frequency of the circuit of Fig. P6.2 is the same as in Example 6.5.2 and:
(a) $V = 1$ V
(b) $V = 4$ V
(c) $V = 10$ V

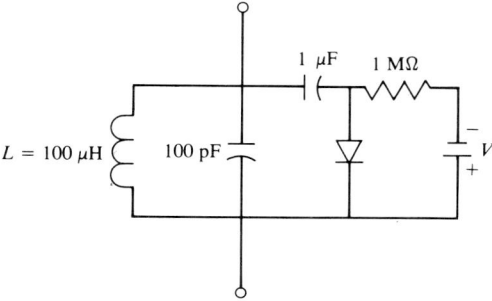

FIGURE P6.21
Resonant circuit with voltage-controlled tuning.

6.22. In a one-sided, abrupt junction in Si at 300 K, the $1/C_d^2$ vs. V_R plot has a slope $S = 4.8 \times 10^{16}$ cm^4/V · F^2 and meets the $1/C_d^2$ axis at $1/C_d^2 = 3.6 \times 10^{16}$ cm^4/F^2.
(a) What are the dopings of the two semiconductors forming the junction?
(b) Is the one-sided hypothesis justified?
(c) What is the contact potential?
(d) What is the width of the depletion region at a bias of -1 V?

6.23. An abrupt, one-sided PN junction in Si at 300 K has $N_D = 4.5 \times 10^{16}$ cm^{-3} and $N_A = 10^{19}$ cm^{-3}.
(a) Compute the reverse bias voltage at which avalanche breakdown occurs.
(b) Compute the maximum field in the depletion region at the onset of the avalanche condition and compare it with published data (see App. 6A).

6.24. Describe the sequence of events following the diffusion of electrons into the P semiconductor of a junction under:
(a) No bias conditions.
(b) Forward bias.
(c) Compare the two and explain why, under forward bias, the diffusing electrons do not simply extend the depletion region by recombination.

6.25. Verify Eq. (6.7.14) solving (6.7.10) under the proper boundary conditions by using the Laplace transform.

6.26. An abrupt PN junction in Si at 300 K has $N_A = 10^{15}$ cm^{-3} and $N_D = 2 \times 10^{15}$ cm^{-3}. $\tau_{eP} = \tau_{hN} = 1$ μs. Assume that in each semiconductor, the total impurity concentration equals the net doping. Compute (and compare with Fig. 6.7.1) the minority carrier concentration distributions vs. x, where $x = 0$ at the physical junction, as in Fig. 6.7.1, for applied bias of:
(a) $V = 0.4$ V.
(b) $V = 0$ V.
(c) $V = -2$ V.
(d) For each case compute the minority carrier diffusion current densities at $x = x_N$ and $x = -x_P$.

6.27. An abrupt Si PN junction at 300 K has a junction cross section $A = 1$ mm, $\tau_{eP} = \tau_{hN} = 0.1$ μs. Assume total doping to equal net doping in each semiconductor. Compute the current for $V = 0.4, 0.5$, and 0.6 V for:
(a) $N_A = 10^{16}$, $N_D = 10^{17}$ cm^{-3}.
(b) $N_A = 10^{17}$, $N_D = 10^{17}$ cm^{-3}.
(c) $N_A = 10^{16}$, $N_D = 10^{18}$ cm^{-3}.
(d) Compare the previous results and conclude whether the most important impurity concentration in determining the diode current is that of the low-doped or that of the high-doped side. Give a reasonable explanation for the result.

6.28. An abrupt PN junction in Si at 300 K has $N_A = 10^{19}$ cm^{-3}, $N_D = 10^{17}$ cm^{-3}, $\tau_{eP} = \tau_{hN} = 0.1$ μs, and junction cross section $A = 1$ mm^2. Assume that the total impurity concentration in each semiconductor equals the net impurity concentration.
(a) Tabulate the I/V characteristic of the device for $-15 < V < 0.75$ V. Do not forget the breakdown phenomenon.
(b) Repeat for the case in which the same net impurity concentration N_D has been obtained by adding As impurities to a substrate with $N_A = 10^{19}$ cm^{-3}.

6.29. Repeat Prob. 6.28 for $T = 400$ K.

6.30. The diode in the circuit of Fig. P6.30 has an abrupt junction in Si with $N_A = 10^{19}$ cm^{-3}, $N_D = 10^{17}$ cm^{-3}, cross section $A = 0.1$ mm^2, and $\tau_{eP} = \tau_{hN} = 0.1$ μs. The temperature is 300 K. Assume that the total impurity concentration in each semiconductor is equal to the net doping.
(a) Compute the current.
(b) Write a computer program to perform the above computations, allowing appropriate choices of conditions and components.

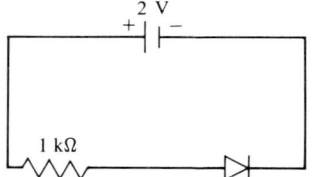

2 V

1 kΩ

FIGURE P6.30

6.31. Assuming the dc voltage drop across each of the forward biased diodes to be 0.5 V, compute the peak-to-peak value of voltage v_2 in Fig. P6.31 for:
(a) $R = 100$ kΩ
(b) $R = 10$ kΩ

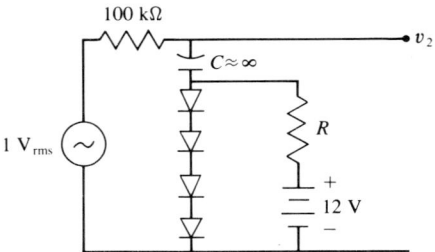

100 kΩ

v_2

$C \approx \infty$

R

1 V$_{rms}$

12 V

FIGURE P6.31
Voltage-controlled signal voltage divider.

6.32. For the diode of Prob. 6.30, assume that the N semiconductor is a cylinder of 2 mm length and 0.1 mm^2 cross section, that the resistance in the external circuit including the P semiconductor is 0 Ω, and that the applied voltage is 14 V reverse bias. Compute the current.

6.33. A PN junction in Si at 300 K has $N_A = 10^{19}$ cm^{-3}, $N_D = 10^{16}$ cm^{-3}, junction cross section 1 mm^2, and $\tau_{eP} = \tau_{hN} = 0.1$ μs. Tabulate power dissipated in the device vs. applied voltage for $0 < V < 0.75$ V.

6.34. A forward bias of 100 V is applied to the series connection of the diode of Prob. 6.33 and a 100-Ω resistor. Using reasonable approximations:
(a) Compute the power dissipated in the diode.
(b) Repeat for the same circuit, but using a diode with the same specifications but host material Ge.

6.35. For an abrupt PN junction in Si at 300 K with $N_A = 10^{19}$ cm^{-3}, $N_D = 2 \times 10^{16}$ cm^{-3}, $\tau_{eP} = \tau_{hN} = 0.1$ μs, and cross section $A = 1$ mm^2, assuming total impurity concentration in each semiconductor to equal the net doping:
(a) Tabulate the dynamic resistance r_d vs. applied voltage for $0.2 < V < 0.75$ V. Compare with Fig. 6.8.2.
(b) Repeat for the data of the diode of Prob. 6.28.

6.36. An abrupt PN junction in Si at 300 K has a cross section of 0.1 mm^2, $N_D = 10^{19}$ cm^{-3}, $N_A = 5 \times 10^{15}$ cm^{-3}, $\tau_{eP} = \tau_{hN} = 10^{-7}$ s. In each semiconductor the total impurity concentration equals the net doping. Compute and compare the depletion and diffusion capacitances under bias voltages of:
(a) $V = -2$ V
(b) $V = 0$ V
(c) $V = 0.5$ V

6.37. The diode of Prob. 6.36 is submitted to forward bias voltages of $V = 0.2$, 0.5, and 0.6 V. Compute and compare the magnitudes of the diffusion and of the recombination currents assuming:
(a) That the recombination time $\tau_r = 10^{-7}$ s, as in Prob. 6.36.
(b) That the addition of appropriate impurities has created efficient trap levels resulting in $\tau_r = 10^{-10}$ s.

6.38. Compare the tunnel diode characteristics with the diffusion current plot in Fig. 6.9.4 and qualitatively indicate the phenomena that occur as the bias varies. Relate these phenomena to the energy band diagrams of Fig. 6.9.3.

Computer Problems

Note: Appropriate choice of subroutines may be useful to permit more convenient programming if more than one problem is to be solved.

6.39. Write a computer program to perform the computations of Prob. 6.8. Interactively allow the user to make choices of N_A, N_D, and T.

6.40. Write a computer program to solve Prob. 6.9, so that a valid solution is obtained not only for parts (a), (b), and (c) but also for part (d).

6.41. Write a computer program to solve Prob. 6.11, interactively allowing the user to choose values of A, D, N_A, N_D, and T.

6.42. Modify the computer program of Prob. 6.39 to solve Prob. 6.17, permitting an interactive choice of reverse bias voltage V.

6.43. Write a computer program to solve problems of the type of Prob. 6.27, interactively allowing choices of N_A, N_D, total doping in each semiconductor, τ_{eP}, τ_{hN}, and T.

6.44. Modify the program of Prob. 6.43 to obtain a graph of the diode V/I characteristic in the forward biased mode, allowing interactive choices of the same parameters as in Prob. 6.43.

6.45. Write a computer program to perform the computations of Prob. 6.35, allowing interactive choices of N_A, N_D, τ_{eP}, and τ_{hN}.

REFERENCES

1. Sze, S. M., and G. Gibbons: "Avalanche Breakdown Voltage of Abrupt and Linearly Grade P-N Junctions in Ge, Si, GaAs and GaP," *Appl. Phys. Lett.*, vol. 8, p. 111, 1966.
2. Grove, A. S.: *Physics and Technology of Semiconductor Devices*, John Wiley & Sons, Inc., New York, 1967.
3. Sze, S. M.: *Semiconductor Devices, Physics and Technology*, John Wiley & Sons, Inc., New York, 1985.
4. Miller, S. L.: "Ionization Rates for Holes and Electrons in Si," *Phys. Rev.*, vol. 105, p. 1246, 1957.

ADDITIONAL READING

Ankrum, Paul D.: *Semiconductor Electronics*, Prentice Hall, Inc., Englewood Cliffs, N.J., 1971.

Gray, P. E., D. De Witt, A. R. Boothroyd, and J. F. Gibbons: *Physical Electronics and Circuit Models of Transistors*, vol. 2, Semiconductor Electronics Education Committee, John Wiley & Sons, Inc., New York, 1964.

Philips, A. B.: *Transistor Engineering*, McGraw-Hill Book Company, Inc., New York, 1962.

Sah, C. T.: "Effects of Surface Recombination and Channel on P-N Junction and Transistor Characteristics," *IRE Trans. Elect. Devices*, vol. ED 9, p. 94, 1962.

Sah, C. T., P. N. Noyce, and W. Shockley: "Carrier Generation and Recombination in P-N Junction and P-N Junction Characteristics," *Proc. IRE*, vol. 45, p. 1228, 1957.

Shockley, William: "The Theory of P-N Junction in Semiconductors and P-N Junction Transistors," *Bell System Tech. J.*, vol. 28, p. 435, 1949.

Shockley, William: *Electrons and Holes in Semiconductors*, Van Nostrand Company, Inc., New York, 1950.

Yang, Edward S.: *Fundamentals of Semiconductor Devices*, McGraw-Hill Book Company, Inc., New York, 1978.

CHAPTER
7

METAL-SEMICONDUCTOR JUNCTION

The previous chapter investigated the exchange of carriers across a junction between two semiconductors of opposite type. As stated, such an exchange is not limited to semiconductors, but, under the proper conditions, can occur between other media.

The analysis of the PN junction suggests that a potential barrier exists across an interface. The relationship between the flow of electrons across a potential barrier and the barrier height determines the performance of several important devices. In this chapter, this relationship will be investigated with particular regard to two specific cases: metal-vacuum and metal-semiconductor junctions.

7.1 METAL-VACUUM INTERFACE— THERMIONIC EFFECT

Figure 7.1.1a shows the energy-position diagram in the immediate vicinity of the interface between a metal and empty space. The material on the left of the interface is clearly a metal, as its Fermi level falls within the conduction band. The diagram for the vacuum consists of the vacuum level line schematically symbolizing the fact that, for an electron to exist in the vacuum, it must be at least at the vacuum energy level arbitrarily chosen as reference, or zero level (cf. Chap. 1).

At the surface of the metal there appears the familiar steep surface potential barrier. Its height, equal to the difference between the vacuum level and the material's Fermi level, is indicated by $q\phi_M$. Remembering that, at moderate temperatures, most conduction electrons are very close to the Fermi level, this height is seen to represent the energy that an electron must acquire to overcome the

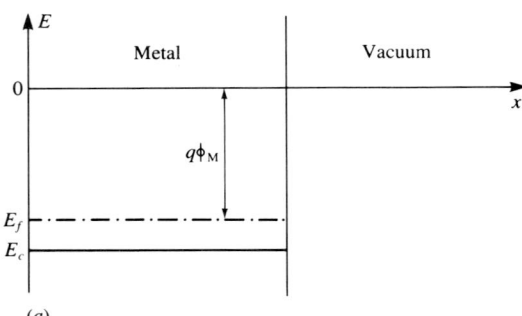

(a)

FIGURE 7.1.1
Metal-vacuum interface. (a) Energy-position band diagram. Notice the potential barrier of energy $q\phi_M$ at the interface. The vacuum level is used as reference ($E = 0$). (b) State density distribution $n_{sc}(E)$, Fermi function $f(E)$, and carrier density distribution $n_d(E)$ at 300 K. The concentration of electrons with energy above the vacuum level is negligible. (c) Same plots as in (b), but for $T = 900$ K. The darkened area measures the concentration of electrons with sufficient energy to escape from the metal.

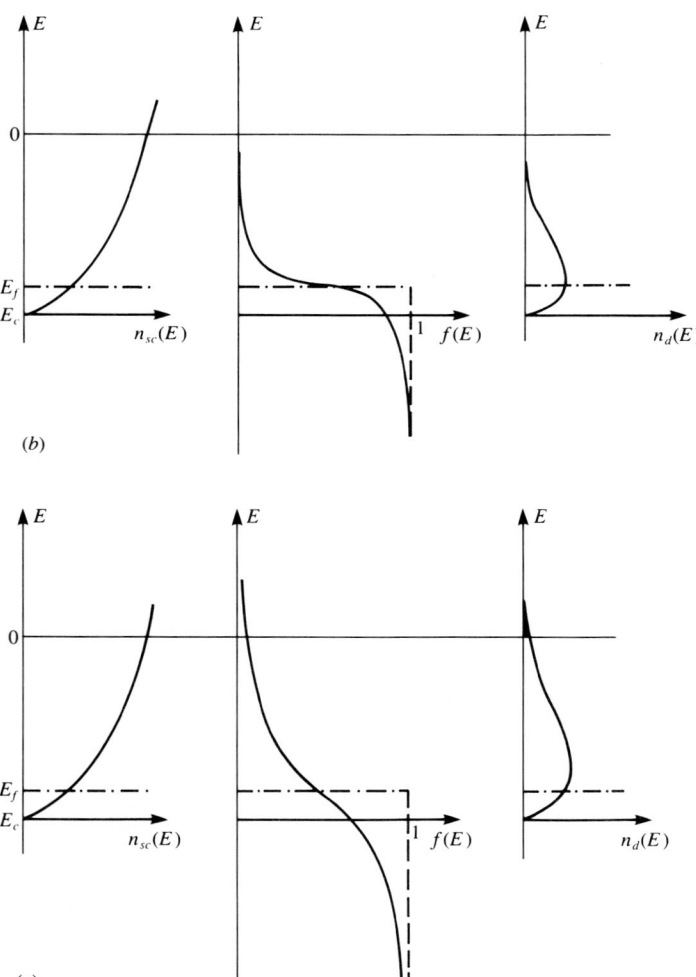

(b)

(c)

TABLE 7.1.1
Work functions and Richardson constants

Element	ϕ_M	R	Element	ϕ_M	R	Element	ϕ_M	R
Oxides	1	0.01	Cs	4.8	16.2	Ta	4.1	60
Polysilicates	3.95		Cu	4.4	65	Th	3.4	60
Al	4.2		Ni	5	30	W	4.5	60
Au	4.7		Pt	5.4	32	W(Tho)	2.6	3

surface voltage barrier and escape from the metal. The *voltage* ϕ_M is a measure of this energy in electronvolts and is known as the *work function*. Work functions for several metals are given, in volts, in Table 7.1.1.

Figure 7.1.1b and c also shows qualitatively the familiar graphs for $n_{sc}(E)$ (the state density distribution in the conduction band), $f(E)$ (the Fermi-Dirac function), and $n_d(E)$ (the electron density distribution). These functions were discussed in Sec. 3.2, where it was shown that the electron density distribution within the conduction band is related to the Fermi function and to the state density distribution by Eq. (3.2.3): $n_d(E) = n_{sc}(E) f(E)$. In Fig. 7.1.1b these graphs are qualitatively drawn for a temperature of 300 K; in Fig. 7.1.1c for $T = 900$ K.

Comparing the electron distribution graph with the zero energy level line (vacuum level) and remembering the properties of the distribution function, it is easy to see that the concentration of the electrons having energy greater than the vacuum level equals the shaded area of the distribution graphs in Fig. 7.1.1c. It is evident that, in this example, at 300 K (Fig. 7.1.1b) essentially no electrons possess enough energy to overcome the surface potential barrier and enter the vacuum. Therefore no current can flow out into the vacuum through the surface of the metal.

If, however, the temperature is raised to 900 K, as qualitatively shown in Fig. 7.1.1c, the shape of the Fermi function is changed, so that now a non-negligible amount of electrons is raised to energy levels greater than $0 = E_f + q\phi_M$. These electrons can overcome the surface barrier and *a current of electrons is emitted from the metal surface into the vacuum*. Because the energy necessary to overcome the surface potential barrier is provided to the electrons by thermal means, the phenomenon is known as *thermionic emission*. Historically, this phenomenon has played a fundamental part in the development of electronics and it is worth while discussing its implications in some detail.

Consider a piece of metal in a hermetically sealed chamber, which has been evacuated by means of a pump, as schematically shown in Fig. 7.1.2a. If the temperature of the metal is high enough, the previous theory predicts that the metal will emit electrons into the surrounding vacuum, as schematically indicated in the figure. In doing so, the metal loses the negative charge of the emitted electrons and becomes positively charged, generating an electric field, which attracts the emitted electrons back toward the metal. The electrons will, therefore, shoot off from the metal, slow down under the electrostatic force, and, eventually,

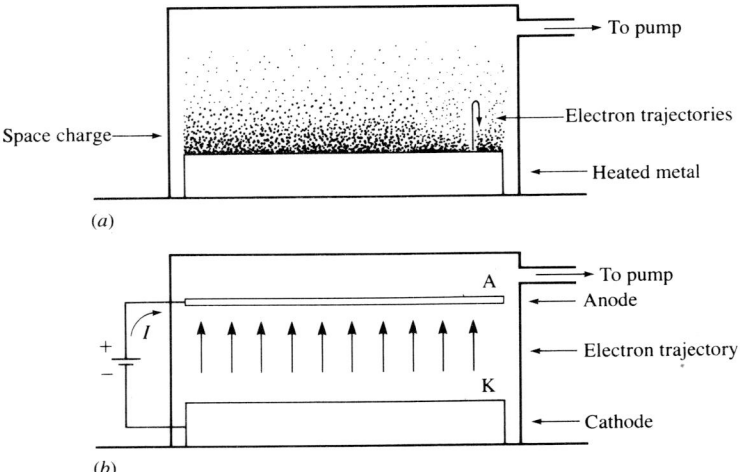

FIGURE 7.1.2
Electron emission into vacuum from heated metal cathode. (a) The emitted electrons form a *space charge*. This negatively charged cloud near the metal surface pushes the emitted electrons back into the metal. (b) A positively charged anode sweeps away the emitted electrons and a current flows through the vacuum.

return to the metal. This is indicated in the figure by the arrow turning back on itself.

Two electron flows are therefore established: (1) an *emission* flow of electrons leaving the metal due to its high temperature and (2) a *drift* of electrons returning to the metal under the pull of the electric field created. It is evident that a condition of dynamic equilibrium will be reached when the two currents equal each other and that no net current will then leave the metal.

At any instant, the region of space in the immediate vicinity of the metal surface contains a cloud of electrons in transit from—and back into—the metal and constituting a negative charge around it. This is known as the *space charge*, and the electric field recalling the emitted electrons is established between the positively charged metal and the space charge.

If, however, a cold electrode A (anode) is added to the system inside the enclosure and a voltage is applied between it and the hot emitting metal K (cathode), as shown in Fig. 7.1.2b, the resulting electric field can overcome the space charge field and draw away the emitted electrons, preventing them from returning to the cathode, so that *a current I is established through the vacuum* between the emitting metal and the new electrode.

The resulting structure is a *vacuum tube*: it contains a hot *cathode* K and a cold *anode* or *plate* A in a vacuum enclosure and the current through it can be controlled by varying the cathode temperature and/or the anode voltage. As stated, the invention of this structure opened up the new field of electronics. Vacuum structures, based on the thermionic emission effect, are still very much in use. One of them, as an example, is the familiar cathode ray tube (CRT).

Figure 7.1.1, illustrating the role played by temperature in raising some of the metal conduction electrons to energy levels permitting thermionic emission phenomena, points out which portion, if any, of the electron distribution density function is above the zero (vacuum) level by comparing the appropriate graphs.

7.2 QUANTITATIVE ANALYSIS OF THE THERMIONIC EFFECT

To compute the emission current, consider the conditions prevailing in a conductor at thermal equilibrium. As described in Sec. 4.1, the conduction electrons are animated by a random distribution of velocities and, statistically, the current through an element of surface δA perpendicular to the x axis is

$$\delta I = -qnv_x \, \delta A$$

where v_x is the average x component of the charge velocity and n the carrier concentration. Notice that these statistical quantities are both computed at the position of δA and averaged over this surface and in its immediate vicinity.

If the surface δA is inside the conductor, as in the case of δA_1 in Fig. 7.2.1, then, because of the random distribution of velocities, the average v_x is zero (cf. Sec. 4.1), as many electrons cross δA_1 in one direction as in the opposite one, and δI is zero. If, however, δA is on the surface of the conductor, as δA_2 in Fig. 7.2.1, then there are conduction electrons only on one side of δA_2, the distribution of the electrons and of their velocities is no longer symmetrical around δA_2, and v_x and δI need no longer be zero.

In this case, however, not all the conduction electrons can contribute to the emission current. To flow through δA_2 an electron must possess at least enough energy to overcome the surface energy barrier ($E \geq E_f + q\phi_M$) and its x component of motion must be directed toward the outside of the conductor (to the right in the figure). Remembering that, due to the randomness of the motion, the second condition is satisfied by only half of the electrons and using the concepts

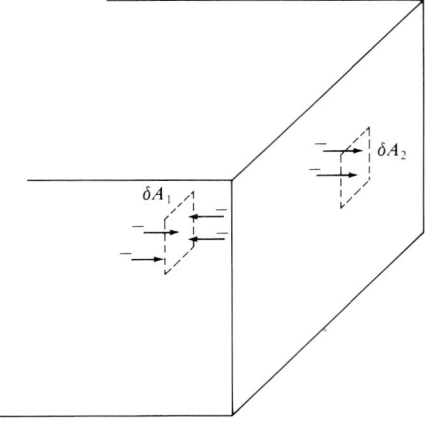

FIGURE 7.2.1
Flow of electrons through elementary surfaces located inside a conductor (δA_1) and on its surface (δA_2).

developed in Chap. 3, δI can be computed as

$$\delta I = q \, \delta A \int_{E_f + q\phi_M}^{\infty} v_{\text{thx}}(E) \, n_{sc}(E) \, \tfrac{1}{2} f(E) \, dE \qquad (7.2.1)$$

in which the symbols have the same meaning as in (3.3.1) and $v_{\text{thx}}(E)$ is the rms value of the x component of *the thermal velocity of the particles at energy level E*. Remembering (4.1.3),

$$v_{\text{thx}}(E) = \sqrt{\frac{2}{3} \frac{E}{m_e}} \qquad (7.2.2)$$

introducing the appropriate values into (7.2.1) [cf. (3.2.1) and (3.2.2)] and extending the computation to the finite surface area A, the emission current can be expressed by the *Richardson-Dushman equation*:

$$I = ART^2 e^{-q\phi_M/(kT)} \qquad (7.2.3)$$

where R is the Richardson constant

$$R = \frac{4\pi q m_e k^2}{h^3} \qquad (7.2.4)$$

with a theoretical value of 120 A/(K$^2 \cdot$ cm^2) (using the free space mass of the electron). In reality, the appropriate equivalent mass of the electron should be used and other phenomena taken into consideration. Measured values of R for several metals are indicated, in A/(K$^2 \cdot$ cm^2), in Table 7.1.1.

It should be kept in mind that, in Eq. (7.2.3), the emission current is computed under the assumption that the electrons emitted are immediately swept far away from the surface. If they are not, then, as previously discussed, they are attracted back toward the emitting body, which is now positively charged, and the steady state is characterized by zero net current and a negatively charged *space charge* in the immediate vicinity of the emitting surface. A current of electrons returning to the emitting body tends to deplete the space charge and balances out the emission current, which tends to replenish it.

Example 7.2.1. A tungsten cathode with a surface of 120 mm^2 is operated at a temperature of 2500 K. Compute the emission current supposing that the electrons are swept away from the emitting surface as fast as they are produced.

Solution. For W, Table 7.1.1 yields $\phi_M = 4.5$ eV and $R = 60$ A/(K$^2 \cdot$ cm^2). Using these values in (7.2.3),

$$I = 1.2 \times 60 \times 6.25 \times 10^6 \, e^{-4.5/0.216} = 0.4 \text{ A}$$

7.3   THE M-S JUNCTION AT EQUILIBRIUM

Figure 7.3.1 shows (at center) the energy band diagrams of a metal (on the left) and a semiconductor (on the right). Figure 7.3.1*a* depicts conditions prevailing before the metal-semiconductor contact is made. The vacuum level (energy level outside the two objects) is shown as a horizontal line indicating that the two are at the same potential.

All energies are in electronvolts so that, numerically, they are the same as the equivalent potentials. All quantities referring to the metal are designated by a subscript M and to the semiconductor by a subscript S (for example, E_{fM} and E_{fS} indicate the Fermi levels of the metal and the semiconductor respectively). The work functions of the metal and the semiconductor are ϕ_M and ϕ_S respectively.

The semiconductor *electron affinity* χ_S is a *voltage* measuring the difference between the vacuum level and the bottom of the conduction band E_{cS} in electronvolts. For semiconductors, this datum, in practice, is more useful than the work function, which varies with doping. Values of electron affinity for some semiconductor materials are listed in Table 3.1.1. The voltage V_{cf} measures the difference between the bottom of the conduction band E_{cS} and the semiconductor Fermi level E_{fS}. Its value can be computed using Eq. (3.4.6).

For the moment, we shall assume that the semiconductor is N-type and that its Fermi level is higher than that of the metal. Both of these conditions are easily recognized in the picture, which also shows the Fermi functions $f(E)$ and the conduction electron density distributions $n_d(E)$ for the metal (on the left) and for the semiconductor (on the right).

Between the metal and the semiconductor there exists a potential barrier, the top of which is at level E_{cS} and is indicated in the figure by the line designated as "top of barrier." Only electrons at this level or higher can be emitted across the interface. The black areas[1] obtained in Fig. 7.3.1a by comparing the density distribution graphs with the top of the barrier level show that, in the case depicted, both metal and semiconductor contain significant concentrations of electrons possessing sufficient energy to migrate across the junction. Therefore, if an intimate contact is established between the metal and the semiconductor, electron emission becomes possible across the junction in both directions. However, a comparison of the black areas for the metal and semiconductor clearly indicates that the emission current I_S from the semiconductor to the metal is much stronger than I_M in the opposite direction, so that *net emission occurs from semiconductor to metal*. This is indicated in the figure by the different thicknesses of the arrows for I_S and I_M (current directions opposite to electron motion).

The migration of negative charge from the semiconductor results in the formation of a *depletion region* of width w at the interface, making the semiconductor positive and so lowering its energy diagram with respect to the metal. This is shown in Fig. 7.3.1b. At equilibrium, a contact potential $V_{bi} = \phi_M - \phi_S$ is established, so that the Fermi levels are aligned and *no net current flows through the interface*. This conclusion is confirmed by noticing that the concentrations of

---

[1] Remember that the black areas of the electron density distributions in the figure represent the concentration of electrons having enough energy to be emitted across the junction.

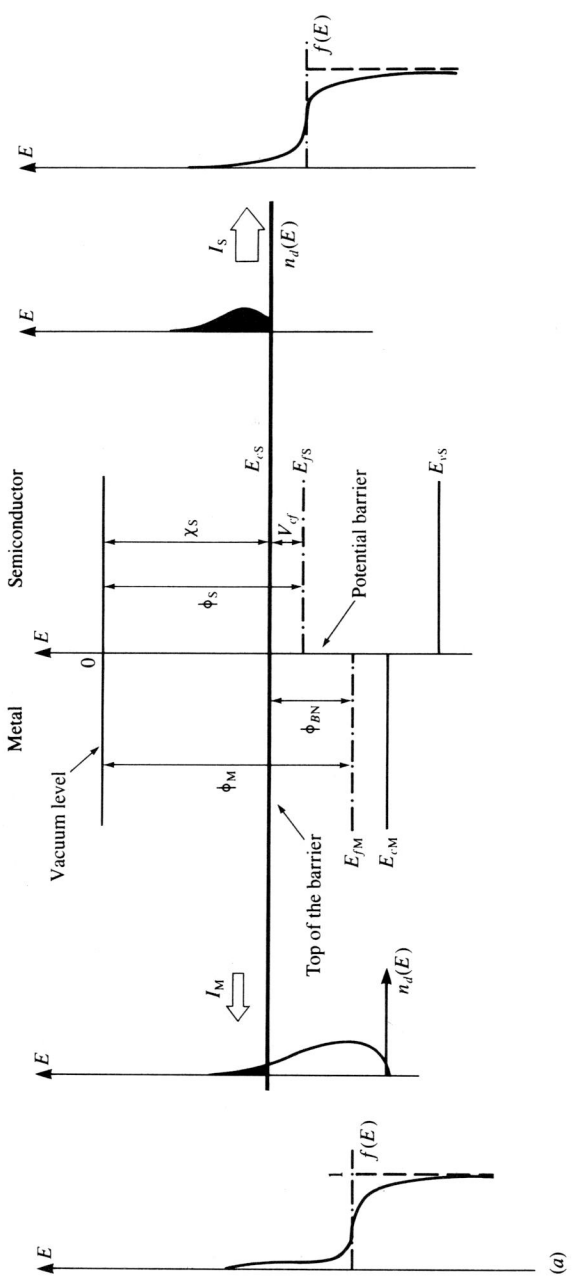

236

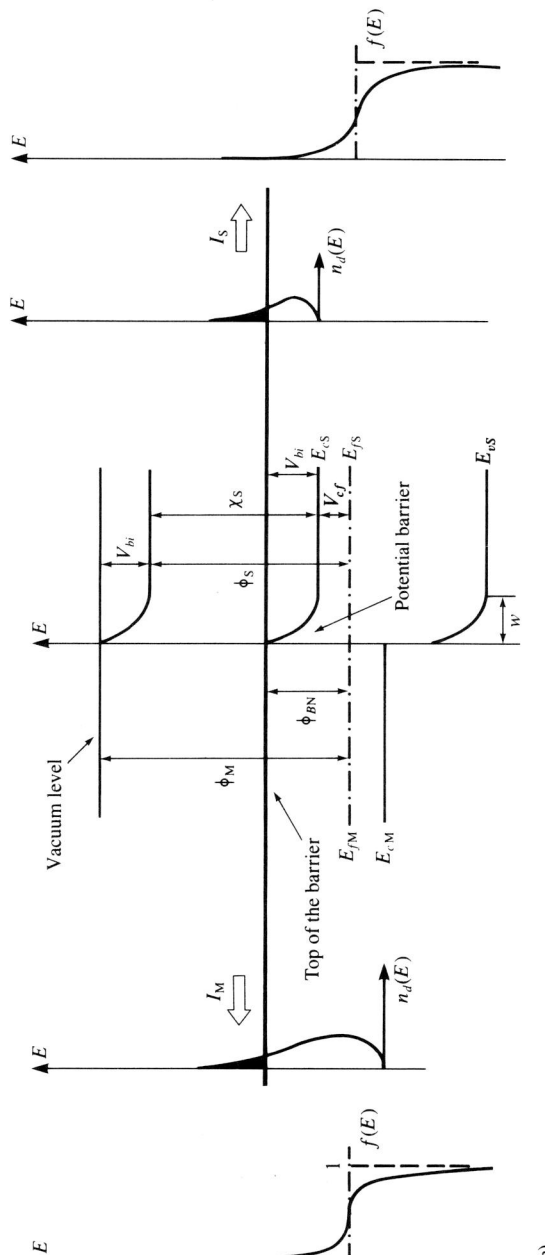

FIGURE 7.3.1

Metal-semiconductor junction. Energy-position diagram showing Fermi function $f(E)$ and carrier distribution density function $n_d(E)$ in the metal (at left) and semiconductor (at right). N-type semiconductor with $E_{fS} > E_{fM}$ is assumed. (*a*) Before the junction is made, the emission current I_S from semiconductor to metal is greater than I_M from metal to semiconductor. (*b*) At equilibrium, the balance of shaded areas of $n_d(E)$ in metal and semiconductor assure equilibrium $I_S = I_M$.

electrons having enough energy to overcome the potential barrier at the interface are essentially the same in the metal and in the semiconductor. These concentrations, represented in Fig. 7.3.1*b* by the black areas of the electron density distribution graphs above the top of the barrier, are balanced, indicating equal emission currents I_S and I_M across the interface in both directions.

The structure just described is often called a *Schottky junction* (or Schottky diode).

Observing Fig. 7.3.1*b*, the alert student will have recognized that the depletion width, charge distribution, electric field, and potential distribution of the Schottky junction are identical to those of a one-sided PN junction. This was to be expected, considering the electrical characteristics of a good conductor. However, in this case, the depletion region is generated exclusively by *migration of majority carriers without any accumulation of minority carriers and recombination phenomena*. As will be seen later on, this significantly affects transient response.

The equilibrium configuration of Fig. 7.3.1*b* shows that an energy barrier (known as the *Schottky barrier*) ϕ_{BN} is raised between the metal and the semiconductor. From the figure,

$$\phi_{BN} = \phi_M - \chi_S = V_{bi} + V_{cf} \tag{7.3.1}$$

If a metal electron is to enter the semiconductor, it must overcome ϕ_{BN}, so that this quantity is of paramount importance in determining the behavior of the device.

Unfortunately, expression (7.3.1) has only theoretical value. Under practical, real-world conditions, when the junction is fabricated, the intimate connection of materials having different crystal structures, such as the metal and the semiconductor, results in lattice distortions at the interface. The quantum configuration is disturbed, extra permissible states are generated at the surface, and finally the value of ϕ_{BN} differs from the predictions of Eq. (7.3.1) and depends on the conditions of junction fabrication.

Some representative values are shown in Table 7.3.1 and in Fig. 7.3.2. Comparison with values in Table 7.1.1 and Table 3.1.1 clearly shows that the influence of the spurious phenomena mentioned here is the determinant factor. Practical values are in excess of those predicted by (7.3.1). Metal–P-type semiconductor

TABLE 7.3.1
M–S barriers ϕ_{BN}

	Si	Ge	GaAs	GaP
Al	0.5–0.77	0.48	0.8	1.05
Au	0.81	0.45	0.9	1.28
Cu	0.69–0.79	0.48	0.82	1.2
Pt	0.9		0.86	1.45
W	0.66		0.8	

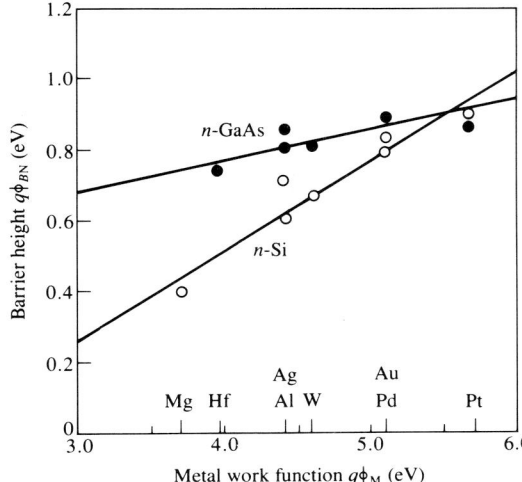

FIGURE 7.3.2
Experimental Schottky barrier values in contacts between different metals and Si or GaAs [1].

barriers can be computed from the metal–N-type semiconductor values by means of Eq. (7.3.2), as indicated later on.

Example 7.3.1. A contact between W and N-type Si with $N_D = 10^{16}$ cm^{-3} is at 300 K. Compute the contact potential, depletion region width, and maximum electric field intensity.

Solution. From Eq. (3.4.6), $V_{cf} = 0.0259 \ln (2.8 \times 10^{19}/10^{16}) = 0.2$ V. From Table 7.3.1, $\phi_{BN} = 0.66$ V, so that, from Eq. (7.3.1), $V_{bi} = 0.66 - 0.2 = 0.46$ V.

As stated, the depletion region width can be computed on the basis of the formula valid for the one-sided PN junction. Imposing the one-sided junction condition on (6.3.21),

$$w = \sqrt{\frac{2 \times 1.04 \times 10^{-12}}{1.6 \times 10^{-19} \times 10^{16}} \, 0.46} = 0.245 \ \mu\text{m}$$

Analogously, from (6.3.15),

$$\mathscr{E}_{max} = \frac{1.6 \times 10^{-19} \times 10^{16}}{1.04 \times 10^{-12}} \, 2.45 \times 10^{-5} = 37.7 \text{ kV/cm}$$

Contact between a metal and a P-type semiconductor, under the assumption that the Fermi level in the metal is higher than in the semiconductor, yields results similar to those just discussed. This case is left to the student as a useful exercise. The mechanism by which equilibrium is achieved is analogous to the previous one. The Schottky barrier is

$$\phi_{BP} = \frac{E_g}{q} - \phi_{BN} = V_{bi} + V_{fv} \tag{7.3.2}$$

where the same observations about the generation of surface states and their effect on the contact potential apply.

Schottky junctions can be forward or reverse biased, just as PN junctions. Metal–N semiconductor junctions are forward biased when the semiconductor is made negative with respect to the metal. The opposite polarity applies to metal–P semiconductor junctions. In both cases, by convention, voltages resulting in forward bias are assigned the positive sign.

7.4 THE M-S JUNCTION UNDER REVERSE BIAS

As in the case of PN junctions, and for the same reasons, negative biasing pulls current carriers away from the junction area, extending the depletion region, and increasing the voltage across it, so that, at steady state, it becomes $V_{bi} - V > V_{bi}$, as shown in Fig. 7.4.1, displaying energy-position and carrier density distribution diagrams for a metal–N semiconductor contact under reverse biased conditions.

Figure 7.4.1 should be compared with Fig. 7.3.1b. To help in this comparison, some of the most significant features of the unbiased equilibrium graphs have been reproduced in Fig. 7.4.1, using dotted lines.

For the semiconductor electrons, the barrier height to be overcome for migration is now increased by $-V$ to $\phi_{BN} - V$ (remember that, by convention,

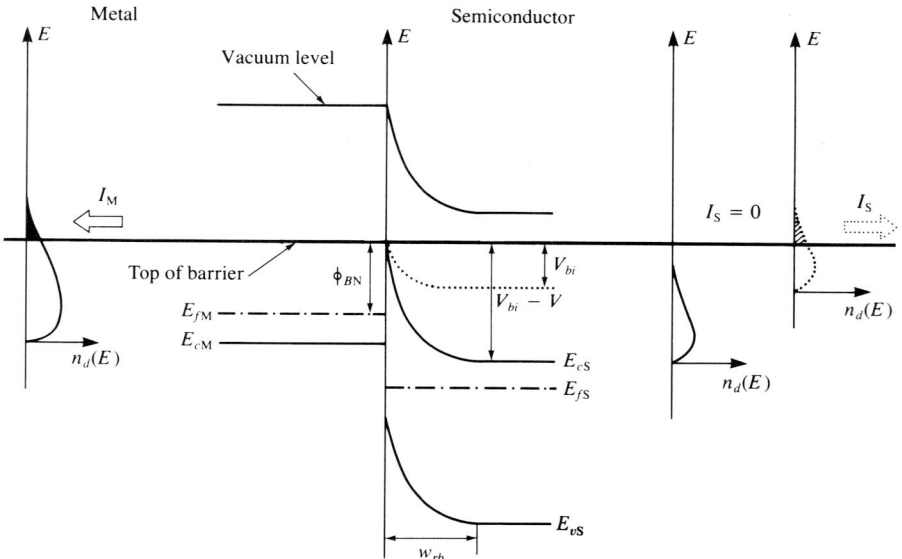

FIGURE 7.4.1
Metal-semiconductor junction under reverse bias. Compare with Fig. 7.3.1b and notice wider w_{rb}, increased barrier for semiconductor electrons, and consequently depressed semiconductor electron distribution density with no electrons above the top of the barrier, reducing I_S to zero. Same assumptions as for Fig. 7.3.1. Dotted lines repeat the unbiased equilibrium conditions for easier reference.

the reverse bias voltage V is negative),[2] *but for the metal electrons it remains the same* (ϕ_{BN}). Indeed, in a metal-semiconductor junction, the depletion region does not extend into the metal, so that application of any bias will affect only the semiconductor depletion region and its barrier potential, while *the barrier potential for metal electrons cannot be affected by the application of an external bias.*

For sufficiently high negative bias, Fig. 7.4.1 indicates that no portion of the graph of the semiconductor electrons distribution density lies above the top of the barrier. Consequently, *the emission current from semiconductor to metal is cut off* and the current through the device equals the emission from metal to semiconductor. For sufficiently high negative bias, this current is saturated. If ϕ_{BN} is large, then this *reverse saturation current* is very small. The student should sketch diagrams similar to Fig. 7.4.1 for different values of the reverse bias voltage to visualize how this voltage influences the significant quantities (such as depletion region width, barrier height, etc.).

As already noticed, in the reverse biased condition, the Schottky junction behavior is very similar to that of the reverse biased one-sided PN junction. The width of the reverse biased depletion region can therefore be computed by

$$w_{rb} = \sqrt{\frac{2\varepsilon}{qN_D}(V_{bi} - V)} \tag{7.4.1}$$

corresponding to a stored charge:

$$Q_s = qN_D w_{rb} = \sqrt{2\varepsilon qN_D(V_{bi} - V)} \tag{7.4.2}$$

and a dynamic depletion capacitance:

$$C_d = \left|\frac{\partial Q_s}{\partial V}\right| = \sqrt{\frac{q\varepsilon N_D}{2(V_{bi} - V)}} A = \frac{\varepsilon A}{w_{rb}} \tag{7.4.3}$$

in complete analogy with the reverse biased one-sided junction. All considerations, concepts, and computation techniques developed for the one-sided reverse biased junction consequently apply, including the diagnostic techniques involving the $1/C_d^2$ vs. V diagram. Such techniques can yield, among other quantities, an experimental value of the contact potential V_{bi}. Comparison of this measured quantity with the theoretical value given by (6.2.9) can shed light on the effect and extent of the lattice distortions mentioned in the previous section, and it is often used in practice to measure experimentally the Schottky barrier height.

Example 7.4.1. A Schottky diode with N-type Si semiconductor at 300 K yields the $1/c_d^2$ vs. V_R plot of Fig. 7.4.2, where c_d is the capacitance per square centimeter of

---

[2] From a perusal of Fig. 7.4.1, this barrier might appear to be $V_{bi} - V$ as there are no conduction electrons below E_{cs}; however, as electron densities are calculated from the Fermi function $f(E)$, which is referred to E_f, the barrier height is best computed from the Fermi level.

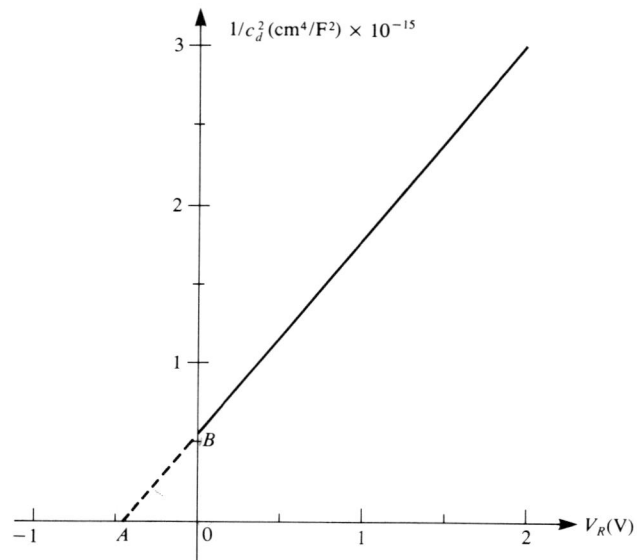

FIGURE 7.4.2
Plot of inverse square junction specific capacitance vs. reverse bias voltage V_R for reverse biased M–S junction of Example 7.4.1.

junction cross section and V_R the reverse bias voltage. Compute the barrier voltage ϕ_{BN}.

Solution. By inspection of Fig. 7.4.2, $V_{bi} = 0.46$ V (point A). Also by inspection (point B) $1/c_d(0)^2 = 0.553 \times 10^{15}$ (cm²/F)², so that, setting $V_R = 0$ in (6.5.6) valid for the one-sided junction, N_D can be computed as

$$N_D = \frac{2 \times 0.46}{1.04 \times 10^{-12} \times 1.6 \times 10^{19} \times 0.553 \times 10^{15}} = 10^{16} \text{ cm}^{-3}$$

Using this value in (3.4.5) and remembering the meaning of V_{cf} :

$$V_{cf} = 0.0259 \ln \frac{2.8 \times 10^{19}}{10^{16}} = 0.2 \text{ V}$$

and finally, from (7.3.1),

$$\phi_{BN} = 0.46 + 0.2 = 0.66 \text{ V}$$

Notice that, as V_{bi} has been computed from what may be assumed to be experimental data, this value is not a purely theoretical one, but should be reliable in practice.

7.5 THE M-S JUNCTION UNDER FORWARD BIAS

Figure 7.5.1 shows the M-S junction energy band diagrams under a forward bias voltage V. For ease of comparison, the main features of the unbiased equilibrium

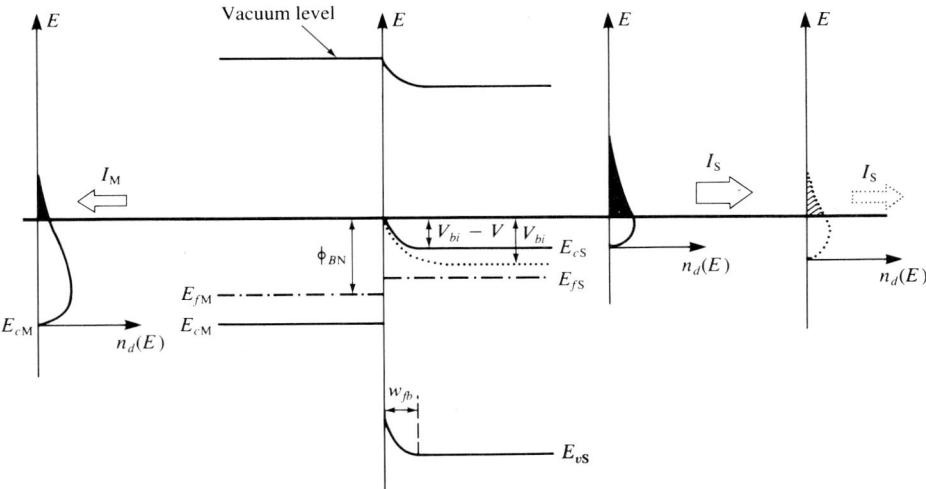

FIGURE 7.5.1
Metal-semiconductor junction under forward bias. Compare with Fig. 7.3.1*b* and notice narrower w_{fb}, decreased barrier for semiconductor electrons, and consequently raised semiconductor electron distribution density with large concentration of electrons above the top of the barrier, increasing I_S so that $I_S > I_M$. Unbiased equilibrium conditions are reproduced in dotted lines for easier reference.

conditions of Fig. 7.3.1*b* are reproduced in dotted lines. As usual, forward bias, pushing carriers toward the junction, replenishes part of the depletion region, decreasing its width and lowering the barrier voltage across it to $\phi_{BN} - V$ (with $V > 0$).

The forward biased semiconductor electron density distribution diagram is raised by V with respect to the equilibrium plot, resulting in a larger concentration of high-energy carriers (black area), so that the semiconductor-metal emission current I_S is increased. The metal carrier density distribution $n_d(e)$ is not affected by bias and the metal-semiconductor emission current I_M remains the same as at equilibrium. This results in a net forward current in the device.

Again the student should sketch similar diagrams, experimenting with different forward bias voltages to visualize the influence of this parameter on device conduction.

There is a similarity between what happens in the forward biased Schottky diode and the analogous conditions in the PN junction: in both cases the lowering of the barrier at the junction disturbs the equilibrium between two currents. However, the currents appearing in the Schottky diode are due to *emission of majority carriers* rather than to *diffusion of minority carriers* as in the case of the PN junction; consequently there are no minority carrier storage effects and so significant differences are to be expected in device behavior.

Even more marked similarities can be noticed by a comparison of Schottky vs. thermionic phenomena. In both cases, net current flow occurs because larger and larger portions of the conduction electron population in the emitting material are raised above the top of the potential barrier which constrains the flow across the interface. The mechanism by which the required energy is imparted to the carriers, however, is different. In the thermionic phenomenon, raising the temperature changes the *shape* of the Fermi function and so of the carrier density distribution, increasing the percentage of high-energy carriers (cf. Fig. 7.1.1*b* and *c*); in the Schottky diode, instead, the forward bias *raises the Fermi level*, shifting the entire carrier density distribution curve upward with respect to the minimum level required for emission, without changing the shape of the distribution (cf. Fig. 7.5.1). These two different mechanisms for increasing the concentration of electrons with sufficient energy for emission should be compared qualitatively, observing the above-mentioned figures in the light of the previous discussion.

The two phenomena differ not only qualitatively (thermal vs. electrical energy) but also quantitatively, because the Schottky barrier is usually about one order of magnitude lower than the metal surface barrier. Notice also that thermionic phenomena, depending on temperature changes, are limited in their speed of response by the thermal inertia of the cathodes, whereas the Schottky emission, requiring only a voltage variation, can be controlled with considerably greater speed.

7.6   QUANTITATIVE ANALYSIS OF THE FORWARD BIASED M-S JUNCTION

The emission currents from the metal and from the semiconductor can be computed on the basis of the same quantitative analysis developed for thermionic emission, as expressed by Eq. (7.2.3). It is only required to substitute the values of the appropriate barrier voltages, ϕ_{BN} and $\phi_{BN} - V$ as displayed in Fig. 7.5.1, in place of the work function of Fig. 7.1.1. As already mentioned, this entails much lower values of the barrier, as easily noticed by comparing values of ϕ_{BN} with those for ϕ_M in Tables 7.1.1 and 7.3.1, and significant current can be obtained at room temperature.

As implied in Fig. 7.5.1, *metal-semiconductor* emission requires the metal electrons to overcome a barrier ϕ_{BN}, so that (7.2.3) becomes

$$I_M = AR^*T^2 e^{-q\phi_{BN}/(kT)} \tag{7.6.1}$$

Conversely, as can also be observed from Fig. 7.5.1, *semiconductor-metal* emission is characterized by a barrier height $\phi_{NB} - V < \phi_{BN}$ (cf. Fig. 7.5.1), so that

$$I_S = AR^*T^2 e^{-q(\phi_{BN} - V)/(kT)} \tag{7.6.2}$$

The constant R^* appearing in (7.6.1) and (7.6.2) is the effective Richardson constant and can be computed by (7.2.4) using the appropriate values of the carrier equivalent mass; usually, however, the appropriate values of R^* are obtained by

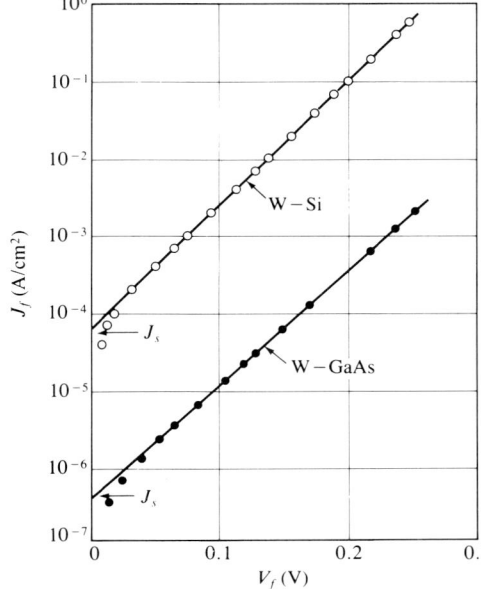

FIGURE 7.6.1

I/V characteristics of Schottky diodes for W–Si and W–GaAs structures. Experimental measurements of current density are compared with the predictions of Eq. (7.6.3) [2].

measurement. Some representative values of R^* in amperes per square kelvin per square centimeter are: electrons in Si, 110; electrons in GaAs, 8; holes in Si, 32; holes in GaAs, 74.

The total net current through the device can be computed as the difference between the two opposing emissions:

$$I = AR^*T^2e^{-q\phi_{BN}/(kT)}(e^{V/V_T} - 1) = AJ_0(e^{V/V_T} - 1) \qquad (7.6.3)$$

A typical plot of this I/V relationship is shown in Fig. 7.6.1 for W–Si and W–GaAs structures. A comparison of Eq. (7.6.3) with Eq. (6.7.20) shows that the I/V characteristics of Schottky diodes display the same type of current to applied voltage dependence as the PN junction, the only difference between the two being the value of the reverse saturation current, which in the Schottky diode is

$$I_0 = AJ_0 = AR^*T^2e^{-q\phi_{BN}/(kT)} \qquad (7.6.4)$$

generally several orders of magnitude larger than the corresponding values for PN diodes.

Typical Schottky diode characteristics together with those of a comparable PN junction diode are displayed for easy comparison in Fig. 7.6.2. Notice the dramatic difference in cut-in voltage.

Example 7.6.1. A Schottky diode between W and Si doped with 10^{16} As atoms/cm^3 has a junction cross section of 1 mm^2.

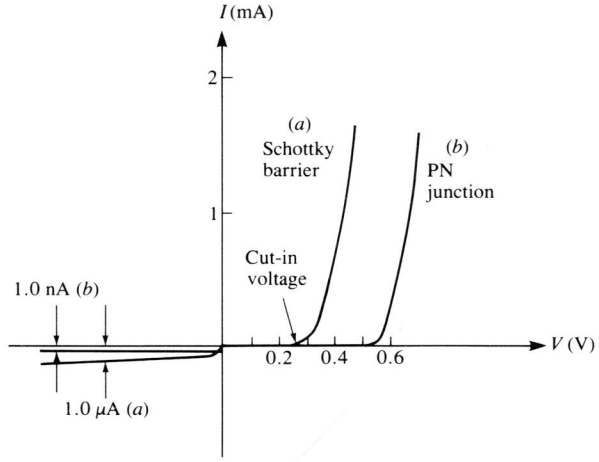

FIGURE 7.6.2
Comparison of Schottky barrier and PN junction diode I/V characteristics showing typical difference in cut-in voltage and reverse saturation current [3].

(a) Compute the current in the diode at $T = 300$ K for a forward bias voltage of 0.25 V.

(b) Consider a Si PN diode at the same temperature, with the same junction cross section, $N_A = 10^{19}$ cm^{-3}, $N_D = 10^{16}$ cm^{-3}, and minority carrier lifetime $\tau_{hN} = 1$ μs. What would be the current for the same forward bias?

(c) What bias is required to bring the current in the PN diode to the same value as that computed in (a) for the Schottky diode?

Solution. From Table 7.1.1, the barrier for a W to Si contact is $\phi_{BN} = 0.66$ V. The effective Richardson constant is $R^* = 110$ A/(K$^2 \cdot$cm^2), so that, from (7.6.4), the inverse saturation current is

$$I_0 = 10^{-2} \times 110 \times 9 \times 10^4 \, e^{-0.66/0.0259} = 8.48 \times 10^{-7} \text{ A}$$

(a) For a forward bias of 0.25 V, from (7.6.3), the current is then

$$I = 8.48 \times 10^{-7} \, e^{0.25/0.0259} = 13.2 \text{ mA}$$

(b) For the PN junction indicated, from (3.3.13), $p_{oN} = 2.25 \times 10^4$ cm^{-3}. From (4.2.9) and (4.3.2), $D_{hN} = 10.5$ cm^2/s and from (6.7.15), $L_{hN} = 3.24 \times 10^{-3}$ cm, so that, from (6.7.20),

$$I_D = 10^{-2} \times 1.6 \times 10^{-19} \, \frac{10.5}{3.24 \times 10^{-3}} \, 2.25 \times 10^4 \, e^{0.25/0.0259} = 1.8 \text{ nA}$$

The student should observe that this forward biased PN diode current is almost three orders of magnitude smaller than the reverse saturation current for the Schottky diode.

(c) Noticing that, from the previous computation, for this PN diode, the reverse saturation current is $I_0 = 0.117$ pA, then using (6.7.20),

$$V \approx 0.0259 \ln \frac{13.2 \times 10^{-3}}{1.17 \times 10^{-13}} = 0.66 \text{ V}$$

significantly larger than the 0.25 V required by the Schottky diode.

7.7 SCHOTTKY DIODE SMALL-SIGNAL PARAMETERS

Remembering the concepts developed in Sec. 6.8, it is easy to use the results of Secs. 7.4 and 7.6 to obtain the approximate small-signal linear equivalent circuit for the Schottky diode shown in Fig. 7.7.1.

As was to be expected in the light of the similarities pointed out in the previous discussions, this circuit is very similar to that of Fig. 6.8.1. The most noticeable difference is the absence of the diffusion capacitance. As conduction in the Schottky diode is due to majority carrier emission, there is no minority carrier charge storage and so no diffusion capacitance. This makes the Schottky diode a *much faster device* than the PN diode.

Another important difference is the much larger value of reverse saturation current. As a result, the forward bias required to obtain significant conduction (cut-in voltage) is much smaller, making the forward biased Schottky diode a better approximation to a short circuit than the PN diode, but the cutoff Schottky diode is not as good an approximation to an open circuit for dc. As a result, the Schottky diode generally gives faster, more efficient performance at high frequencies and as a switch, and is therefore often preferred to the conventional diode for these applications. Some of these quantitative differences are illustrated in Example 7.7.1.

Under equivalent operating conditions, both r_d and C_d are computed using the same algorithms as for the PN junction.

Example 7.7.1. Compute the small-signal equivalent parameters for both the Schottky and the PN diodes of Example 7.6.1, draw the two equivalent circuits at work points corresponding to the same current 13.2 mA, and compare the ac performances.

Solution. For the Schottky diode, from (3.4.6), $V_{cf} = 0.0259$ ln $(2.8 \times 10^{19}/ 10^{16}) = 0.2$ V, so that, from (7.3.1), $V_{bi} = 0.66 - 0.2 = 0.46$ V. Therefore, from (7.4.3),

$$C_d = 10^{-2} \sqrt{\frac{1.6 \times 10^{-19} \times 1.04 \times 10^{-12} \times 10^{16}}{2 \times 0.21}} = 0.629 \text{ nF}$$

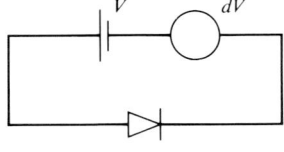
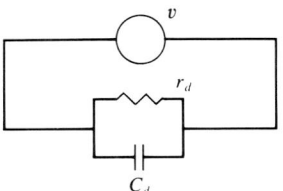

FIGURE 7.7.1
Small-signal dV applied to forward biased Schottky diode: circuit schematic and equivalent small-signal schematic.

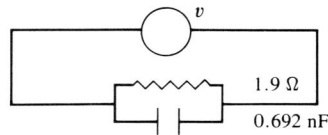
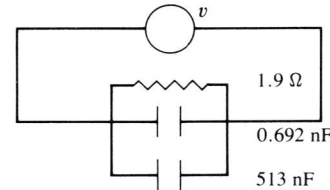

FIGURE 7.7.2
Equivalent circuits for Example 7.7.1.

For the dynamic resistance, using the approximate expression valid for PN junctions (cf. Sec. 6.8),

$$r_d = (40 \times 0.0132)^{-1} = 1.9 \ \Omega$$

For the PN junction at 13.2 mA, r_d and C_d are the same as for the Schottky diode [as the student can check by computing V_{bi} from (6.2.9) and C_d from (6.5.3)]. The diffusion capacitance can be computed from (6.8.3),

$$C_{\text{diff}} = \frac{10^{-2} \times 1.6 \times 10^{-19} \times 2.25 \times 10^4 \ \sqrt{10^{-5}}}{0.0259} \ e^{0.66/0.0259} = 513 \ \text{nF}$$

The equivalent circuit schematic is as shown in Fig. 7.7.2. At this work point the parameters are: $r_d = 1.9 \ \Omega$ for both diodes; $C_d = 629$ pF for the Schottky, $C_d = 514$ nF for the PN diode. The time constants are: Schottky, $2 \times 6.29 \times 10^{-10} = 1.26$ ns; PN diode, $2 \times (6.29 \times 10^{-10} + 513 \times 10^{-9})$ or 1.027 μs, which is almost 1000 times larger than for the Schottky. This is due to the very large value of the diffusion capacitance.

7.8 THE OHMIC CONTACT

In the previous sections the M-S junction was shown to behave similarly to a diode. However, semiconductor devices need electrodes, which are generally obtained by metallization, and these metal-semiconductor contacts should not behave as rectifying diodes, but instead as low-resistance ohmic contacts, assuring easy conduction in both directions.

A contact is said to be ohmic when large currents can flow through it in both directions under negligibly low voltage drops. This section describes two different ways to implement these conditions in metal-semiconductor contacts.

As an example of the first of these techniques consider a metal–N semiconductor junction, but, for this application, assume that the materials constituting the junction have been chosen so that the work function of the metal is lower than that of the N semiconductor:

$$\phi_M < \phi_S$$

so that *the metal Fermi level is above the semiconductor Fermi level.*

The energy band diagram of such a junction at the instant at which contact is made (before carrier exchanges have occurred) is shown in Fig. 7.8.1a. The figure also depicts the top of the barrier between metal and semiconductor (at level E_{cM}) together with the carrier density distributions $n_d(E)$ in the metal (at left) and in the semiconductor (at right), indicating as usual by black areas the con-

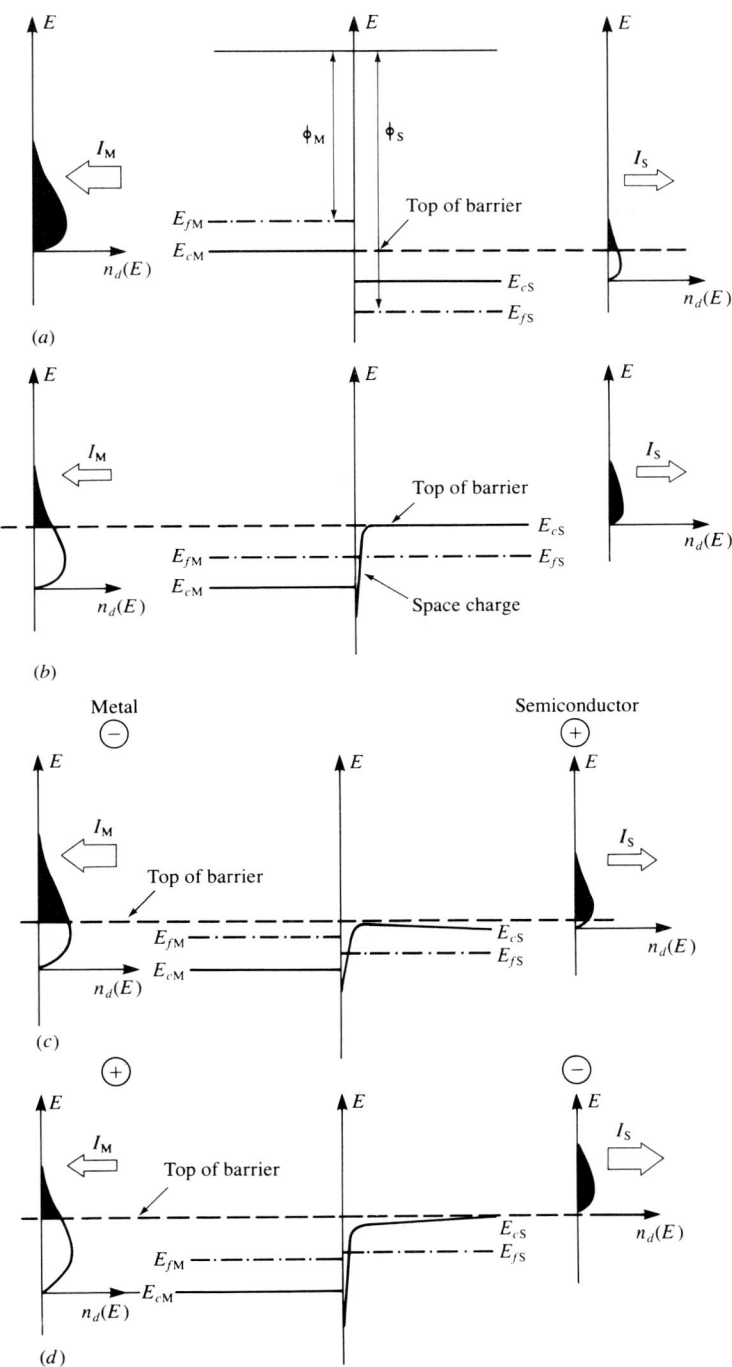

FIGURE 7.8.1
Metal–N semiconductor ohmic contact. Band diagram. It is assumed that $E_{fS} < E_{fM}$. (a) Before the contact is fabricated. (b) Nonbiased equilibrium condition. (c) Reverse bias. (d) Forward bias.

centrations of carriers of sufficient energy for emission across the junction. It appears evident that the conditions for electron emission are more favorable in the metal-semiconductor direction than in the opposite one. The resulting transient electron emission transfers a negative charge into the semiconductor, raising its energy band diagram, aligning the Fermi levels, and yielding the equilibrium band diagram of Fig. 7.8.1b. In the new equilibrium condition, the graph of the semiconductor electron density distribution is raised, so that the increased electron emission into the metal balances the emission into the semiconductor.

It is easily proved that the condition shown in Fig. 7.8.1b implies good ohmic contact. Indeed, consider a voltage making the semiconductor positive. As shown in Fig. 7.8.1c, such a voltage decreases the energy of the semiconductor electrons, lowering the barrier and the semiconductor density distribution diagram, finally resulting in net electron emission from the metal. Conversely, a voltage making the semiconductor negative raises the corresponding density distribution diagram as shown in Fig. 7.8.1d, finally resulting in net electron emission from the semiconductor.

The important fact is that both these currents are due to majority carrier emission and therefore can be quite large and that the barrier to be overcome in both cases is very small, so that the voltage drop at the contact is negligible, meeting the conditions for low-resistance ohmic contact.

The electronic charge making the semiconductor negative at equilibrium consists of majority carriers; consequently it does not generate recombination phenomena and is therefore not stored in a depletion region. It is instead located in a narrow layer at the surface of the semiconductor in the form of a space charge, rather than a depletion region (cf. Fig. 7.8.1b through d).

In the reverse biased direction (metal negative, semiconductor positive), the electrons coming from the metal must overcome the small barrier due to the space charge and conduction is at first limited by this space charge. As the reverse bias is increased, finally the space charge is depleted as soon as it is formed and from then on the current is limited only by the bulk resistance of the metal in series with that of the semiconductor. In the forward direction the current is always bulk resistance limited. These phenomena are apparent by the slope of the ohmic contact I/V curve, shown qualitatively in Fig. 7.8.2.

Similar conditions apply for the junction of a metal to a P-type semiconductor, *provided the semiconductor's Fermi level is above that of the metal* ($\phi_M > \phi_S$).

A second technique to generate ohmic contacts takes advantage of the tunneling effect. In a metal-semiconductor junction, if the semiconductor is *degenerately doped*, then the width of the depletion region is correspondingly very small (cf. Sec. 6.3), so that, under any circumstances, carriers can easily *tunnel* under the narrow barrier, as mentioned in Secs. 1.4 and 2.5 and discussed in App. 1A.2. This makes conduction easy in both directions, irrespective of the barrier height.

In conclusion, ohmic contacts can be obtained either by choosing contact metals with $\phi_M > \phi_S$ for making contact with N semiconductors, $\phi_M < \phi_S$ for P semiconductor contacts, or simply by degenerately doping the semiconductor next to the contact region.

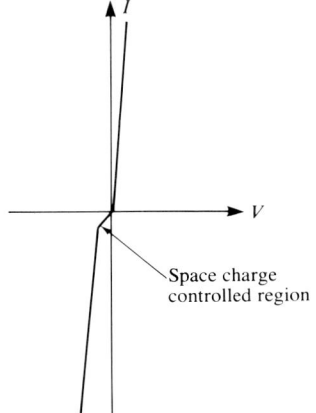

Space charge
controlled region

FIGURE 7.8.2
M-S ohmic contact V/I characteristic, qualitatively showing the
space charge controlled region (exaggerated in the drawing).

7.9 SUMMARY

Conduction electrons are kept within the boundaries of a metal by a surface barrier. To escape, they must acquire an amount of energy equal to the height of the barrier, measured in electronvolts by the metal's *work function* ϕ_M.

This can be achieved by raising the temperature of the metal. The phenomenon is called *thermionic emission.*

The emission current is expressed by the *Richardson-Dushman equation* (7.2.3), in which there appears the *Richardson constant R*, with a theoretical value of 120 A/(K$^2 \cdot$ cm^2) [Eq. (7.2.4)] and practical values shown in Table 7.1.1.

When intimate contact is made between an N semiconductor and a metal, if the semiconductor's Fermi level is above that of the metal, then a transient net emission of electrons occurs from semiconductor to metal, generating a depletion region similar to that of a one-sided PN junction, but involving no minority carriers.

The *barrier for metal electrons* is $\phi_{BN} = V_{bi} + V_{cf}$, where V_{bi} is usually affected by lattice distortions at the interface. Experimental values of ϕ_{BN} are given in Table 7.3.1 and in Fig. 7.3.2.

Just as in a one-sided PN junction, *negative biasing extends the depletion region*, raising the potential barrier for semiconductor electrons but not affecting the metal conditions. For sufficiently high bias this cuts off semiconductor-metal emission.

The remaining unchanged metal-semiconductor emission now constitutes a net *reverse saturation current*, generally much larger than in PN junction diodes.

A *depletion capacitance* is also created in complete analogy to the one-sided PN junction case.

Forward bias increases semiconductor-metal emission, without significantly changing metal emission, so that large *forward currents* can flow.

The total *Schottky diode net current* can be computed from Eq. (7.6.3), the reverse saturation current from Eq. (7.6.4). This is usually several orders of magnitude larger than the corresponding PN diode reverse saturation currents.

Schottky diode characteristics are similar to **PN** diode I/V curves, but the cut-in voltage is smaller.

Because in Schottky diodes there is no minority carrier storage, the device is *faster than conventional diodes* and is preferable as a high-speed switch.

Ohmic contacts can be obtained by using metals with $\phi_M < \phi_S$ for contacts with N semiconductors and $\phi_M > \phi_S$ for P semiconductor junctions. Another commonly used artifice to obtain ohmic contacts consists of degenerately doping the semiconductor in correspondence with the contact region. In this case, conduction across the interface is mediated by tunneling.

PROBLEMS

7.1. A thoriated cathode is at a temperature $T = 1000$ K.
  (*a*) What percentage of the permissible states is occupied at the lowest energy level allowing emission?
  (*b*) Consider an electron near the surface and moving toward it with a kinetic energy of 4.8×10^{-19} J. Is it emitted? If so, what is its residual kinetic energy after emission (emission energy)?

7.2. In the arrangement of Fig. 7.1.2*b*, the voltage V is slowly raised from zero to higher positive values. Qualitatively predict what happens. Assume that the anode is far from the cathode (several times the thickness of the space charge).
  Hint: Consider the superposition of the field between the cathode and the space charge on the field between the anode and the cathode.

7.3. Electron emission from a Th sample is to be obtained by illuminating its surface with electromagnetic radiation. What is the maximum wavelength to be used?
  Hint: The photon energy is $E_{ph} = hf$.

7.4. The cathode of a vacuum diode is made of thoriated tungsten, has a length of 3 cm and a diameter of 1 mm, and is at a temperature of 1800 K. If the positive plate voltage is such that the anode current is 200 mA:
  (*a*) What percentage of the emitted current returns to the cathode?
  (*b*) What prevents this current from reaching the anode?

7.5. Figure P7.5 shows the energy-position diagram of a metal and a semiconductor

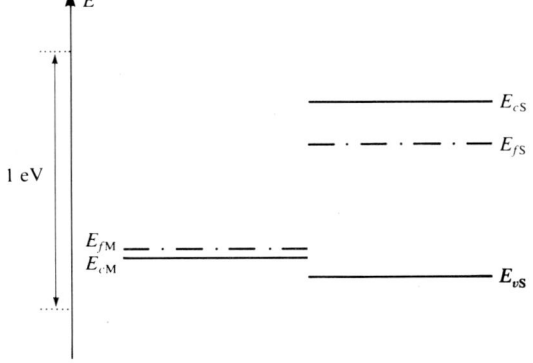

FIGURE P7.5
The figure is drawn to scale. The energy axis scale is indicated at the left of the figure.

before a junction is made. Energies are shown to scale in electronvolts. It is assumed that the temperature is 300 K, that the two specimens are at the same potential, and that the metal is Au.

(a) Compute the semiconductor electron affinity χ_{eS}.

(b) For the semiconductor specimen determine the host material, the type (P, N, or intrinsic), and the impurity concentration.

(c) What is the theoretical M-S barrier?

(d) What is the actual value of the M-S barrier and why?

7.6. At 300 K, for a contact between Al and N-type Si with $N_D = 10^{16}$ cm^{-3}, determine the theoretical and actual barrier heights (use the data of Fig. 7.3.2 and Table 3.1.1 or of Table 7.3.1).

(a) Compare the two data and give reasons for the discrepancy.

(b) Compute the actual V_{bi}, w, and $\mathscr{E}_{max}$.

7.7. A contact between W and P-type Si with $N_A = 10^{16}$ cm^{-3} is at 300 K.

(a) Draw the energy band diagrams of the two materials before the junction is made, assuming them to be at the same potential (as in Fig. 7.3.1a).

(b) Draw the energy band diagram of the junction in equilibrium, assuming that no distortion of the lattice with creation of spurious states has occurred.

(c) Compute ϕ_{BP}, w, and $\mathscr{E}_{max}$ under the same assumption.

(d) Repeat parts (b) and (c) for the practical case, including the effect of lattice distortion at the interface.

7.8. In a reverse biased metal-semiconductor junction the Fermi levels are not aligned. As a consequence, some current should flow. Compare the energy band diagram of a metal–N-type semiconductor junction at reverse bias with the unbiased diagram. Analyze the currents flowing in the two cases and explain what happens.

7.9. For a Au–N-type Si junction with $N_D = 3 \times 10^{15}$ cm^{-3} and a 0.5 mm^2 cross section at 300 K:

(a) Compute the depletion capacitance for a bias $V = -4$ V, using the value of ϕ_{BN} from Table 7.3.1.

(b) Assuming uniform doping, draw the $1/c_d^2$ vs. V_R curve.

(c) Repeat using the theoretical value of V_{bi} obtained from the data of Tables 7.1.1 and 3.1.1 and compare the two diagrams. Comment.

7.10. For the metal-semiconductor junction of Prob. 7.9, biased at $V = -4$ V, compute the average capacitance to a finite step of voltage of amplitude $\Delta V = 0.05$ V, tending to increase the reverse bias.

7.11. (a) Compute and tabulate the I/V characteristic of a Schottky diode between W and N-type Si with $N_D = 10^{17}$ cm^{-3} at 300 K, with a junction cross section $A = 1$ mm^2. Extend the tabulation from 0 V to the point in the forward biased region where the current is 100 mA (use values from Table 7.3.1).

(b) Compare with the analogous values for the PN junction of Prob. 6.28a and comment.

7.12. Compute and tabulate the I/V characteristic of the Schottky diode of Prob. 7.11, using the theoretical value of ϕ_{BN} obtained from the data of Tables 7.1.1 and 3.1.1. Compare with the result of Prob. 7.11.

7.13. Would you expect the diffusion capacitance of a Schottky diode to be larger or smaller than that of an equivalent PN junction? Why? What do you expect the ratio of the two diffusion capacitances to be?

7.14. For a W–N-type Si junction with $N_D = 2 \times 10^{16}$ cm^{-3} at 300 K with junction cross section $A = 0.1$ mm^2:

(a) Tabulate the small-signal dynamic resistance r_d vs. the forward bias V.

(b) Compare this tabulation with that of Prob. 6.35 for a comparable PN junction.

7.15. A metal with work function $\phi_M = 4$ V forms a junction with an N-type semiconductor with $\chi_S = 3.6$ V and $E_g = 1.4$ eV. It is assumed that no spurious states are created at the junction and that the temperature is 300 K.

(a) What is the maximum impurity concentration N_D required to ensure an ohmic contact?

(b) At $N_D = 10^{15}$ cm^{-3}, compute the height of the barrier to be overcome by the metal electrons in order to migrate into the semiconductor.

7.16. The polysilicon contact technology, mentioned in Chap. 5, makes use of heavily doped polycrystalline material in lieu of a metal for the fabrication of special electrodes. This technology will be discussed in greater detail in connection with MOSFETs (cf. Chap. 9). On the basis of the concepts developed up to now, consider the contact between a polysilicon electrode degenerately doped with 10^{19} As atoms/cm^3 and an N-type Si semiconductor with $N_D = 10^{15}$ cm^{-3}.

(a) Is the contact ohmic?

(b) Sketch the energy band diagram of the above system before the junction is fabricated and at equilibrium with the junction in place. Compute the barrier voltage ϕ_{BN}. How is the barrier generated?

(c) Roughly estimate the bias range corresponding to the space charge limited conduction.

Computer Problems

7.17. Write a computer program to solve Prob. 7.11a. Present the results in tabular form.

7.18. Write a computer program to solve Prob. 7.12. Present the results in tabular form.

REFERENCES

1. Cowley, A. M., and S. M. Sze: "Surface States and Barrier Height of Metal-Semiconductor Systems," *J. Appl. Phys.*, vol. 36, p. 3212, 1965.
2. Crowell, C. R., J. C. Sarace, and S. M. Sze: "Tungsten-Semiconductor Schottky Barrier Diodes," *Trans. Met. Soc. AIME*, vol. 233, p. 478, 1965.
3. Yang, Edward S.: *Fundamentals of Semiconductor Devices*, McGraw-Hill Book Company, Inc., New York, 1978.

ADDITIONAL READING

Mead, C. A.: "Metal-Semiconductor Surface Barriers," *Solid State Elect.*, vol. 9, p. 1023, 1966.
Rhoderick, E. H.: "Comments on the Conduction Mechanism in Schottky Diodes," *J. Phys. D., Appl. Phys.*, vol. 5, p. 1920, 1972.

CHAPTER

8

JUNCTION FIELD-EFFECT TRANSISTORS

8.1 J-FET STRUCTURE

The mechanism of operation of the junction field-effect transistor (J-FET) will be analyzed using the schematic configuration of Fig. 8.1.1. Figure 8.1.2 represents the same structure in the planar configuration.

Both configurations are in very simplified form and are grossly distorted geometrically to help in graphically describing and analyzing the basic features of J-FET operation. The structure of actual devices usually incorporates several additional details to improve performance and/or fabrication and depends on the specific technology used (e.g., single or integrated devices, etc.). Two types of J-FET are produced: the N- and the P-channel types. The structure shown in the figures depicts an N-channel J-FET.

Two electrodes, designated as source (S) and drain (D), are connected[1] to the two ends of a bar of N-type semiconductor, usually in the form of a thin layer, represented by the less heavily darkened region in the figure, establishing a

---

[1] In practice, the metal–N semiconductor contact is obtained through two degenerately doped n^+ regions, to ensure ohmic contact (cf. Sec. 7.8). This will be discussed in Sec. 8.6 (fabrication).

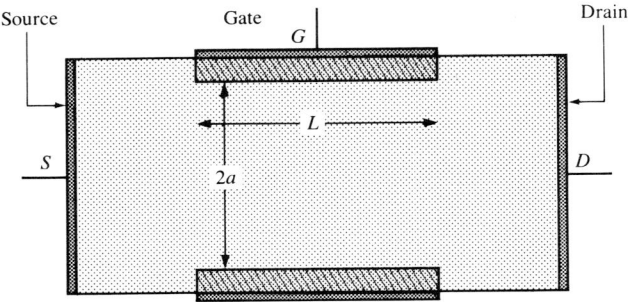

FIGURE 8.1.1
Basic structure of a J-FET. Upper and lower gate electrodes are connected together (G).

continuous conducting path between them. In both figures the thickness of this semiconductor path has been grossly magnified to permit proper labeling. As shown, a portion of this layer, called the *channel*, of length L and width $2a$, is sandwiched between two heavily doped p^+ regions (darkened in the figure), called the *gate*, with which it forms PN junctions. Typically the channel is very narrow ($2a$ is usually of the order of a micrometer) and $L \gg 2a$. The thickness of the device is uniform and designated by z.

In Fig. 8.1.2, one of these p^+ regions is the substrate wafer on which the N-type semiconductor layer is epitaxially grown; the other p^+ region is obtained by diffusion into the N semiconductor. In both figures, metallic contacts are connected to the surfaces of the two p^+ regions and are electrically connected together to form one electrode, labeled G (for gate), providing electrical access to this part of the device.

From now on, unless otherwise specified, the source electrode will be assumed to be at reference potential (indicated as ground in the figures to follow),

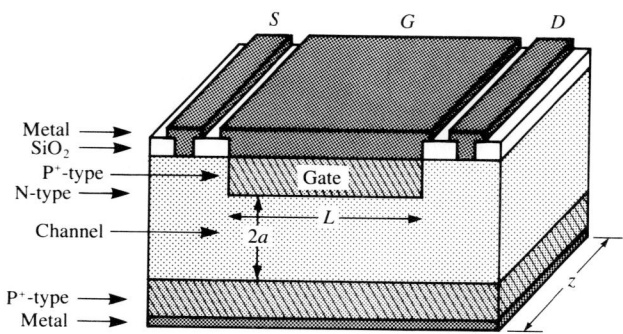

FIGURE 8.1.2
Essential structure of a planar J-FET.

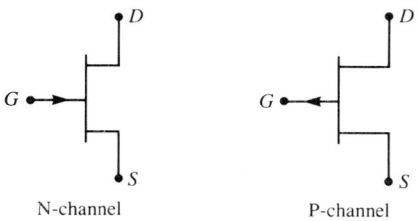

FIGURE 8.1.3
J-FET circuit symbols for N-channel and P-channel devices. Arrows indicate direction from P to N material.

and the gate and drain potentials with respect to this reference will be designated by V_G and V_D respectively.

As previously stated, the example shown depicts an N-channel J-FET. In P-channel J-FETs, the substrate and gate are N-type, the channel is P-type.

Figure 8.1.3 shows the circuit symbol for the J-FET, indicating source, drain, and gate terminals. The type of J-FET represented by each symbol is specified by the direction of the arrow in the gate connection. Consistently with the conventions adopted to indicate the polarity of the PN diode constituting the gate of the J-FET, this direction is from the P to N material.

A typical sequence of steps used in the fabrication of a J-FET will be discussed in Sec. 8.6.

8.2 J-FET PRINCIPLE OF OPERATION

Reference is made to Fig. 8.2.1, reproducing the configuration of Fig. 8.1.1. The N semiconductor constitutes an ohmic conduction path, uninterrupted by junctions, between the source and the drain, so that conduction is possible in the two directions. In this path, the resistance of the two comparatively bulky regions next to the contacts is negligible compared to that of the narrow channel, so that the drain to source resistance should be $R_d \approx L/(\sigma z \, 2a)$.

However, even if both V_G and V_D are zero (G and D shorted to S), as shown in Fig. 8.2.1a, a contact potential V_{bi} is developed across the PN junctions and two depletion regions are formed. They are indicated by the two clear regions near the junctions in Fig. 8.2.1a, where it should be remembered that the upper and lower gate electrodes are electrically connected to each other.

The width $w(V_{bi})$ of these depletion regions is determined by the contact potential and can be computed as discussed in Secs. 6.2 and 6.3 [Eq. (6.2.9)]. Notice that these regions, being depleted of their current carriers, are insulators and so do not contribute to the conductance of the channel, which is therefore divided into a nonconducting, depleted portion (clear in the figure) and a conductive, *active* part (darkened in the figure).

The figure further shows that the depletion regions extend only into the N-type channel, because the junctions are one-sided, due to the strong doping of the p^+ regions. It can therefore finally be concluded that the active channel width

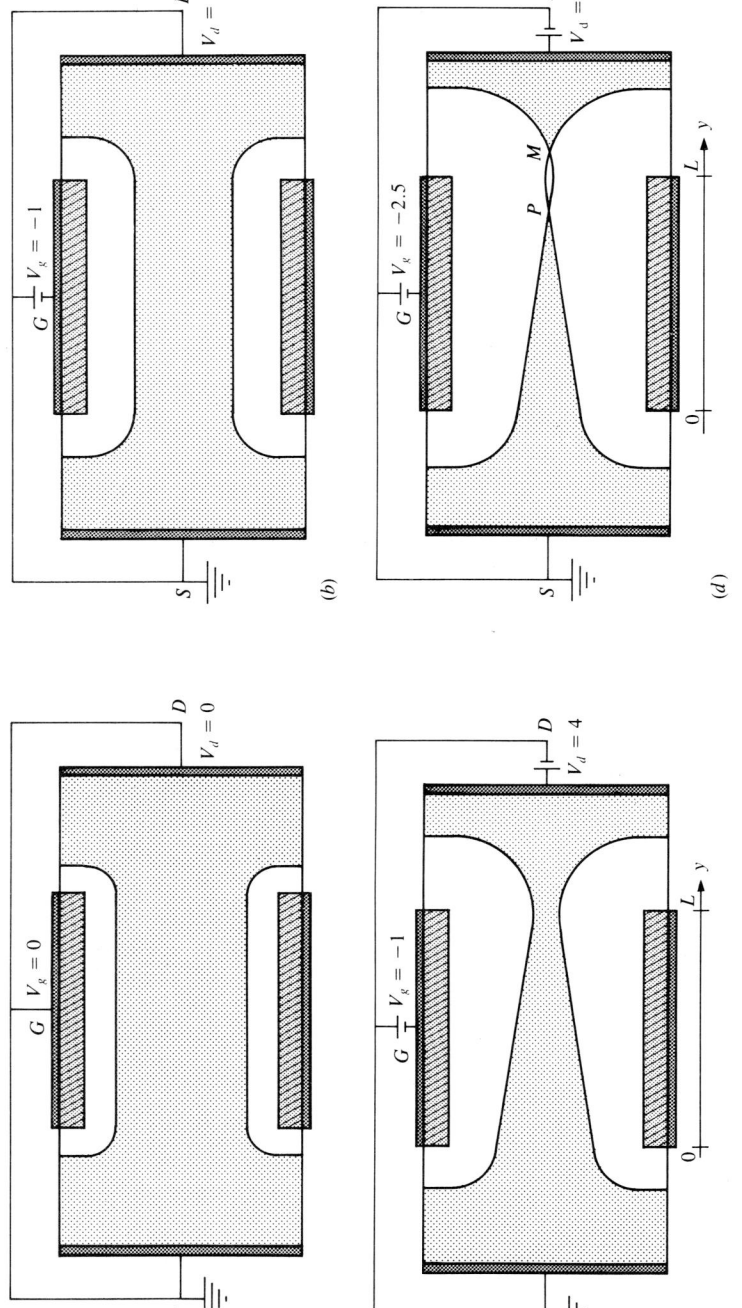

FIGURE 8.2.1

Schematic representation of J-FET operation. Dark grey areas represent metal electrodes. Darkened areas indicate P semiconductors, less heavily darkened areas N semiconductors, clear areas symbolize depleted regions. Upper and lower gate contacts are electrically connected together and constitute one electrode. (*a*) Unbiased condition. (*b*) Negatively biased gate, unbiased drain. Notice wider depletion regions. (*c*) Negatively biased gate, positively biased drain, below pinchoff. The active channel width varies with the position along the *y* axis, indicated at the bottom of the picture. (*d*) Bias conditions beyond pinchoff.

258

is reduced from $2a$ to $2[a - w(V_{bi})]$ and the channel resistance is correspondingly increased.

In the following, for the student's convenience, some simple calculations of depletion region and active channel widths will be carried out, as an example, using the theory developed in Chap. 6. A more complete quantitative analysis will be offered in Sec. 8.3. For the purposes of this example, in the computations the following data will be assumed to define the structure:

$$N_A = 10^{19} \text{ cm}^{-3}; \qquad N_D = 10^{16} \text{ cm}^{-3}; \qquad 2a = 2 \ \mu\text{m}$$

Then, under the unbiased conditions of Fig. 8.2.1a, from Eq. (6.2.9), $V_{bi} = 0.87$ V and, from Eq. (6.3.20) or (6.3.21), $w(V_{bi}) = 0.34 \ \mu$m, so that the active channel width becomes $2(a - w) = 1.32 \ \mu$m.

If V_G is made negative, as shown by Fig. 8.2.1b, then the two junctions are reverse biased, the voltage across them is raised to $V_{bi} - V_G$ (remember that V_G is negative), the depletion region width is increased to $w(V_{bi} - V_G)$, and the junction is further constricted, so that its resistance becomes even greater.

Under the data described, if $V_G = -1$ V, then $V_{bi} - V_G = 1.87$ V, so that the reverse biased depletion region widths become $w_{rb} = 0.5 \ \mu$m each and the active channel width $\approx 1 \ \mu$m. These values correspond to Fig. 8.2.1b.

If now a positive voltage V_D is applied to the drain, the conducting channel acts as a voltage divider between drain and source so that the potential along the channel becomes a function $V(y)$ of the coordinate y, i.e., of the position along the length of the channel. In Fig. 8.2.1c, $V(y)$ varies from 0 near the source to V_D near the drain. At an arbitrary point of abscissa y, then the voltage across the junction is $V_{bi} + V(y) - V_G$, where $V(y)$ is a voltage between zero and V_D. The width of the depletion regions correspondingly becomes a function of y and the width of the active channel decreases as we move from the end of the channel near the source in the direction of the drain, where it reaches a minimum. Here the voltage across the junction is maximum and equal to $V_{bi} - V_G + V_D$. This is schematically indicated in Fig. 8.2.1c.

With $V_G = -1$ V and $V_D = 4$ V, as shown in Fig. 8.2.1c, and with the assumed data, then, near the source, where $V(y) = 0$, the channel to gate voltage $V(y) + V_{bi} - V_G$ is 1.87 V and the reverse biased depletion region width $w_{rb} = 0.5$ μm, with an active channel 1 μm wide, just as in the previous computation. However, near the drain, where $V(y) = 4$ V, the channel to gate voltage rises to 5.87 V, the depletion region width becomes 0.88 μm, and the active channel is reduced to its minimum width of 0.24 μm. Figure 8.2.1c also shows that, moving along the channel from source to drain, the conductive active portion becomes narrower and narrower.

If V_D keeps increasing, the minimum active channel width will finally be reduced to zero. The channel is then said to be *pinched off*. This happens when the junction voltage reaches a value V_{po} such that the corresponding width of each depletion region becomes equal to a, i.e., half the geometrical width of the channel, or $w(V_{po}) = a$. This *pinchoff voltage* V_{po} is an important characteristic property of the device and can be computed from the device fabrication data, as discussed later [Eq. (8.6.3)]. Under the conditions adopted for Fig. 8.2.1, $V_{po} = 7.66$ V. For a given V_{bi} and V_G, the drain voltage at which pinchoff begins is $V_{D, po} = V_{po} - V_{bi} + V_G$.

If the biasing voltages are increased to the point where the channel to gate voltage exceeds V_{po}, as depicted in Fig. 8.2.1d, then the point at which the channel is pinched off (P in the figure) moves to the position where the junction voltage equals V_{po} and the depletion region extends a little toward the drain to point M, so that a short length of the channel becomes part of the depletion region.

In Fig. 8.2.1d, the maximum channel to gate voltage is $V_D + V_{bi} - V_G = 5 + 0.87 + 2.5 = 8.73$ V. This is greater than the previously computed value of 7.66 V for V_{po}. These conditions prevail at point M, where the depletion region width is 1.05 μm and so greater than the geometric channel's half-width a. Notice that point M lies in the bulky region of the N semiconductor next to the drain, outside of the narrow channel. Point P, instead, is within the channel, where the channel potential is down to $V_{D, po} = 7.66 - 0.87 - 2.5 = 4.29$ V and so the junction potential equals V_{po}.

Between point P and point M, depletion has turned the channel material into an insulator. In this region, the figure also shows the configuration of the mathematically computed edges of the depletion regions, showing that, between P and M, the two depletion regions would overlap and the active channel width would become negative.

Figure 8.2.1d shows that the depletion region width at different points is a function of the local junction voltage. Consider now a channel biased well into pinchoff, as in Fig. 8.2.1d. Between the source and the pinchoff point P the channel is uninterrupted and there is a voltage $V_{D, po}$ across it, so a current must flow. This current cannot end abruptly at the pinchoff point, thanks to Kirchoff's principle, so *a stream of carriers is injected into the depletion region* at the pinchoff point P. These carriers are swept across the depletion region in the direction of the channel axis by the strong electric field in it and finally reach the drain electrode, so *the drain current equals the current flowing between the source and the pinchoff point P*.

As the drain voltage V_D increases, point P moves to the left, but only by a minute amount, so that, as a rough first approximation, the geometric configuration of the channel between the source and point P remains practically unchanged, and so does its resistance. As the voltage at point P remains, by definition, $V_{D, po}$, then the current from S to P, which is determined by this voltage and the channel resistance, remains the same; therefore, finally, *in a J-FET in pinchoff, the current is essentially independent of the drain voltage* and remains at the constant value I_{DS} reached at the onset of pinchoff. Notice that this *pinchoff saturation current*, although independent of V_D, is a function of V_G.

The mechanism of conduction described above makes it clear that, throughout the device, current is mediated by the channel majority carriers (electrons for N-channel, holes for P-channel), showing that *the J-FET is a unipolar device*. When the channel is made of N-type material, the majority carriers mediating the current are electrons and the device is properly named N-channel. In P-channel devices the current is mediated by holes, in accordance with the nomenclature.

8.3 QUANTITATIVE ANALYSIS OF THE J-FET

As usual, the desired quantitative relationships can be obtained by simply translating the qualitative considerations of Sec. 8.2 into the language of mathematics.

Figure 8.3.1 represents a section taken along the length of the channel, qualitatively showing the dependence of the depletion region width on the position y. The active channel width never reaches zero, so *the figure applies below pinchoff.*

An unknown current I_D flows along the active channel. To compute it, calculate the ohmic voltage drop dV generated by this current over the infinitesimal displacement dy shown in the figure. Using (4.2.8) and with evident notation,

$$dV = I_D \, dR = I_D \frac{dy}{\sigma A} = \frac{dy}{2qN_D\mu_e z[a - w(y)]} I_D \tag{8.3.1}$$

in which, by (6.3.21), remembering Secs. 6.5 and 8.2,

$$w(y) = \sqrt{\frac{2\varepsilon}{qN_D}\left[V(y) - V_G + V_{bi}\right]} \tag{8.3.2}$$

substituting (8.3.2) into (8.3.1), and reordering:

$$I_D \, dy = 2qN_D\mu_e z\left\{a - \sqrt{\frac{2\varepsilon}{qN_D}\left[V(y) - V_G + V_{bi}\right]}\right\} dV \tag{8.3.3}$$

The above expression can now be integrated *under the assumption that pinchoff has not been reached.* As previously noted, below pinchoff, Fig. 8.3.1 applies, showing that, as y varies from 0 to L, $V(y)$ varies from 0 to V_D. Consequently, if the left side is integrated from 0 to L, integration of the right side should extend

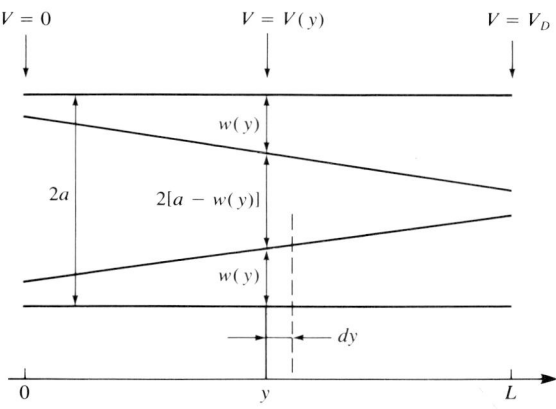

$V = 0$ $\qquad$ $V = V(y)$ $\qquad$ $V = V_D$

$w(y)$

$2a$ $\qquad$ $2[a - w(y)]$

$w(y)$

dy

0 $\qquad$ y $\qquad$ L $\qquad\rightarrow y$

FIGURE 8.3.1
Depletion region and active channel widths together with potential distribution along the channel length, below pinchoff.

from 0 to V_D. The left side is easily integrated, remembering that at steady state I_D is independent of y (Kirchoff's law). Evaluating both definite integrals and dividing both sides by L,

$$I_D = \frac{2qN_D\mu_e z}{L} \left\{ aV_D - \frac{2}{3}\sqrt{\frac{2\varepsilon}{qN_D}} \left[(V_D - V_G + V_{bi})^{3/2} - (V_{bi} - V_G)^{3/2} \right] \right\} \quad (8.3.4)$$

Notice that the above expression is valid only below pinchoff, as stated in determining the limits of the integrals taken. Remembering the definition of V_{po} of Sec. 8.2 and then using (6.3.21) for a one-sided junction,

$$w(V_{po}) = \sqrt{\frac{2\varepsilon}{qN_D}} V_{po} = a \quad (8.3.5)$$

from which V_{po} can be computed as a function of the device fabrication data:

$$V_{po} = \frac{qN_D a^2}{2\varepsilon} \quad (8.3.6)$$

Comparing this expression with (8.3.4), after some algebra and introducing the admittance parameter,

$$G = \frac{2qN_D\mu_e za}{L} \quad \text{mhos} \quad (8.3.7)$$

Equation (8.3.4) can then be rewritten as

$$I_D = G\left\{ V_D - \frac{2}{3}\frac{1}{\sqrt{V_{po}}} \left[(V_D + V_{bi} - V_G)^{3/2} - (V_{bi} - V_G)^{3/2} \right] \right\} \quad (8.3.8)$$

In accordance with Sec. 8.2, pinchoff is reached when V_D reaches the value $V_{D,po} = V_{po} - (V_{bi} - V_G)$. Substituting this value into (8.3.8) and remembering the arguments presented in Sec. 8.2, after some algebra it can be concluded that, beyond pinchoff, the saturation value of the drain current is

$$I_{Ds} = G\left[\frac{V_{po}}{3} - (V_{bi} - V_G)\left(1 - \frac{2}{3}\sqrt{\frac{V_{bi} - V_G}{V_{po}}} \right) \right] \quad (8.3.9)$$

A useful approximate form of (8.3.9) is

$$I_{Ds} = I_{DD}\left(1 + \frac{V_G}{V_{po}} \right)^2 \quad (8.3.10)$$

in which I_{DD} is the saturation current for zero gate bias and is computed by setting $V_G = 0$ in (8.3.9):

$$I_{DD} = G\left[\frac{V_{po}}{3} - V_{bi}\left(1 - \frac{2}{3}\sqrt{\frac{V_{bi}}{V_{po}}} \right) \right] \quad (8.3.11)$$

The student should keep in mind that, in the equations of this section, including (8.3.2), V_G is always a negative quantity.

In the previous computation of the depletion region width [expression (8.3.2)], it was implied that the field across the depletion region is perpendicular to the gate surface. Similarly, in computing the current through the channel [expression (8.3.3)] it was implied that the field producing the current is in the y direction, i.e., perpendicular to the previous field.

These simplifying assumptions make all computations very easy. Unfortunately, however, they are not valid except in the special case in which the length of the channel is much longer than its width: $L \gg 2a$. This assumption, necessary to make the above formulas valid for design, is known as the *long channel assumption*.

With the help of (8.3.8) and (8.3.9) it is possible to draw by points the I/V characteristics of the J-FET from the device design parameters. An example of such computation is given below.

Example 8.3.1. An N-channel Si J-FET at 300 K has thickness $z = 1$ mm, geometric channel half-width $a = 1$ μm, length $L = 25$ μm, and doping $N_D = 10^{16}$ cm^{-3} for the channel and $N_A = 10^{19}$ cm^{-3} for the p^+ regions. The total impurity concentration in the channel is also 10^{16} cm^{-3}. Compute: (a) the contact potential, (b) pinchoff voltage, (c) current at $V_G = -2$ V, $V_D = 3$ V, and (d) saturation current at $V_G = -2$ V using both the theoretical and approximate expressions. Compare these two values.

Solution

(a) From (6.2.9),

$$V_{bi} = 0.0259 \ln \frac{10^{16} \times 10^{19}}{2.25 \times 10^{20}} = 0.87 \text{ V}$$

(b) From (8.3.6),

$$V_{po} = \frac{1.6 \times 10^{-19} \times 10^{16} \times 10^{-8}}{2 \times 1.04 \times 10^{-12}} = 7.66 \text{ V}$$

(c) From (4.2.9), $\mu_e = 1123$, so that, from (8.3.7),

$$G = \frac{2 \times 1.6 \times 10^{-19} \times 10^{16} \times 1123 \times 10^{-5}}{25 \times 10^{-4}} = 0.0144 \text{ mho}$$

Comparing the maximum junction voltage $3 + 0.87 + 2 = 5.87$ with the pinchoff voltage V_{po}, the J-FET is not in pinchoff, so (8.3.8) applies and

$$I_D(-2, 3) = 0.0144 \left\{ 3 - \frac{2}{3\sqrt{7.66}} [(3 + 0.87 + 2)^{3/2} - (0.87 + 2)^{3/2}] \right\}$$

$$= 10.703 \text{ mA}$$

(d) From (8.3.9),

$$I_{Ds}(-2) = 0.0144\left[\frac{7.66}{3} - (0.87 + 2)\left(1 - \frac{2}{3}\sqrt{\frac{0.87 + 2}{7.66}}\right)\right] = 12.3 \text{ mA}$$

To compute the same quantity by the approximate formula, first compute the saturation current at zero gate bias I_{DD} from (8.3.11):

$$I_{DD} = 0.0144\left[\frac{7.66}{3} - 0.87\left(1 - \sqrt{\frac{0.87}{7.66}}\right)\right] = 27 \text{ mA}$$

Then, from (8.3.10),

$$I_{Ds} = 27\left(1 - \frac{2}{7.66}\right)^2 = 14.7 \text{ mA}$$

which is within 20 percent of the theoretical value.

Figure 8.3.2 illustrates the above example, showing a set of characteristic I/V curves for the device.

As already noted, the above theory was developed under a set of simplifying assumptions. In many cases, however, it can yield satisfactory predictions, as can be verified by inspection of Fig. 8.3.3 [1], displaying in (b) a set of I/V characteristics, theoretically computed on the basis of Eqs. (8.3.8) and (8.3.9),[2] and in (a) a set of experimentally obtained curves. It should be noted that the device represented by Fig. 8.3.3 complies with the most limiting of the assumptions of the theory: the long channel assumption of Sec. 8.3. For other devices the theory requires modification. In the case of short channel devices, the electric field configuration may become rather complex, requiring graduate-level considerations, beyond the scope of this text. Some deviations from the elementary theory will be discussed in Sec. 8.5.

Observation of Fig. 8.3.3 shows that the largest discrepancies between theoretical and measured performance occur at high currents. Equations (8.3.8) and (8.3.9) make no allowance for the series resistances of the bulk material (cf. Sec. 8.5). If appropriate values for these elements of the equivalent circuit are assumed and their effect on the current for given bias voltages is calculated, then the theoretically computed curves approximate experimental values much more closely.

8.4   SIGNAL TRANSFER—GAIN

The above discussion of current-voltage relationships within a J-FET implies some very important potentialities of the device as a circuit element.

---

[2] Appropriate data are: $N_A = N_D = N_{tot} = 2.5 \times 10^{15}$ cm^{-3}, $z = 0.25$ cm, $L = 0.0025$ cm, $a = 0.000162$ cm.

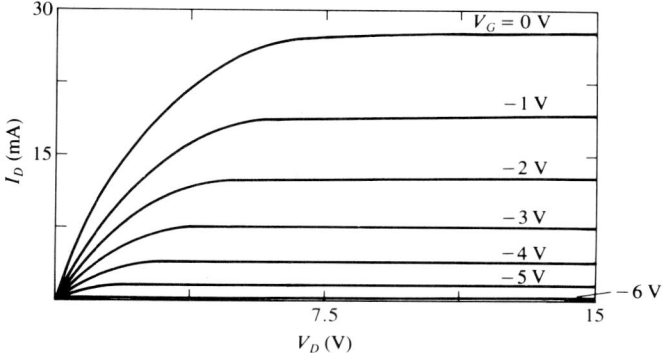

FIGURE 8.3.2
J-FET characteristics theoretically computed for the conditions of Example 8.3.1.

Consider the circuit of Fig. 8.4.1 in which a J-FET is shown together with the voltages applied to it. Similarly to what was done in Fig. 6.8.1, the gate voltage consists of two components: a *dc bias voltage* V_{GG} and an *ac signal* v_{in}.

Two separate circuits can be identified in the figure: a *gate or input circuit* containing the signal generator and a *drain or output circuit* containing the *load* R_L. The fundamental implication of the previous discussion is that *the current in*

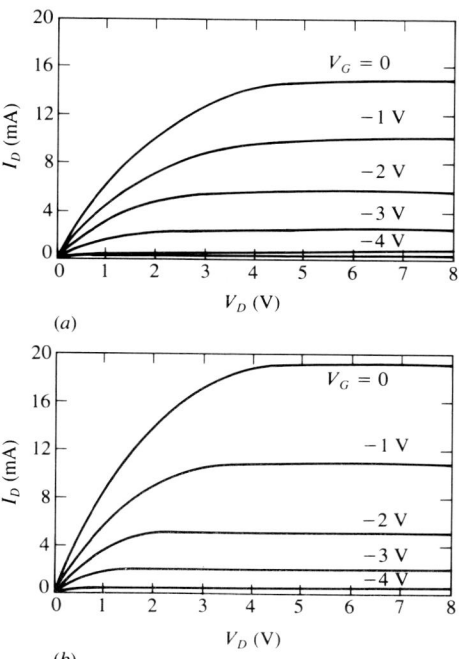

FIGURE 8.3.3
I/V characteristics of N-channel J-FET [1]. (*a*) Experimentally determined. (*b*) Computed from Eqs. (8.3.8) and (8.3.9).

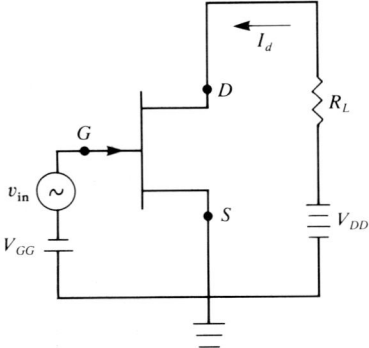

FIGURE 8.4.1
J-FET in amplifier circuit.

the output circuit can be controlled by varying the voltage applied to the input circuit.

This implies that, in the circuit of the figure, the output current I_D is composed of a dc and an ac component, the latter being controlled by the waveshape of the signal generator v_{in}, thereby implementing a *signal transfer* from the input to the output circuit.

The input circuit contains the reverse biased gate diode of the J-FET, so that, to a first approximation, the current in it is negligible and the signal generator can control the gate voltage without having to spend any appreciable amount of power in driving the input. On the other hand, the output current I_D may contain an appreciable ac component, so that *a significant amount of signal power is delivered to the load.* In conclusion, the transfer of signal from the input to the output is accompanied by a *power gain.*

As the student has already learned in active circuit courses, the ability to achieve gain with no more than moderate distortion is one of the most important properties of semiconductor devices and the design engineer must learn to analyze and determine the optimal conditions to achieve this end.

8.5 SMALL-SIGNAL LINEAR EQUIVALENT CIRCUIT

An important special case of signal transfer occurs when the amplitude of the signal is very small. Over this very limited range of voltage variations, the relationships between voltages and currents obtained in Sec. 8.3 can be approximated by linear expressions, easily computed by considering only the first-order terms of their Taylor series expansion.

Within such an approximation, the device can then be represented by a *small-signal equivalent linear circuit* and its behavior can be determined by the familiar methods of linear circuit analysis. The appropriate values of the equivalent circuit elements depend on the dc bias voltages, i.e., on the dc work point, and can be obtained by computing the first-order coefficients of the Taylor

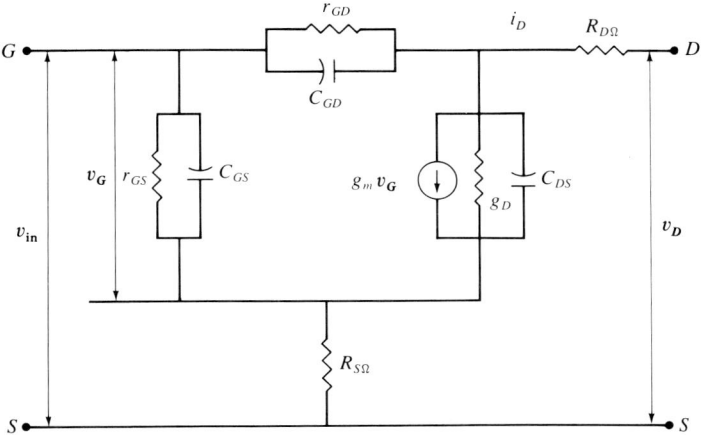

FIGURE 8.5.1
Small-signal linear equivalent circuit for J-FET.

expansion of the expressions for the I/V characteristics around the designated dc work point.

The above remarks are meant as a capsule summary of small-signal equivalent linear circuits theory, as a reminder for the student, who is assumed to have already been exposed to this material in previous active circuit courses [2, 3]. Only the most important definitions will be recalled here (also cf. Sec. 6.8).

A small-signal linear equivalent circuit for the J-FET is shown in Fig. 8.5.1. Approximate values for its circuit components will be determined on the basis of the theories presented up to now. The most important parameters are:

Mutual Conductance (or Transconductance)

This is defined as the ratio of an infinitesimal variation of output current to the infinitesimal variation of input voltage required to produce it under constant output voltage conditions. For the J-FET, the output current is the drain current and the input voltage is the gate voltage, so that

$$g_m = \left[\frac{\partial I_D}{\partial V_G}\right]_{V_D = \text{constant}} \tag{8.5.1}$$

Output Admittance

This is defined as the ratio of an infinitesimal output current to the infinitesimal output voltage required to produce it under constant input voltage conditions:

$$g_D = \left[\frac{\partial I_D}{\partial V_D}\right]_{V_G = \text{constant}} \tag{8.5.2}$$

These parameters depend on the choice of work point around which the small signal is applied. For J-FETs, the most useful conditions of linear operation are the *saturation mode* (beyond pinchoff) and the *linear mode*, in which operation is limited to very small values of the drain voltage.

Saturation Region Parameters

From (8.3.9), using (8.5.1),

$$g_{ms} = \frac{\partial I_{Ds}}{\partial V_G} = G\left(1 - \sqrt{\frac{V_{bi} - V_G}{V_{po}}}\right) \tag{8.5.3}$$

Applying (8.5.2) to (8.3.9),

$$g_{Ds} \approx 0 \tag{8.5.4}$$

in agreement with the statement of Sec. 8.2 that, beyond pinchoff, the saturation current is independent of the drain voltage. This first approximation value will be revised later on.

Linear Region Parameters

In the linear region the total drain voltage $V_D \ll V_{bi} - V_G$, so that, introducing this condition into (8.3.8) and using a first-order Taylor series approximation,[3]

$$I_{DL} \approx G\left(1 - \sqrt{\frac{V_{bi} - V_G}{V_{po}}}\right) V_D \tag{8.5.5}$$

Using this approximate expression, from (8.5.1),

$$g_{mL} = \frac{G}{2} V_D \frac{1}{\sqrt{V_{po}(V_{bi} - V_G)}} \tag{8.5.6}$$

and from (8.5.2),

$$g_{DL} = G\left(1 - \sqrt{\frac{V_{bi} - V_G}{V_{po}}}\right) \tag{8.5.7}$$

This value is identical to the value for g_{ms} of (8.5.3). This does not happen to be a coincidence.

---

[3] Expanding the expression $(x + \delta x)^{3/2}$ into a Taylor series and retaining only the first-order term in δx (supposed very small),

$$(x + \delta x)^{3/2} \approx x^{3/2} + \tfrac{3}{2}\sqrt{x}\,\delta x$$

as the student can easily verify.

Capacitances

The capacitances appearing in the equivalent circuit represent the depletion capacitances existing between the different electrodes because of the reverse biased junctions.

Their theoretical computation may prove very laborious and their influence on the response of the device depends very strongly on the circuit in which the device is used. For instance, if the device is providing a large voltage gain, then the gate to drain capacitance becomes the most important element in determining the high-frequency response (because of the Miller effect; cf. Refs. 2 and 3).

In usual industrial practice, the parameter provided by the manufacturer to indicate the ability of the device to handle high frequencies is the *cutoff frequency* f_{co}, defined as the frequency at which the current gain with shorted output becomes unity. From the equivalent circuit, at shorted output, the small-signal input current becomes

$$i_{in} \approx v_G \, 2\pi f_{co} \, C_G \qquad (8.5.8)$$

while the short-circuit output current is of the order of magnitude:

$$i_o \approx g_m v_G \qquad (8.5.9)$$

where $C_G = C_{GS} + C_{GD}$. Equating these two currents, the cutoff frequency is seen to be

$$f_{co} \approx \frac{g_m}{2\pi C_G} \qquad (8.5.10)$$

This quantity can be roughly estimated by adopting for g_m its maximum value $g_m \leq G$[cf. (8.5.3)]. As for C_G, its computation may require graduate-level considerations. In most cases, remembering that there are two depletion regions in parallel and that each of them extends to the entire area beneath the gate, the expression $C_G \approx 2\varepsilon Lz/W$ is adopted, in which W is assumed to be some quantity related to the geometric half-width of the channel (such as $a/2$). Using (8.3.7), f_{co} is commonly computed as

$$f_{co} = \frac{qN_D a^2 \mu_e}{\pi \varepsilon L^2} \, \alpha \qquad (8.5.11)$$

where α varies between $\frac{1}{4}$ and 2. In practice, the values of f_{co} provided by the manufacturers are obtained experimentally by measurements on the fabricated device. However, (8.5.11) is of value to the designer, indicating as it does that improved high-frequency performance can be obtained by using high-mobility semiconductors and keeping L as small as possible. This, however, may invalidate the long channel condition, making the previous design equations invalid and requiring more sophisticated expressions.

For the circuit design engineer, the f_{co} parameter can provide useful information, but must be used with caution, exercising much engineering common sense and, in the light of design experience, mostly as a " ball-park " indication of the hf limitations of the device.

The other equivalent circuit components appearing in Fig. 8.5.1 can be considered as introducing second-order variations to the device performance.

Leakage Resistances

Indicated in the equivalent circuit by r_{GD} and r_{GS}, these resistances represent leakage currents in the reverse biased PN junction at the gate. As such they are generally extremely large resistors (cf. Chap. 6), especially in the small-signal ac mode. They can generally be omitted except when manufacturing difficulties result in high surface leakage conditions.

Source and Drain Ohmic Resistances

Indicated in the schematic as $R_{S\Omega}$ and $R_{D\Omega}$, these resistors represent the ohmic resistance of the electrode contacts and of the bulk material between the ends of the channel and the electrodes. Their value does not depend on the applied voltages, as does the channel resistance, and they are usually very small. Their influence is felt when g_m or g_D become large. It is possible to eliminate them from the equivalent circuit by modifying the values of g_m and g_D to take into account their effect on device performance.

From the equivalent circuit

$$i_D = g_m v_G + g_D[v_D - i_D(R_{S\Omega} + R_{D\Omega})] \tag{8.5.12}$$

and
$$v_G = v_{in} - i_D R_{S\Omega} \tag{8.5.13}$$

Substituting (8.5.13) into (8.5.12) and solving for i_D gives

$$i_D = \frac{g_m v_{in} + g_D v_D}{1 + g_m R_{S\Omega} + g_D(R_{S\Omega} + R_{D\Omega})} = g'_m v_{in} + g'_D v_D \tag{8.5.14}$$

implicitly defining the modified admittances g'_m and g'_D. Incorporating the above modifications, the equivalent schematic takes on the simple and useful form shown in Fig. 8.5.2.

The modified mutual admittance of Eq. (8.5.14) is illustrated by Fig. 8.5.3, plotting the mutual conductance at saturation vs. gate voltage. The solid line

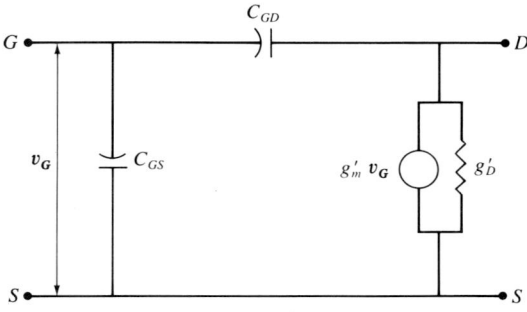

FIGURE 8.5.2
Approximate small-signal linear equivalent circuit.

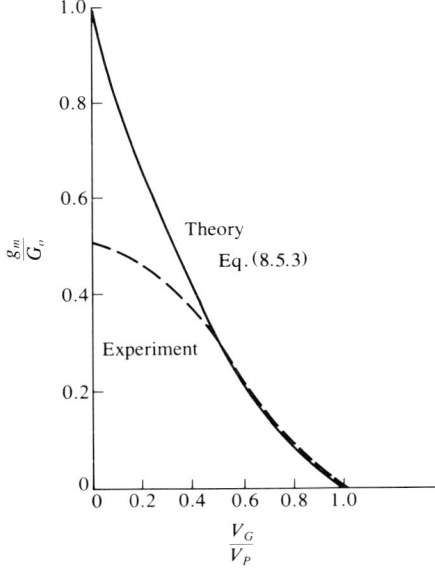

FIGURE 8.5.3
Saturation transconductance vs. gate voltage, theoretically computed [Eq. (8.5.3), solid line] and measured (dashed line). The discrepancy at high values of g_{ms} can be minimized by taking into account the series source resistance [Eq. (8.5.14)] [4].

shows g_{ms} as computed from (8.5.3), the dashed line shows experimental values. Notice the large deviation at high values of g_{ms}, as predicted by (8.5.14).

The reader familiar with feedback theory will easily recognize that the above parameter variations are the consequence of internal feedback produced within the transistor by resistors $R_{S\Omega}$ and $R_{D\Omega}$ of Fig. 8.5.1.

Channel Length Modulation

As a first approximation, assume the drain current in the saturation region to be independent of drain voltage. Then, in the saturation region, the drain behaves as an ideal current generator [cf. Eq. (8.5.4)]. In Sec. 8.2 this approximation was adopted by noticing that, beyond pinchoff ($V_D > V_{D,\,po}$), the effective length of the channel varies only by a small amount as the drain voltage is increased. In Fig. 8.2.1d, the potential at point P is $V_{D,\,po}$ and at M it is V_D. The distance PM then represents the width of a depletion region across which there is a voltage difference $V_D - V_{D,\,po}$. We shall designate the distance PM by $2\delta L$. As a first approximation, then, the channel length variation can be assumed to be one-half of that distance or

$$\delta L = \frac{1}{2}\sqrt{\frac{2\varepsilon}{qN_D}(V_D - V_{D,\,po})} = \sqrt{\frac{\varepsilon}{2qN_D}[V_D - (V_{po} + V_G - V_{bi})]} \quad (8.5.15)$$

in which use was made of (6.3.21), appropriately modified, and of the expression for $V_{D,\,po}$ obtained in Sec. 8.2.

A variation of channel length affects the saturation drain current only through the conductance factor G, so that, with evident notation, the saturation current at V_D is

$$I_{Ds}(V_D) = I_{Ds}(V_{D,\,po}) \frac{L}{L - \delta L} \tag{8.5.16}$$

The slope of the I_{Ds} vs. V curve (i.e., the average drain admittance) between drain voltages V_{D1} and V_{D2} is evidently

$$g_{Ds} = \frac{I_{Ds2} - I_{Ds1}}{V_{D2} - V_{D1}} \tag{8.5.17}$$

or, using (8.5.15) and after some algebra,

$$g_{Ds} = \frac{I_{Ds}}{V_{D2} - V_{D1}} \frac{\delta L_2 - \delta L_1}{1 - (\delta L_2 - \delta L_1) + \delta L_1 \, \delta L_2 / L} \tag{8.5.18}$$

in which the last addend in the denominator is usually so small that it can be neglected.

Example 8.5.1. For the J-FET of Example 8.3.1, compute: (a) g_{DL} for $V_G = -2$, (b) the saturation cutoff frequency f_{co} (use $\alpha = 0.25$). For the work point $V_D = 6$ V, $V_G = -2$ V, compute: (c) g_m, (d) g_D, (e) the modified transfer and output admittances assuming $R_{S\Omega} = R_{D\Omega} = 45\ \Omega$ and draw the equivalent circuit.

Solution

(a) From (8.5.7), using the results of Example 8.3.1,

$$g_{DL} = 0.114\left(1 - \sqrt{\frac{0.87 + 2}{7.66}}\right) = 5.58 \text{ mmho}$$

(b) f_{co} can be obtained directly from (8.5.11), or, remembering its definition, computing $C_G \approx 2 \times 1.04 \times 10^{-12} \times 25 \times 10^{-4} \times 0.1/0.5 \times 10^{-4} = 10.4$ pF (in which the average depletion region width has been assumed equal to $a/2$, in agreement with a choice of $\alpha = 0.25$). Using this value,

$$f_{co} = \frac{0.0144}{2\pi \times 10.4 \times 10^{-12}} = 219 \text{ MHz}$$

(c) At the given work point: $V_{D,\,po} = 7.66 - (0.87 + 2) = 4.76$ V. As $V_D = 6 > 4.76$, at this work point the J-FET is in the saturation region and

$$g_m = g_{ms} = g_{DL} = 5.58 \text{ mmho}$$

(d) Computing g_{Ds} at $V_G = -2$ V and between $V_{D1} = 5.9$ V and $V_{D2} = 6.1$ V, from (8.5.15),

$$\delta L_1 = \sqrt{\frac{1.04 \times 10^{-12}}{2 \times 1.6 \times 10^{-19} \times 10^{16}} (5.9 - 4.79)}$$

$$= \sqrt{3.25 \times 10^{-10} \times 1.11} = 1.9 \times 10^{-5} \text{ cm}$$

$$\delta L_2 = \sqrt{3.25 \times 10^{-10}(6.1 - 4.79)} = 2.06 \times 10^{-5} \text{ cm}$$

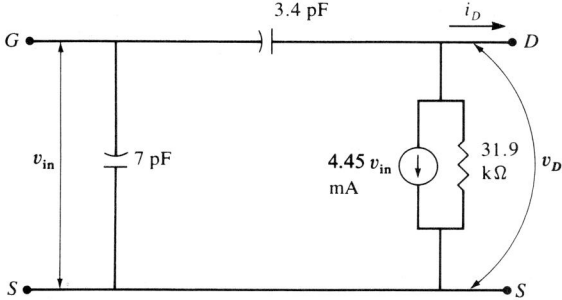

FIGURE 8.5.4
Approximate linear equivalent circuit for Example 8.5.1.

so that, from (8.5.18), using the suggested approximation,

$$g_{Ds} \approx \frac{12.3}{0.2} \frac{(2.06 - 1.9) \times 10^{-5}}{25 \times 10^{-4} - (2.06 + 1.9) \times 10^{-5}} = 40 \ \mu\text{mho}$$

(e) From (8.5.14),

$$g'_{ms} = \frac{5.58 \times 10^{-3}}{1 + 5.58 \times 10^{-3} \times 45 + 40 \times 10^{-6} \times 90} = \frac{5.58 \times 10^{-3}}{1.2547} = 4.45 \ \text{mmho}$$

$$g'_{Ds} = \frac{40 \times 10^{-6}}{1.2547} = 31.9 \ \mu\text{mho}$$

The reduced equivalent circuit is shown in Fig. 8.5.4, where the total gate capacitance C_G has been divided into a gate to drain and a gate to source capacitance, in the approximate ratio of 1 to 2.

8.6 J-FET FABRICATION

The sequence of steps actually employed in the fabrication of a J-FET device varies greatly from device to device and from manufacturer to manufacturer. The following outline illustrates one of the possible implementations. In order to keep the amount of detail within the scope of this text, some simplifications have been introduced in the sequence outlined.

The process described uses four masks, which are schematically shown in Fig. 8.6.1. They follow the geometry shown in Figs. 8.6.2 through 8.6.5 and refer to one device only. These masks assume that the photoresist used is of the negative type, except mask #4, which is designed for positive photoresist. In a real industrial process, their geometry would display a greater degree of sophistication and, of course, each mask would contain many repeated patterns, one for each device to be manufactured on the same wafer. With the help of the following outline, it should be easy to realize how critically important is the accurate positioning (registration) of these masks during the process.

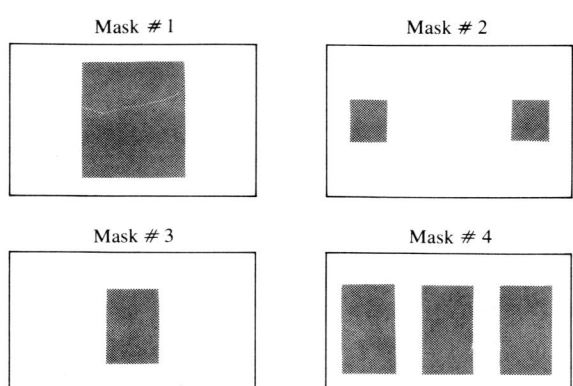

Mask # 1

Mask # 2

Mask # 3

Mask # 4

FIGURE 8.6.1
Idealized sketches of the four masks used in the J-FET fabrication sequence outlined in the text.

Fabrication Sequence Outline

1. A thin layer of N-type material (of the order of 1 μm) is deposited on a p^+ substrate either epitaxially or by a diffusion process.
2. The surface of the N-type layer is oxidized, usually by the wet process to obtain a SiO_2 layer suitable for masking purposes.
3. By a photostep, a suitable well is etched into the oxide to expose the N-type layer in the region where the p^+ gate will be fabricated. This step uses mask #1.
4. The p^+ gate is either diffused or deposited by implantation into the N-type material. This step is very carefully controlled to ensure proper characteristics in the channel. Figure 8.6.2 shows a cross section of the device at this stage of the fabrication. The geometrical dimensions are not shown to scale, but are distorted for better visualization of the various elements. Notice how the diffusion process determines the geometrical width of the channel. The importance of controlling this width to close tolerances is evidenced by the role played by the channel half-width a in Eqs. (8.3.6) through (8.3.9).
5. Next the oxide layer is stripped off and a new layer is grown.

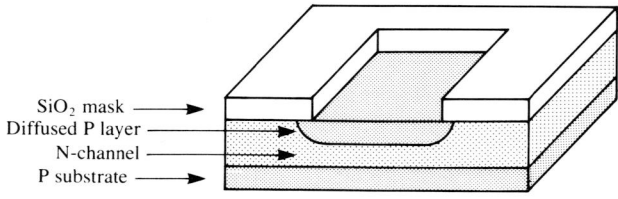

SiO$_2$ mask
Diffused P layer
N-channel
P substrate

FIGURE 8.6.2
Cross section schematically showing the structure of the device after step 4 of the fabrication sequence outlined in the text.

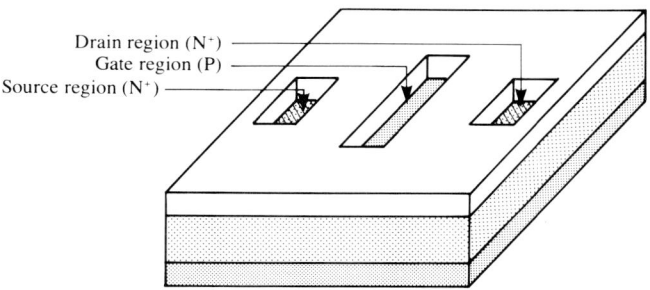

Drain region (N⁺)
Gate region (P)
Source region (N⁺)

FIGURE 8.6.3
Device structure just before metallization (after step 7). Notice the wells where the gate and the n^+ contact material are exposed by the etched oxide.

6. By a photostep, using mask #2, two wells are produced in the oxide and two degenerately doped layers of n^+ material are diffused into the N-type layer. These will be used to assure ohmic contact with the drain and source electrodes (cf. Sec. 7.8).

7. Through another photostep, using mask #3, a third well is etched into the oxide, at the location of the diffused p^+ gate layer, so now access to the underlying material is available for the source, drain, and gate electrodes. Figure 8.6.3 shows the specimen at this stage of the process. Actually, as described, this step is not very practical. Modifications to it will be described later on.

8. A metallization step deposits the metal for the electrodes over the whole face of the device.

9. A final photostep, using mask #4, etches the deposited metal layer separating the various electrodes. Figure 8.6.4 shows a cross section of the device at this stage. The numbers correspond to the various steps of this sequence and indicate at what stage each part of the structure was fabricated.

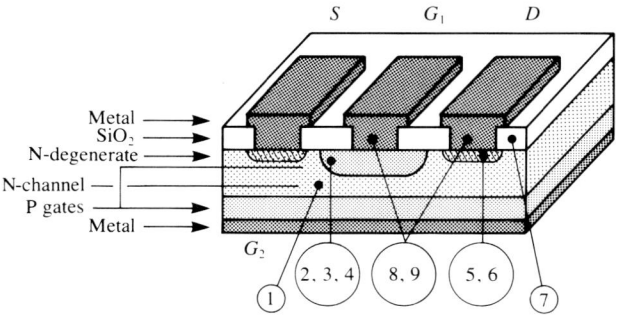

Metal
SiO₂
N-degenerate
N-channel
P gates
Metal

FIGURE 8.6.4
Cross section showing the internal structure of the completed device. Notice that the geometrical proportions of the different parts are grossly distorted. The numbers within the circles refer to the number of the step at which the specific feature was fabricated in the sequence outlined in the text.

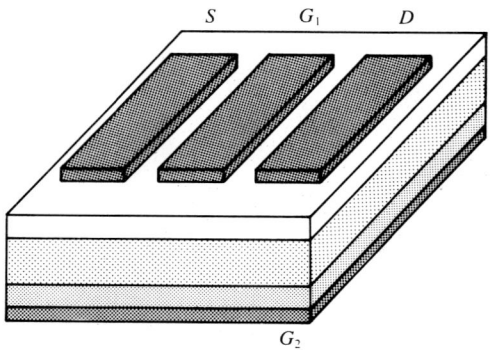

S G_1 D

G_2

FIGURE 8.6.5
Schematic three-dimensional sketch of the external structure of the completed device, showing the electrodes. Gates G_1 and G_2 are electrically connected together to form one electrode.

Several steps, as described in the above fabrication sequence, are grossly oversimplified and would prove impractical under real fabrication conditions.

In particular, step 5 seldom involves stripping of the mask oxide layer deposited in step 2. Analogously to step 5 of the PN diode fabrication sequence of Sec. 6.1, a reoxidation is often performed here without stripping the oxide. The resulting insulating oxide layer is of essentially uniform thickness for the same reasons outlined in Sec. 6.1.

If step 7 were conducted as described in the outline, the photoresist layer constituting the mask would be directly in contact with the exposed n^+ source and drain contact wells. This would seriously contaminate the semiconductor surfaces. In practice, a reoxidation is often performed at this stage, generating a continuous oxide layer of essentially uniform thickness, without any windows. By a photostep, three contact windows for source, drain, and gate are then etched in this continuous oxide layer. Notice that, in this case, mask #3 does not conform with the sketch of Fig. 8.6.1, but consists of three windows, instead of one, as etching of the source and drain windows must now be repeated.

At step 8, the bottom surface of the support wafer (G_2) is usually not metallized. Electrical connection to the p^+ material is instead obtained by bonding the bottom of the wafer to a gold-plated support (Au–Si bonding) by means of a suitable bonding agent.

Finally, Fig. 8.6.5 shows the outer appearance of the completed device with its electrodes. In actual structures, the shape of the electrodes is usually more complex to facilitate external connections. This is particularly true in integrated circuits, where the electrodes also provide the circuit connections to the other devices on the same chip.

8.7   THE MESFET

The gate of the J-FET, as previously described, is a PN junction in which variations of the bias control the width of the depletion region, thereby modulating the width of the active channel and so, ultimately, its resistance.

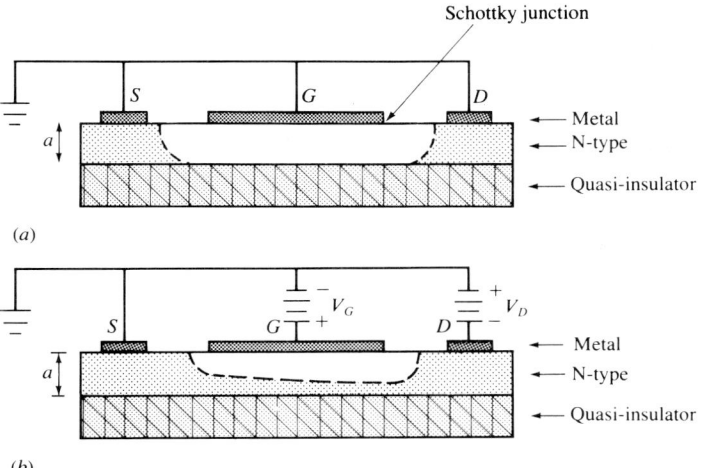

FIGURE 8.7.1

MESFET structure. The clear area is the depletion region. (a) Zero bias on the gate and drain. The MESFET is cut off, because of small channel width a and low channel doping: MESFET normally off. (b) Positive V_G. Narrower depletion region. The MESFET is not cut off. For V_D sufficiently lower than V_G, it may even not be pinched off.

In Chap. 7 it was repeatedly stated that the metal-semiconductor Schottky diode is a much faster device than an equivalent PN junction, so that faster FETs can be obtained by substituting a metal-semiconductor junction as the gate. The resulting device is a MESFET (metal-semiconductor field-effect transistor).

In order to achieve maximum speed, MESFET structures usually differ in many important details from their junction counterpart, the J-FET. These differences can best be appreciated by a comparison of a schematic representation of a typical MESFET, such as Fig. 8.7.1, with Fig. 8.2.1 showing a J-FET structure.

The gate of the MESFET of Fig. 8.7.1 consists of a single metal-semiconductor junction, instead of the two PN junctions sandwiching the channel in the J-FET of Fig. 8.2.1. The second of these PN junctions is substituted by a semi-insulating substrate, so that only one depletion region is formed in the channel. Consequently, the gate capacitance, consisting of one junction instead of the two depletion capacitances in parallel discussed in Sec. 8.5, equals about half of the J-FET input capacitance.

The MESFET channel, epitaxially grown on the semi-insulating substrate, usually consists of very high mobility material, such as a low-doping III–V, N-type semiconductor (as an example, compare the electron mobilities of GaAs with those of Si and Ge in Tables 4.2.1, 4.2.2, and 4.2.3), significantly contributing to device speed. Equation (8.5.11) shows that f_{co} increases proportionally to μ_e and so confirms this conclusion.

This epitaxial N-type layer is often extremely thin, so that, even at zero bias, the channel is pinched off, because the depletion region resulting from the

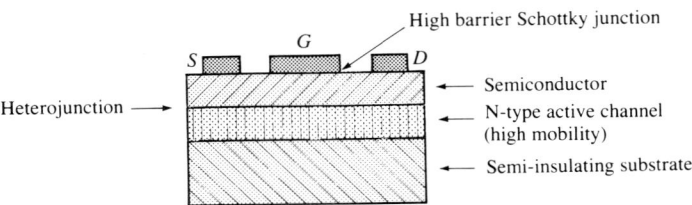

FIGURE 8.7.2
Heterojunction MESFET.

contact potential is wider than the geometric channel thickness. This is shown in Fig. 8.7.1*a*. Under these conditions the MESFET is *normally off*. To bring such a device out of cutoff, the gate must be made positive (forward bias), as shown in Fig. 8.7.1*b*.

Notice that such bias, in a J-FET, would imply the injection of minority carriers into the channel and their accumulation near the junction, with the attendant storage delay and diffusion capacitance discussed in Chap. 6. In the MESFET, the unipolar nature of the Schottky junction avoids such delays, as discussed in Sec. 7.7.

The dc gate current in MESFETs is significantly larger than in J-FETs, even at reverse bias [cf. Eq. (7.6.4)], but as this reverse current is saturated it does not influence the small-signal input impedance. However, for dc and for large swings of the input signal, the J-FET gate is usually a better approximation to an open circuit than the MESFET input.

To minimize the reverse dc gate current of the MESFET, a high Schottky barrier is desirable; however, as previously stated, the choice of channel material is usually dictated by the requirements of very high mobility, suggesting low-doping III–V semiconductors as a suitable choice. Unfortunately this choice often implies low barrier heights and so high reverse gate currents [cf. Eq. (7.6.4)]. An elegant solution is offered by the heterojunction technique.

On the usual semi-insulating substrate, a suitable layer of the high-mobility channel material is epitaxially grown; then a layer of a different semiconductor, forming a heterojunction, is grown between it and the gate electrode, as shown in Fig. 8.7.2. This new material is chosen so that it results in a high barrier Schottky junction with the gate metal. The heterojunction confines conduction electrons to the high-mobility material, which becomes the active channel. The active channel

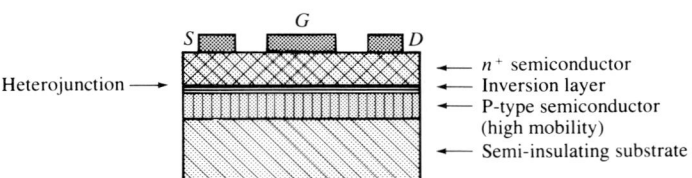

FIGURE 8.7.3
Heterojunction, inversion layer, normally on MESFET.

width can then be controlled by the gate voltage. The result is a MESFET with high Schottky barrier and high-mobility active channel material, permitting high-speed operation.

Another heterojunction MESFET device is suggested by the possibility, indicated in Sec. 6.10, of producing inversion layers by means of heterojunctions. In the device depicted in Fig. 8.7.3 such an inversion layer is created by the heterojunction in the high-mobility P-type material. This high-conductance layer constitutes the active channel, which can be modulated by the gate voltage. The result is a normally-on fast MESFET.

8.8 SUMMARY

The structure and circuit symbol of the J-FET are shown in Figs. 8.1.1 through 8.1.3. J-FETs are *unipolar devices*, they can be N- or P-channel, depending on whether conduction is mediated by electrons or holes. N-channel J-FETS can be as much as three times faster than the equivalent P-channel structures.

In normal operation, the N-channel J-FET is biased with a negative V_G and a positive V_D. The gate PN junction is therefore reverse biased and a depletion region extends into the geometric width $2a$ of the channel. The width of the active channel is consequently constricted, decreasing from the source to the drain end, so that *the channel resistance depends on both the drain and gate voltages.*

The drain current can therefore be controlled by means of the gate voltage, resulting in the possibility of *gain.*

For a given gate voltage V_G, as the drain voltage is increased, the current I_D also increases, *but not quite proportionally to the voltage applied* because of the increment in resistance.

Where the voltage across the junction is greater than the *pinchoff voltage* V_{po}, the width of the active channel goes to zero—the channel is said to be pinched off. Under these conditions the current ceases to increase with the drain voltage, but becomes essentially independent from it and equal to a constant value I_{Ds} dependent only on V_G.

The pinchoff voltage can be computed by means of Eq. (8.6.3). Before pinchoff, the dependence of the drain current from gate and drain voltages is expressed by Eq. (8.3.8). Within the pinchoff region, the saturation current is computed by Eq. (8.3.9).

In the small-signal approximation, the J-FET behaves linearly, in accordance with the linear small-signal equivalent circuit of Fig. 8.5.1, the parameters of which can be computed from the device fabrication data by Eqs. (8.5.3) and (8.5.4) for the saturated region (after pinchoff). In the linear region ($V_D \ll$) the analogous parameters are given by Eqs. 8.5.6 and 8.5.7.

The leakage resistors can in general be considered negligibly large, except when fabrication conditions produce uncommonly large surface leakage currents.

The junction capacitances are usually represented by an experimentally measured parameter: the cutoff frequency f_{co}, defined as the frequency at which the small-signal short-circuit current gain of the device becomes unity. An order-of-magnitude theoretical expression for this parameter is given by Eq. (8.5.11).

The feedback effect of the series ohmic bulk resistances $R_{SΩ}$ and $R_{DΩ}$ can be approximated by introducing the modified transfer and output admittances implicitly defined by Eq. (8.5.14). Then the simplified equivalent circuit of Fig. 8.5.2 can be used.

Substituting a metal-semiconductor junction in place of the gate PN junction of the J-FET results in a MESFET structure. Because of the characteristics of Schottky diodes and of other design features, MESFETs can be much faster devices than J-FETs.

PROBLEMS

8.1. (a) Draw the structure of a P-channel J-FET, analogous to Fig. 8.1.1, indicating the type of each semiconductor and identifying each significant part of the structure.

(b) By suitable sketches indicate and qualitatively comment on the conditions of the channel and its resistance under bias conditions equivalent to those of Fig. 8.2.1a, b, c, and d. In all cases indicate the bias polarities.

8.2. A P-channel J-FET is characterized by $N_D = 10^{19}$ cm^{-3}, $N_A = 3 \times 10^{16}$ cm^{-3}, $2a = 1$ μm. Compute the active channel width under the following conditions (if the width is not uniform, compute maximum and minimum values):

(a) $V_G = V_D = 0$
(b) $V_G = 0.5$ V, $V_D = 0$
(c) $V_G = 0.5$ V, $V_D = -2$ V
(d) $V_G = 0.5$ V, $V_D = -5$ V

8.3. Repeat Prob. 8.2 for an N-channel J-FET with $N_A = 10^{19}$ cm^{-3} and $N_D = 10^{17}$ cm^{-3}, using appropriate values for V_G and V_D equivalent to those in (a), (b), (c), and (d). Compare with the results of Prob. 8.2.

8.4. For the J-FET of Prob. 8.3:

(a) Compute the pinchoff voltage V_{po}.
(b) Assuming $V_G = -0.5$ V, compute the drain pinchoff voltage $V_{D, po}$.
(c) Compare with the results of Prob. 8.3.

8.5. For the case presented by Fig. 8.2.1d, roughly compute the distance PM.

Hint: What is the difference of potential between P and M?

8.6. Find an expression for the value of V_G that cuts off a J-FET irrespective of V_D.

8.7. A J-FET has the following data: $T = 300$ K, $N_A = 10^{19}$ cm^{-3}, $N_D = 8 \times 10^{15}$ cm^{-3}, $z = 1$ mm, $a = 0.75$ μm, $L = 20$ μm.

(a) What is the gate cutoff voltage?
(b) Tabulate the drain current for V_D varying from 0 to 3 V in steps of 0.5 V and $V_G = 0, -1, -2, -3$ V.
(c) For the given gate voltages compare the predictions of (8.3.9) with those of (8.3.10).

8.8. Qualitatively predict what happens to V_{bi}, V_{po}, $V_{G, co}$, $V_{D, po}$, and I_{Ds}, as computed in Prob. 8.7, if:

(a) N_D is increased.
(b) N_A is increased.
(c) a is decreased.
(d) L is increased.

8.9. What is the physical meaning of the constant G?

8.10. All formulas presented are subject to the long-channel assumption. Discuss what simplifying consequences of this assumption were used in deriving the formulas and what simplifications they permitted. Consider the effect of abandoning the assumption. What mathematical problems would arise?

8.11. The N-channel J-FET of Fig. P8.11 has the design parameters of the device of Prob. 8.7.

(a) Determine all currents and voltages at the dc work point.

Hint: Assume pinchoff operation and then verify your assumption.

(b) What is the ac signal across the 1000 Ω resistor?

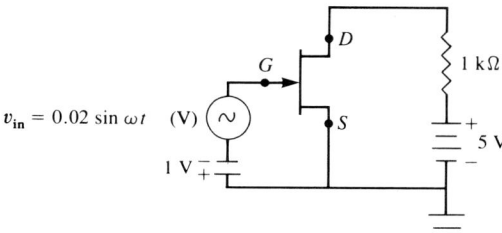

FIGURE P8.11

8.12. The N-channel J-FET of Fig. P8.12 has the following data: $T = 300$ K, $N_A = 2 \times 10^{19}$ cm^{-3}, $N_D = 1.6 \times 10^{16}$ cm^{-3}, $a = 0.6$ μm, $z = 0.5$ mm, and $L = 25$ μm. At the desired work point $V_G = -0.5$ V and the device should be in pinchoff with a gain of 6.

(a) Compute appropriate values for the resistors.

(b) Compute the dc voltage from drain to ground and the drain to source voltage and check that the device is in pinchoff.

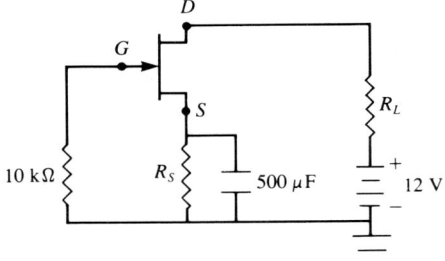

FIGURE P8.12

8.13. The J-FET of Prob. 8.12 is used in the circuit of Fig. P8.13, in which the gate bias is varied by means of the potentiometer from 0 to -2 V in steps of -0.5 V.

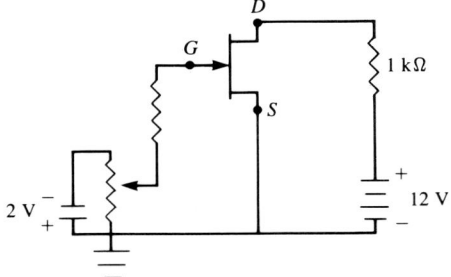

FIGURE P8.13

(a) Tabulate the small-signal gain of the stage vs. gate bias voltage.

(b) Qualitatively indicate what happens if the load is changed to 2000 Ω.

8.14. A P-type J-FET has fabrication data equivalent to those of Prob. 8.7 (but with $N_D = 10^{19}$ cm^{-3} and $N_A = 8 \times 10^{15}$ cm^{-3}). Solve Prob. 8.11 using this transistor and appropriately modifying Fig. P8.11. Compare with the results of Prob. 8.11.

8.15. Draw a small-signal linear equivalent circuit for the amplifier of Fig. P8.12, using as input a small-signal current generator between gate and ground. Use any reasonable approximation *but state it.*

8.16. For the J-FET of Prob. 8.12 operating in the circuit of Fig. P8.12:

(a) Compute the approximate cutoff frequency f_{co}.

(b) Using the experimental value published by the manufacturer of $f_{co} = 100$ MHz, compute the corresponding C_G.

Express the high-frequency 3-dB point in terms of f_{co}, assuming that the input is provided by a current generator between the gate and ground and that:

(c) C_G is all between the gate and the source.

(d) C_G is all between the gate and the drain.

8.17. For the J-FET of Prob. 8.12 compute the output resistance.

8.18. Draw a simplified small-signal linear equivalent circuit (similar to that of Fig. 8.5.2) for the amplifier of Fig. P8.12, assuming that the transistor has $R_{S\Omega} = R_{D\Omega} = 25$ Ω, $g_{ms} = 2.82$ mmho, $g_{Ds} = 2 \times 10^{-5}$ $\mho$, $C_G = 4.5$ pF, of which two-thirds appear between gate and source and one-third between gate and drain.

Computer Problems

8.19. Write a computer program to solve problems of the type of Probs. 8.2 and 8.3, interactively allowing the user to make appropriate choices of N_D, N_A, a, V_G, and V_D.

8.20. Write a computer program to solve Prob. 8.7, making provisions to allow the user interactively to modify fabrication and operation parameters.

8.21. Write a computer program to solve Prob. 8.11.

8.22. Write a computer program to solve Prob. 8.12.

8.23. Write a computer program to solve Prob. 8.13.

8.24. Write a computer program to solve Prob. 8.17, permitting appropriate interactive choices of parameters.

REFERENCES

1. Grove, A. S.: *Physics and Technology of Semiconductor Devices*, John Wiley & Sons, Inc., New York, 1967.
2. Millman, Jacob, and C. C. Halkias: *Integrated Electronics, Analog and Digital Circuits and Systems*, McGraw-Hill Book Company, Inc., New York, 1972.
3. Gray, Paul, and C. L. Searle: *Electronics Principles, Physics, Models and Circuits*, John Wiley & Sons, Inc., New York, 1969.
4. Yang, Edward S.: *Fundamentals of Semiconductor Devices*, McGraw-Hill Book Company, Inc., New York, 1978.

ADDITIONAL READING

Shockley, William: "A Unipolar Field Effect Transistor," *Proc. IEEE*, vol. 54, p. 307, 1966.
Sze, S. M.: *Physics of Semiconductor Devices*, John Wiley & Sons, Inc., New York, 1982.

Sze, S. M.: *Semiconductor Devices, Physics and Technology*, John Wiley & Sons, Inc., New York, 1985.

Todd C. D.: *Junction Field Effect Transistors*, John Wiley & Sons, Inc., New York, 1968.

Yang, Edward S.: "Current Saturation Mechanisms in J-FET," *Adv. Electr. Phys.*, vol. 31, p. 247, 1972.

CHAPTER
9

THE MOSFET

9.1 MOSFET STRUCTURE

There are two types of metal oxide–semiconductor field-effect transistors (MOSFET): the *enhancement* and the *depletion MOSFET*. Each of these types can be *P- or N-channel* (P-MOS or N-MOS). Figure 9.1.1 shows the circuit diagram symbols of an N-MOS and a P-MOS; the difference between them is indicated by the direction of the arrow. The device has four electrodes: source (*S*), drain (*D*), gate (*G*), and substrate (*B*). In operation the substrate is almost always directly connected to the source.

The basic structure of a P-channel MOSFET is shown in Fig. 9.1.2. The P-MOS and N-MOS structures are *complementary*. In a P-MOS the substrate is N-type and the source and drain are P-type, as shown in Fig. 9.1.2, whereas in an N-MOS the substrate is P-type and the source and drain are N-type. The structure of an N-MOS is shown in Fig. 9.4.1.

N-channel MOSFETs can be as much as three times faster than equivalent P-channel devices and are often preferred, especially for high-speed applications. Their fabrication, however, presents some difficulties, so that, historically, the first MOSFETs to be fabricated were of the P-channel type.

A somewhat simplified sequence of fabrication steps will now be described. This outline refers to P-channel MOSFETs, because of their simpler fabrication cycle. The fabrication of N-channel MOSFETs will be illustrated later on, with the introduction of CMOS and polysilicon gate technology (cf. Secs. 9.7 and 9.8), because of the significant role played by these technologies in the development of successful techniques for the production of N-channel MOSFETs.

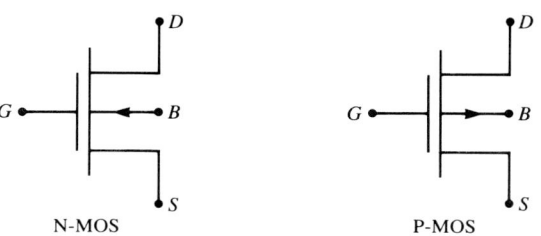

FIGURE 9.1.1
Circuit symbols for N-MOS and P-MOS showing source (S), gate (G), drain (D), and substrate (B) terminals. The device type is designated by the orientation of the arrows in the substrate connection.

In following the fabrication outline, frequent reference should be made to Fig. 9.1.2, where the numbers within the circles refer to the step number at which the indicated structural element was fabricated.

The state of the device after the most important steps of its fabrication sequence is illustrated in Fig. 9.1.4. Notice that, in all these representations, to make the figures legible, the geometric proportions of the drawings are grossly distorted and do not accurately represent the relative dimensions of the different structural elements. The set of masks used in the process is shown, in a highly schematized form, in Fig. 9.1.3, assuming that the photoresist used in each photo-step is of the negative type.

Fabrication Outline

1. The surface of an N-type wafer (the substrate) is oxidized, usually by the wet process. This oxide layer will be used not only during fabrication, as a mask for diffusion processes (masking oxide), but will also remain, as *field oxide*, in the finished product. It must therefore be thick enough to act as an efficient diffusion mask (cf. Secs. 5.4 and 5.5), but it must also be thin and pliable enough not to cause reliability problems. Its thickness is therefore chosen by compromise between these conflicting requirements.

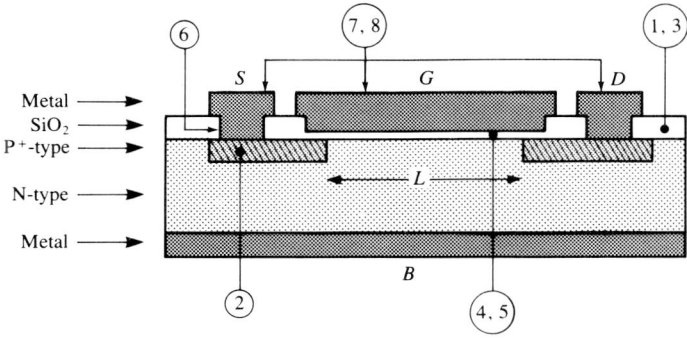

FIGURE 9.1.2
Schematic representation of a P-channel MOS structure. The relative dimensions of the elements of the structure are grossly distorted. The numbers enclosed in the circles refer to the step numbers of the fabrication outline in the text.

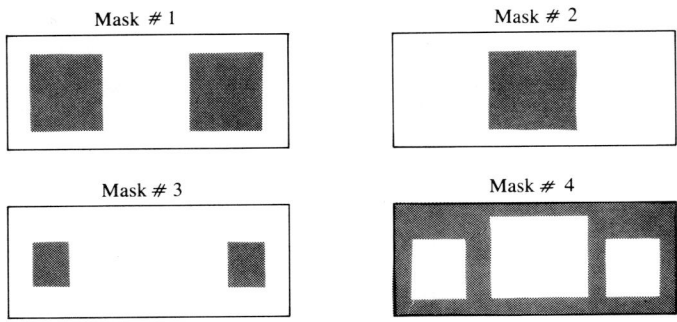

FIGURE 9.1.3
Masks used in the fabrication sequence of Sec. 9.1. Actual production masks are usually of a much more complex design.

2. By a photostep, using mask #1 and subsequent etching, two windows are opened in the oxide layer, exposing two areas of the underlying semiconductor surface. Then, through the windows, a p^+ layer is diffused into each of these areas of the substrate. These p^+ regions will later become the source and drain of the MOSFET. The distance between these two layers is an important parameter of the device. Figure 9.1.4*a* shows the structure of the device after this step. Notice the darker p^+ material showing through the windows and the lighter N-type material of the substrate.

   In Fig. 9.1.4*b* the device has been cut so that the internal structure can be observed in the cross section. The important distance between the darker p^+ source and drain regions is designated by L. Notice that the p^+ material also diffuses laterally over a short length underneath the edges of the mask oxide.

3. The device is now submitted to a second oxidation process resulting in an oxide layer that covers the entire surface. Although this oxidation is performed without stripping off the existing mask oxide, the thickness of the resulting *field oxide* is essentially uniform, because of the faster oxidation rate in the regions where the substrate is exposed (cf. Secs. 5.3 and 6.1).

4. By a photostep, using mask #2, a well is etched in the oxide between the source and drain regions. It is of fundamental importance for the future operation of the device that this well should expose not only all of the length L of the substrate between the source and drain but also some portion of these two regions, as shown in cross section in Fig. 9.1.4*c*.

5. Next, a thin film of high-quality oxide of accurately controlled thickness x_o (the *gate oxide*) is grown over the semiconductor surfaces exposed by this well. Because of the strict quality and dimensional requirements, the dry process is almost universally used in this step (cf. Sec. 5.3). This process also negligibly increases the thickness of the field oxide. The device structure at this stage of fabrication is shown by the cross section of Fig. 9.1.4*d*. The thin gate oxide is seen to overlap part of both the source and drain regions, as was also shown in Fig. 9.1.2.

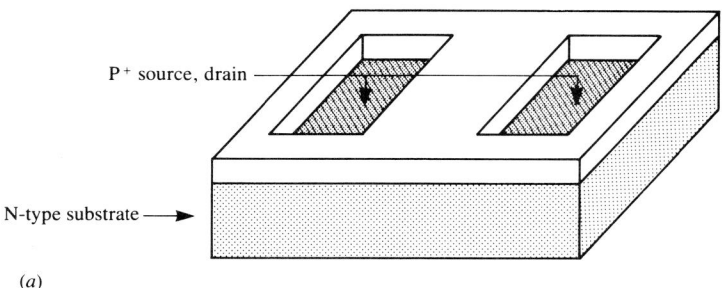

(a)

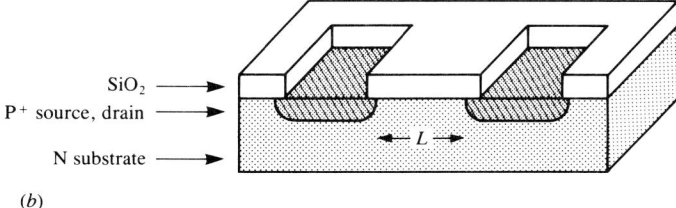

(b)

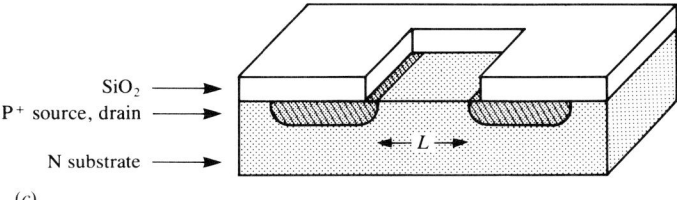

(c)

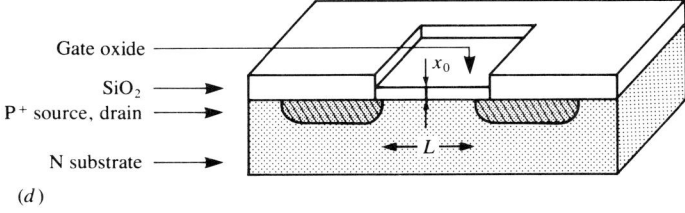

(d)

FIGURE 9.1.4

Structure of the device at different stages of the fabrication process. (a) Three-dimensional view of the device after step 2. Drain and source windows expose the darker underlying p^+ regions diffused into the substrate. (b) Cross section after step 2, showing the distance L between the darker p^+ source and drain regions. (c) Cross section after step 4. The window exposes the edges of the p^+ source and drain regions and the substrate of length L between them. (d) Cross section after step 5. The gate oxide of thickness x_0 covers the channel region and the edges of the source and drain semiconductors. (e) After step 6 the p^+ semiconductors are exposed at the source and drain locations. The device is ready for metallization. (f) Three-dimensional view of the completed device showing the electrodes. Often the substrate is electrically shorted to the source. (g) Cross section of the completed device.

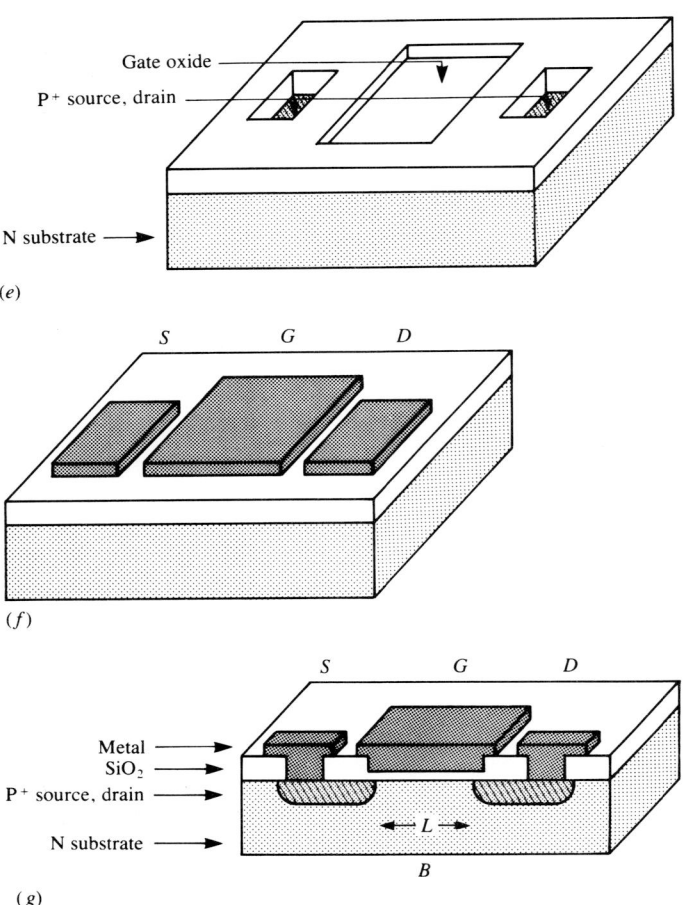

FIGURE 9.1.4
Continued

6. Using mask #3 and a photostep, two windows are etched in the field oxide to expose the two p^+ regions (the source and the drain), as shown in Fig. 9.1.4e, so that ohmic contacts with these semiconductors can be fabricated.

7. To obtain these contacts, the whole wafer surface is covered with a metallic layer by a metallization process step.

8. Using mask #4 and a final photostep, the metal layer is etched to form the source, drain, and gate electrodes. A three-dimensional view of the completed device is shown in Fig. 9.1.4f. Notice that, at this time, the substrate is often electrically connected to the source terminal.

   The completed device structure is displayed in cross section in Fig. 9.1.4g, where some of the most important structural components are indicated by the arrows.

The above sequence of steps and the various parts of Fig. 9.1.4 should be compared with Fig. 9.1.2 to obtain a clear idea of the actual sequence of operations and of the shape and location of the different masks. Particular notice should be taken of the strict requirements for registration of the sequence of masks, especially the very critical mask #2. Many variations and refinements of this fabrication sequence are used in practice.

It should be mentioned here that a technological breakthrough of particular importance in MOSFET fabrication was achieved with the introduction of the polysilicon gate technology, in which layers of doped, conductive polycrystalline silicon (polysilicon) are used for the gate electrode instead of metal. This technique, and the advantages offered by the polysilicon electrode, will be discussed in Sec. 9.8.

As for the mechanism of operation of the device, examining the structure of Fig. 9.1.2, it is noticed that, between the source and drain terminals, two p^+ regions are separated by an N-type semiconductor, resulting in *two back-to-back diodes*. It is therefore evident that, if a voltage is applied between these electrodes, *no current can flow*.

In the following it will be shown that application of proper gate bias will permit current flow through a narrow *channel* of length L, where the electric field generated by the gate potential *locally turns the N semiconductor into a P-type* (field effect), providing an uninterrupted P-type path between the source and the drain.

9.2 MOS CAPACITOR BAND DIAGRAMS

Figure 9.2.1 shows the energy band diagrams of a metal oxide–semiconductor (MOS) structure. It should be clear that the MOS structure comprises only a part

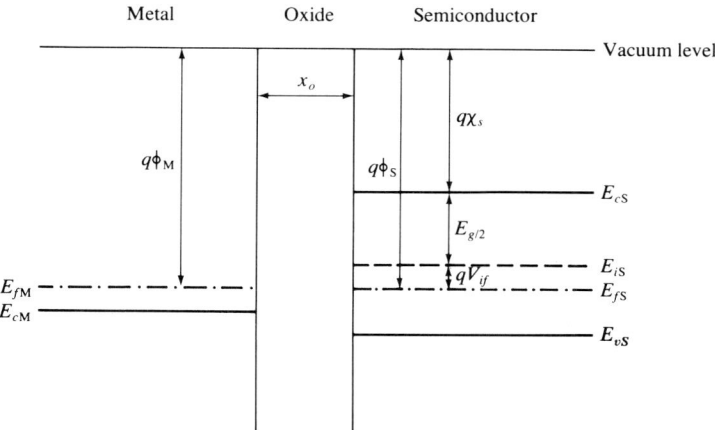

FIGURE 9.2.1
MOS capacitor unbiased band diagram. The semiconductor is assumed to be P-type with $\phi_S = \phi_M$.

of the complete MOSFET: the source and drain regions are missing and only the capacitor-like portion under the gate electrode is considered.

In the description of Sec. 9.1, the substrate was assumed to be N-type, because the fabrication of devices of this type (corresponding to P-channel MOSFETs) is simpler. In the following analysis, instead, the substrate is assumed to be a P-type semiconductor.[1] For the moment, to simplify the discussion, it is assumed that $\phi_M = \phi_S$, i.e., that the Fermi levels are aligned when the semiconductor and the metal are at the same potential as shown in Fig. 9.2.1. This simplifying assumption will be removed later on (cf. Sec. 9.6).

The figure also displays the voltage V_{if}, measuring, in electronvolts, the important energy difference $E_i - E_f$ in the bulk of the P-type semiconductor. This quantity has been shown [see Eq. (3.4.4)] to determine the carrier equilibrium concentrations. For a given impurity concentration, it can be computed using (3.4.9).

The structure evidently constitutes a capacitor, with a SiO_2 dielectric of thickness x_o, while, of the two armatures, one is metallic as usual but the other is made of a semiconductor material.

Because of this semiconductor armature the MOS capacitor exhibits a peculiar behavior under bias. In order to analyze this behavior it is expedient to define two quantities:

1. The *oxide capacitance per unit area* of interface surface:

$$C_o = \frac{\varepsilon_{ox}}{x_o} \tag{9.2.1}$$

is expressed in farads per square centimeter, where ε_{ox} is the permittivity of the dielectric ($\varepsilon_{ox} = 3.9\varepsilon_o = 3.45 \times 10^{-13}$ F/cm for SiO_2; $\varepsilon_{ox} = 6\varepsilon_o = 5.3 \times 10^{-13}$ F/cm for $Si_3 N_4$).

2. The *charge per unit area* of interface surface: Q_s is the total net charge stored in the semiconductor per square centimeter of oxide-semiconductor interface.

During device operation, voltages are applied across the MOS capacitor. As already remarked, because one of the MOS capacitor plates is made of semiconductor material, the MOS structure's behavior differs from that of a conventional capacitor with metallic armatures and depends on the polarity and magnitude of the applied voltage.

Three conditions of operation can be observed.

---

[1] This corresponds to an N-channel MOSFET structure. As already stated, the N-MOS, because of its superior speed, is particularly interesting in practice and because of this we shall analyze its behavior in preference to the P-channel device.

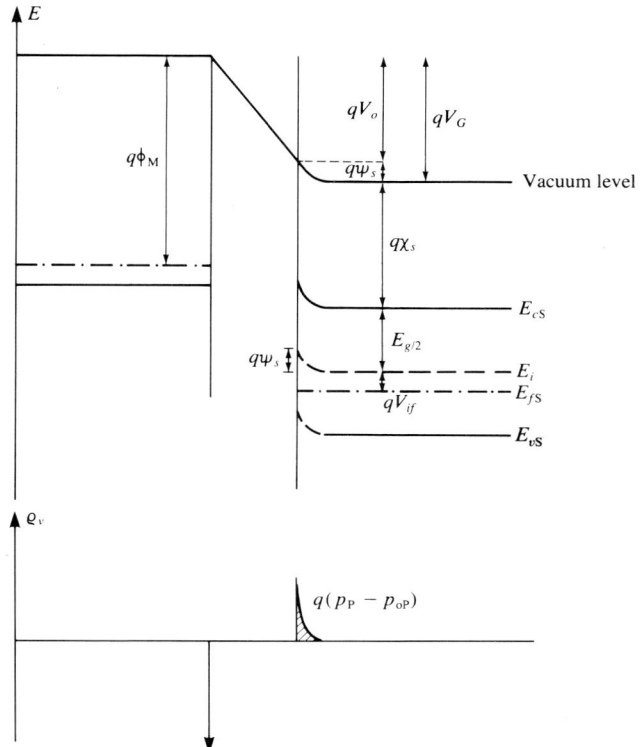

FIGURE 9.2.2
MOS capacitor under negative bias. Enhancement mode. Accumulation of majority carriers near the oxide interface enhances the extrinsic nature of the semiconductor within this layer.

Accumulation (or Enhancement)

A bias voltage V_G making *the metal negative with respect to the semiconductor* is applied between the gate and the source of the MOS structure. By convention, when the bias voltage has this polarity, V_G is assigned the negative sign.

In order to establish a voltage across the dielectric, the two armatures must be charged, so a transient of current is drawn from the bias generator. In the semiconductor, this current produces an accumulation of holes near the oxide interface.[2] The resultant net charge density distribution ρ_v is plotted in Fig. 9.2.2, showing that charges have piled up near the oxide interface (positive charges in the semiconductor, negative on the metal side). In the metal, of course, the charge

---

[2] From a different, but useful, point of view, the electrostatic field resulting from the application of the bias attracts the semiconductor majority carriers toward the oxide-semiconductor interface, creating the accumulation condition.

is all at the surface, because no electric field can be set up within the almost perfect conductor. However, in the semiconductor the charge is distributed over a finite depth.

In accordance with Poisson's theorem, this net charge distribution ρ_v results in a voltage drop ψ_s within a finite layer of the semiconductor near the oxide interface. Because the charge density can be quite high in this case, this layer is rather thin.

The band diagram of the MOS structure, as usual, displays potential distributions in terms of a displacement and distortion of the plot. Figure 9.2.2 shows that most, but not all, of the bias voltage (V_G) appears across the dielectric (V_o), and only a small voltage ψ_s is dropped near the semiconductor surface. Within this layer the hole concentration p_P is greater than the thermal equilibrium value. In the notation of Chap. 4, $p_P > p_{oP}$, so that, thanks to (3.4.4), $E_i - E_f$ must increase in this region. This can be observed in Fig. 9.2.2, where all the lines representing the characteristic levels are bent but *the Fermi level line remains horizontal*. This is necessary, because, at steady state, no current can flow in the MOS structure due to the presence of the dielectric. Because of the distortion of the energy levels, then, near the interface, $E_i - E_f > qV_{if}$. We conclude that, because of the applied voltage, a thin layer of the semiconductor near the oxide interface behaves as if it were more heavily doped. In this region, the extrinsic P-type nature of the semiconductor is *enhanced*.

Depletion

If the bias voltage makes the *metal positive*, then in accordance with the convention adopted, $V_G > 0$. Now the charging current transient is in the opposite direction and holes move away from the oxide-semiconductor junction, so that a depletion region is created.

The charge density $-qN_A$ in the depletion region is plotted in Fig. 9.2.3 and, in accordance with Poisson's law, results in the potential distribution shown by the distorted band energy diagrams of the same figure, indicating that the semiconductor surface at the oxide interface is at the *surface potential ψ_s*.[3] This happens because a portion of the bias voltage V_G is dropped across the depletion region of width w. Because in the depletion region the charge density cannot be greater than $-qN_A$, the width of this region is not negligible. This width can be computed from (6.3.16) as

$$w = \sqrt{\frac{2\varepsilon}{qN_A}}\,\psi_s \qquad (9.2.2)$$

---

[3] The potential of the bulk of the semiconductor (corresponding to the substrate electrode, which, as stated, is usually shorted to the source) is taken as the reference (zero) potential.

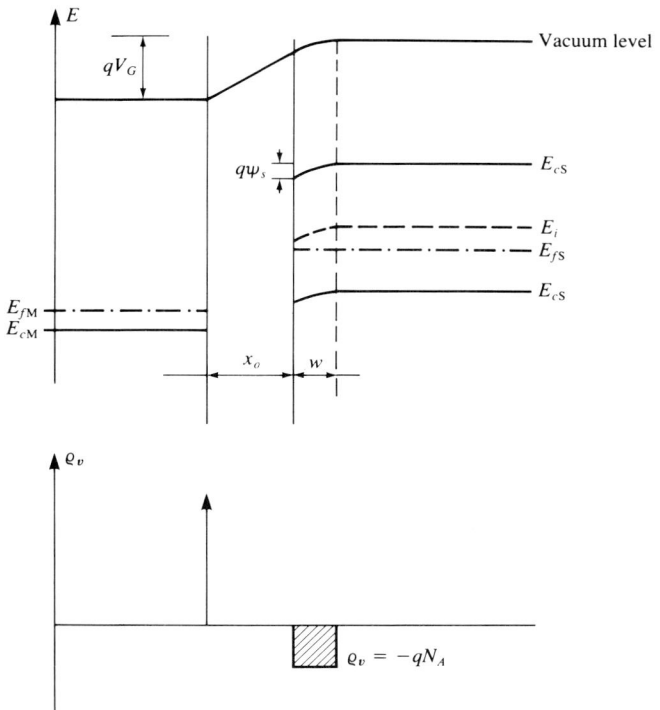

FIGURE 9.2.3
MOS capacitor under small positive bias. Depletion mode. The voltage ψ_s is dropped across a depletion region with charge density $\rho_v = -qN_A$.

The charge stored per unit interface area is consequently

$$Q_s = -qN_A w = Q_B \qquad (9.2.3)$$

where the symbol Q_B indicates that this charge, being stored in the depletion region, is *bound* to the lattice.

Again, the decreased carrier concentration near the oxide-semiconductor interface corresponds to a decrement of $E_i - E_f$, in accord with (3.4.4). This can be observed in Fig. 9.2.3.

Inversion

If the positive bias is increased, the surface potential ψ_s increases and the band energy diagram becomes more and more distorted. When ψ_s becomes so large that, in the distorted graph, the intrinsic Fermi level line bends *below the Fermi level* E_f, as shown in Fig. 9.2.4, then, very close to the oxide-semiconductor interface, $E_i - E_f < 0$ and, as shown by Eqs. (3.4.3) and (3.4.4), in that region the P semiconductor becomes N-type. An N-type *inversion layer* is formed in the semi-

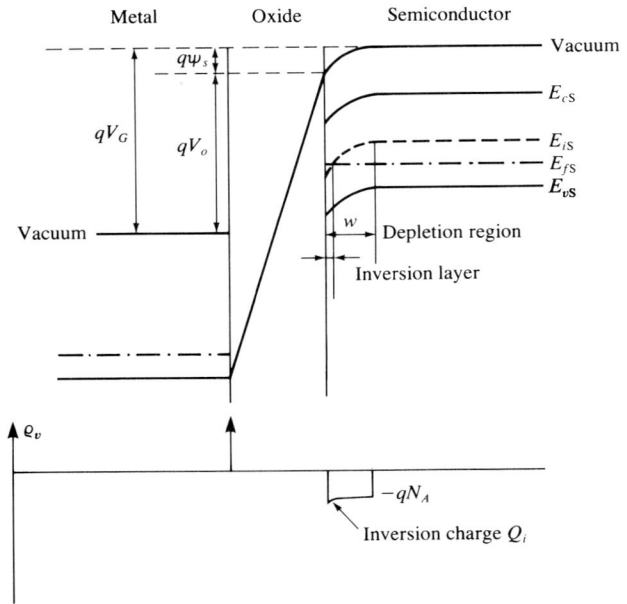

FIGURE 9.2.4
MOS capacitor biased at the onset of weak inversion. $E_i - E_f < 0$ next to the oxide interface, where the semiconductor becomes N-type.

conductor in the immediate vicinity of the oxide interface. This *inversion of the semiconductor type* occurs when $\psi_s > V_{if}$.

To understand by what mechanism inversion can occur, consider the electron-hole pairs continuously being generated by thermal energy within the semiconductor. Under thermal equilibrium conditions, of course, they move around at random and, after some time, recombine and disappear. If, however, they are generated within the depletion region, then the electric field sweeps them away at high speed and in opposite directions. When the field is strong enough, the electrons will be swept all the way to the edge of the oxide layer before they have a chance to recombine with any of the holes, which, instead, are swept away from the oxide layer into the bulk of the semiconductor. This accumulation of electrons near the oxide results in a net *inversion charge* Q_i per unit area of interface and is responsible for the inversion of the semiconductor type.

The net charge density distribution at the onset of the inversion phenomenon is plotted in Fig. 9.2.4. Over most of the depletion region the charge density is $-qN_A$, corresponding to the density of the immovable impurity ion charges. In the immediate vicinity of the oxide interface, however, the pileup of free electrons increases the charge density above the depletion value over a very thin inversion layer, constituting the mobile inversion charge Q_i. To understand the function of the inversion layer in MOSFET operation, it must be borne in mind that the inversion electrons are free conduction electrons, so that the inversion layer is a conductive N-type semiconductor.

As long as $\psi_s - V_{if}$ is small, the electron concentration in the inversion layer is also small [cf. Eq. (3.4.3)] and conductivity is correspondingly small, but if ψ_s is increased to the point where $\psi_s - V_{if} \approx V_{if}$, then (3.4.3) shows that the concentration of electrons in the inverted region is comparable to that of the holes in the bulk of the semiconductor. Remembering (4.2.8), this shows that the conductivity of the inverted layer is significant with respect to the conductivity of the bulk semiconductor. This condition is designated as *strong inversion* and is depicted in Fig. 9.2.5.

When the gate bias is large enough to generate inversion, then, in Fig. 9.1.2, there appears an inverted *channel* at the semiconductor surface immediately below the gate insulation oxide. The semiconductor in this channel behaves as N-type, providing an uninterrupted N-type semiconductor path from the source to the drain, so that the back-to-back PN junctions disappear along this path and conduction becomes possible. In the following, the inverted channel conductivity will be considered negligible below the strong inversion limit. *Inversion will be assumed to be synonymous with strong inversion* and will be said to occur when appreciable conduction starts, i.e., when

$$\psi_s \geq \psi_{si} = 2V_{if} \tag{9.2.4}$$

in accordance with the previous definition of the onset of strong inversion.

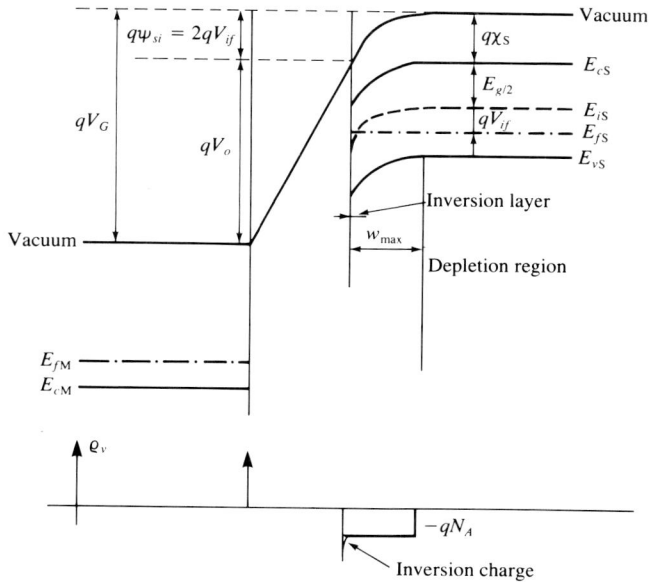

FIGURE 9.2.5
MOS capacitor biased at the onset of strong inversion. Near the interface $E_i - E_f = V_{if}$ and the conductivity of the inversion layer is comparable to that of the bulk semiconductor, $\psi_s = \psi_{si} = 2V_{if}$.

As shown in Figs. 9.2.4 and 9.2.5, the net charge density in most of the depletion region is due, as usual, to the charges of the ionized impurity atoms and is equal to $-qN_A$. However, in the narrow inversion layer in the immediate vicinity of the oxide interface, charge density is strongly increased by the presence of the inversion electrons.

The concentration of these inversion electrons follows (3.4.7), in which, however, $E_f - E_i$ must correspond to the conditions prevailing in the inversion layer (depicted in Fig. 9.2.5), yielding

$$n_\mathrm{P} = n_i\, e^{q(\psi_s - V_{if})/(kT)} \geq n_i\, e^{q(\psi_{si} - V_{if})/(kT)} = p_\mathrm{oP} = N_A \qquad (9.2.5)$$

At this point the student is advised to compare the various equalities and inequalities expressed by Eq. (9.2.5) with the definitions of ψ_s, ψ_{si}, V_{if}, and strong inversion introduced in this section and to verify and clarify their meaning and validity. This clarification can be helped by observation of Fig. 9.2.5 where some of these quantities are indicated on the energy band diagram.

If now the bias voltage is increased beyond the limit that generates strong inversion, then two phenomena occur:

1. The depletion region width (and with it the bound charge Q_B stored in the impurity ions) tends to increase with $\psi_s^{1/2}$ [cf. (6.3.19)].

2. The charge Q_i stored in the electron inversion layer increases *exponentially* with ψ_s, in accordance with Eq. (9.2.5).

The result is represented in Fig. 9.2.6. The *increment in stored charge necessary to balance the increased bias is practically all stored in the inversion layer.* Indeed:

1. The very large increment in Q_i requires only a negligibly small variation of ψ_s, which therefore *remains essentially unchanged* at $\psi_s \approx \psi_{si} = 2V_{if}$. Consequently:

2. The extension of the depletion region width is negligible and w retains the value:

$$w_{\max} = \sqrt{\frac{2\varepsilon}{qN_A}\, 2V_{if}} \qquad (9.2.6)$$

so that:

3. The bound ionic charge stored in the depletion region remains practically constant at $Q_B = -qN_A w_{\max}$ [cf. Eq. (9.2.3)]. The graphs of charge density and potential distribution at the onset of strong inversion (Fig. 9.2.5) should be compared with the corresponding graphs of Fig. 9.2.6, depicting conditions well beyond the strong inversion limit, and the relationships among ψ_s, ψ_{si}, w, $w_{\max}$, and Q_i should be observed and logically justified.

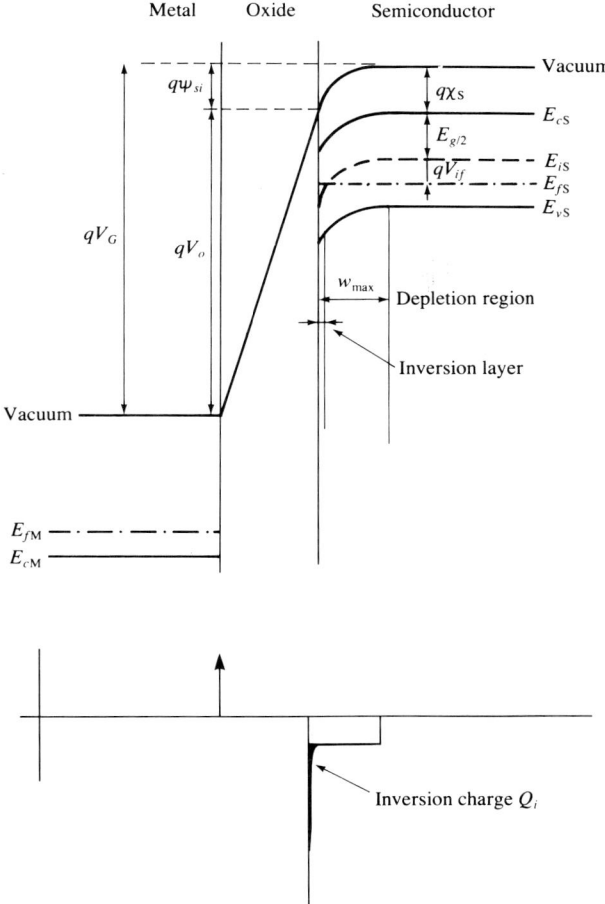

FIGURE 9.2.6

MOS biased beyond the strong inversion limit. The negligible variation of ψ_s beyond ψ_{si} results in a negligible increment of w above w_{max}, but in a large increment of Q_i in Fig. 9.2.5.

Summary

In conclusion, the phenomena accompanying the biasing of the MOS capacitor can be summarized as follows:

1. Negative bias ($V_G < 0$). Majority carriers accumulate near the oxide-semiconductor interface. This results in an *enhancement* of the P nature of the semiconductor at the interface. No appreciable conduction can occur between the source and drain in the MOSFET (cf. Fig. 9.2.2).

2. Low positive bias ($\psi_s < \psi_{si}$). A *depletion* region is formed. There is no appreciable conduction in the MOSFET (cf. Fig. 9.2.3).

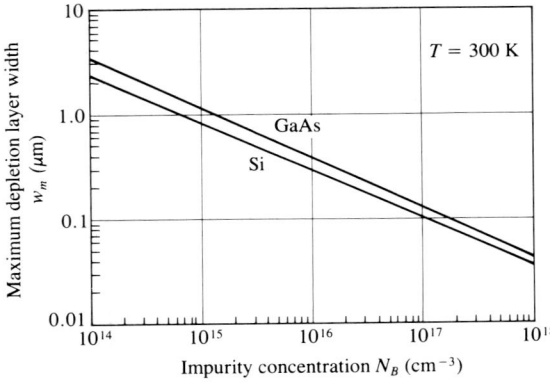

FIGURE 9.2.7
Plot of depletion region width at strong inversion (w_{max}) vs. impurity concentration for Si and GaAs semiconductors [1].

3. Large positive bias ($\psi_s \geq \psi_{si}$). An *inversion* layer is generated. This *N-type channel* permits significant conduction in the MOSFET (cf. Figs. 9.2.5 and 9.2.6).

Once inversion has been achieved, if the bias is further increased, the inversion charge per unit interface area Q_i increases rapidly, and the voltage V_o across the gate oxide layer increases with it, as we shall soon see. Instead, the depletion region width (and with it the bound charge Q_B and the surface potential ψ_s) do not appreciably change any more with increments of bias and remain at essentially constant maximum values w_{max}, Q_B, and ψ_{si} respectively. Figure 9.2.7 plots the maximum depletion region width w_{max} for Si and GaAs as a function of impurity concentration N_A.

The condition $\psi_s \geq \psi_{si} = 2V_{if}$ relates the onset of inversion to the surface potential ψ_s. However, determining and measuring ψ_s may pose some difficulty. The threshold of inversion should be expressed in terms of a more accessible quantity, such as the bias voltage V_G. From Fig. 9.2.4, under all conditions,

$$V_G = V_o + \psi_s \tag{9.2.7}$$

where V_o is the voltage across the dielectric. This in turn is related to the total stored charge per unit surface $Q_s = Q_B + Q_i$:

$$V_o = -\frac{Q_s}{C_o} = -\frac{Q_B + Q_i}{C_o} \tag{9.2.8}$$

where C_o is the oxide capacitance per unit area of oxide interface, defined in Eq. (9.2.1) and the negative sign is due to the convention adopted for the bias sign (+ for positive metal). However, at the onset of inversion, Q_i is as yet negligible so, from (9.2.6) and (9.2.7), for inversion to occur the bias voltage must at least equal the *threshold value* for inversion V_{th}:

$$V_G = V_{th} = -\frac{Q_B}{C_o} + 2V_{if} \tag{9.2.9}$$

where, remembering (9.2.6), Q_B is

$$Q_B = -qN_A w_{max} = -\sqrt{2\varepsilon q N_A (2V_{if})} \qquad (9.2.10)$$

Example 9.2.1. A Si MOS structure at 300 K is characterized by $N_A = 4 \times 10^{14}$ cm^{-3}, $x_o = 0.1$ μm. The assumption $\phi_M = \phi_S$ holds true. If the metal is brought to a potential of $+1$ V with respect to the semiconductor, (a) does inversion occur? (b) Compute V_o.

Solution. From (3.4.9),

$$V_{if} = 0.0259 \ln \left(\frac{4 \times 10^{14}}{1.5 \times 10^{10}} \right) = 0.264 \text{ V}$$

From (9.2.6) or Fig. 9.2.7,

$$w_{max} = \sqrt{\frac{2 \times 1.04 \times 10^{-12}}{1.6 \times 10^{-19} \times 4 \times 10^{14}} \, 0.528} = 1.3 \times 10^{-4} \text{ cm}$$

From (9.2.3),

$$Q_B = -1.6 \times 10^{-19} \times 4 \times 10^{14} \times 1.3 \times 10^{-4} = -8.39 \times 10^{-9} \text{ C/cm}^2$$

From (9.2.1),

$$C_o = 3.45 \times 10^{-13}/10^{-5} = 3.45 \times 10^{-8} \text{ F/cm}^2$$

so that finally, from (9.2.9),

$$V_{th} = \left(\frac{8.39 \times 10^{-9}}{3.45 \times 10^{-8}} \right) + 0.528 = 0.77 \text{ V}$$

(a) As $V_{th} = 0.77 < 1$: yes, inversion occurs.
(b) From (9.2.7), remembering that, at inversion, $\psi_s = 2V_{if}$, then, in our case,

$$V_o = 1 - 0.528 = 0.472 \text{ V}$$

There now remains to analyze quantitatively how this behavior of the MOS structure affects the performance of the MOSFET device. Two important characteristics will be considered: the gate capacitance and the drain current.

9.3 QUANTITATIVE ANALYSIS OF THE MOS GATE CAPACITANCE

As discussed in the previous section, application of a bias to the MOS structure is accompanied by the storage of charges at or near the structure interfaces. This, of course, implies that a capacitance is associated to the structure. Its value depends on the mode of operation.

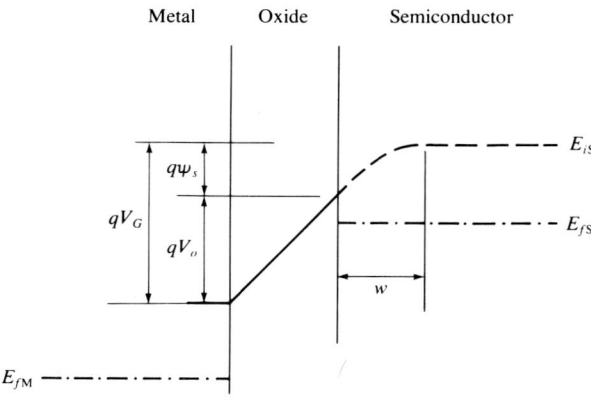

Metal   Oxide   Semiconductor

FIGURE 9.3.1
Potential distribution and relationships among pertinent voltages and energy levels in MOS capacitor structure biased in the depletion mode.

Accumulation

In the majority carrier accumulation mode (negative gate bias), the charge is stored at the oxide-semiconductor interface, so the associated *capacitance per unit interface area* equals the oxide capacitance given by (9.2.1). This capacitance does not vary with the gate bias voltage.

Depletion

The potential distribution in the MOS structure in the depletion mode is shown in Fig. 9.3.1. The structure is seen to be equivalent to the series connection of two capacitors: the oxide capacitor [cf. Eq. (9.2.1)] and the depletion region capacitor:

$$C_d = \frac{\varepsilon}{w} \tag{9.3.1}$$

where ε is the permittivity of the semiconductor material and w the width of the depletion region. From (9.2.7), using (9.2.8) and remembering that, in the depletion mode, $Q_s = Q_B = -qN_A w$ and that $\psi_s = qN_A w^2/(2\varepsilon)$ [in accordance with (6.3.16)], then

$$\frac{qN_A}{2\varepsilon} w^2 + \frac{qN_A}{C_o} w = V_G \tag{9.3.2}$$

which yields the depletion region width corresponding to the bias V_G:

$$w = \frac{\varepsilon}{C_o}\left(-1 + \sqrt{1 + \frac{2C_o^2}{qN_A\varepsilon} V_G}\right) \tag{9.3.3}$$

Using this value in (9.3.1), the series connection of C_o and C_d becomes (cf. Prob. 9.9)

$$C = \frac{C_o}{\sqrt{1 + 2(C_o^2/qN_A\varepsilon)V_G}} \qquad (9.3.4)$$

and the MOS system capacitance in the depletion mode is seen to vary with the gate bias voltage V_G.

Inversion

As noticed in Sec. 9.2, in the inversion mode, the width of the depletion region remains at the constant value w_{max} given by (9.2.6). Using this value in (9.3.1) and computing the structure's capacitance per unit surface as the series connection of C_o and C_d, then, at inversion,

$$C = \frac{C_o}{1 + C_o\sqrt{(2/qN_A\varepsilon)2V_{if}}} \qquad (9.3.5)$$

a constant capacitance per unit surface. Notice that this value implicitly supposes that any charge added across the structure is deposited at the edge of the depletion region. This contradicts the statement of Sec. 9.2, according to which, at inversion, any additional charge is stored in the inversion layer *next to the oxide-semiconductor interface*.

Actually two cases must be considered:

1. When the signal applied to the gate is of *low frequency*, the charge deposited has time to migrate through the depletion region and settle in the immediate neighborhood of the oxide-semiconductor interface. The structure's capacitance then equals C_o.
2. If the applied signal is a *high-frequency* one, then, during one cycle, there is not enough time for the charge to migrate and the extra charge remains at the edge of the depletion region, resulting in the capacitance per unit interface area, expressed by (9.3.5).

In the inversion mode, therefore, the gate capacitance does not depend on the bias voltage V_G, but may vary with the frequency of the applied signal.

In conclusion, the MOS structure capacitance varies with the bias V_G as shown in Fig. 9.3.2, where the solid lines follow the predictions of Eqs. (9.2.1), (9.3.4), and (9.3.5). In the light of the above theory, it is easy to identify the three regions of operation analyzed in the text and described by the three equations. The dots represent computer-generated capacitance values, following a slightly more sophisticated theory [2]. Measured small-signal capacitance values as a function of bias V_G are plotted in Fig. 9.3.3 for several different frequencies of the ac signal. In the depletion and inversion modes, because of the frequency dependence of the capacitance, the graph splits into a family of curves for different signal frequencies.

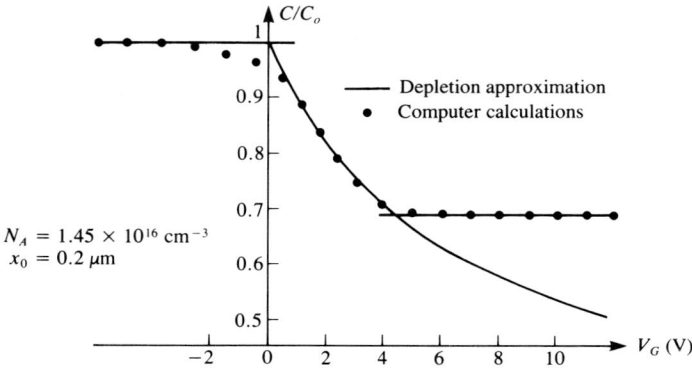

FIGURE 9.3.2
Theoretically computed dependence of the MOS capacitance vs. bias voltage V_G in a metal–SiO$_2$–P-type semiconductor structure [2].

Example 9.3.1. Compute the capacitance per square centimeter of the MOS structure of Example 9.2.1 under a bias of (a) -1 V; (b) 0.5 V; (c) $+1$ V.

Solution

(a) For $V_G = -1 < 0$, the structure is in the *accumulation mode*, so from Example 9.2.1,

$$C = C_o = 34.5 \text{ nF/cm}^2$$

(b) From Example 9.2.1, $V_{th} = 0.77$ V. As $V_G > 0$, but $V_G < V_{th}$, the structure is in the *depletion mode* and, from (9.3.4), at high frequencies,

$$C = \frac{3.45 \times 10^{-8}}{\sqrt{1 + 2(3.45 \times 10^{-8})^2/(1.6 \times 10^{-19} \times 1.04 \times 10^{-12} \times 4 \times 10^{14})}}$$

$$= 7.94 \text{ nF/cm}^2$$

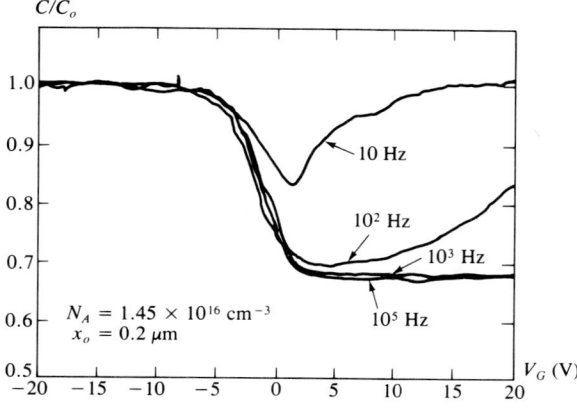

FIGURE 9.3.3
Experimental plot of MOS structure capacitance vs. bias voltage at different frequencies [3].

at low frequencies, $C = C_o = 34.5$ nF/cm^2 with intermediate values at intermediate frequencies.

(c) As $V_G = 1 > 0.77 = V_{th}$, the structure is in the *inversion mode*. The capacitance depends on the frequency of the applied signal:

At *low frequencies*: $$C = C_o = 34.5 \text{ nF/cm}^2$$

At *high frequencies*, from (9.3.5),

$$C = \frac{3.45 \times 10^{-8}}{1 + 3.45 \times 10^{-8}\sqrt{2 \times 0.528/(6.6 \times 10^{-17})}} = 6.45 \text{ nF/cm}^2$$

with intermediate values at intermediate frequencies.

9.4 QUANTITATIVE ANALYSIS OF THE MOSFET I/V CHARACTERISTICS

Figure 9.4.1 represents the pertinent features of an N-channel enhancement MOSFET structure referenced to a set of spatial xy coordinates. Positive bias voltages are applied to both gate (V_G) and drain (V_D) electrodes. The bias is assumed large enough to insure operation in the inversion mode. The operating

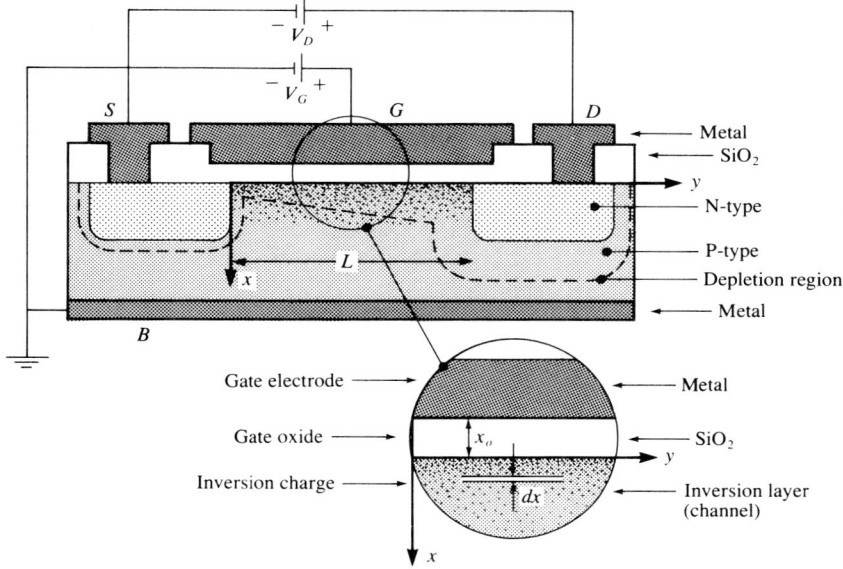

FIGURE 9.4.1

Structure of a biased N-channel MOSFET referred to xy axes. The inversion charge distribution is symbolized by dots of decreasing concentration, the edges of the depletion region by the dashed line. The insert is a magnified reproduction of the region within the circle, showing some of the geometrical elements used in the mathematical analysis.

conditions of the biased MOSFET are more complex than those of the MOS structure considered in Sec. 9.3 because of the effect of the application of the drain voltage V_D.

Threshold Voltage

If no voltage is applied to the drain, then the channel is equipotential, the voltage across the depletion region (surface potential of the semiconductor at the oxide interface) is $\psi_s = \psi_{si} = 2V_{if}$, Q_B is given by (9.2.9), and both V_{th} and Q_B are the same for all points of the channel.

However, under a drain voltage $V_D > 0$, the inversion layer is no longer equipotential; instead *the voltage V between each point and the substrate electrode depends on the position along the length of the channel*. The surface potential becomes

$$\psi_{si}(y) = 2V_{if} + V \tag{9.4.1}$$

a function of position, and so the depletion region width (and with it Q_B) vary from point to point. In Fig. 9.4.1 this is shown by the dotted line, qualitatively indicating the edge of the depletion region:

$$Q_B = -qN_A w_{max} = -\sqrt{2\varepsilon q N_A(2V_{if} + V)} \tag{9.4.2}$$

Using this value in the threshold condition (9.2.9),

$$V_{th} = -\frac{Q_B}{C_o} + 2V_{if} = \frac{\sqrt{2\varepsilon q N_A(2V_{if} + V)}}{C_o} + 2V_{if} \tag{9.4.3}$$

it is seen that *the threshold voltage depends on the channel potential V and therefore varies from point to point along the length of the channel*: $V_{th} = V_{th}(V) = V_{th}(y)$.

Notice that the surface potential ψ_{si} is now equal to $2V_{fi} + V$ because of the added voltage between the surface and substrate electrode due to the effect of the drain bias.

At each point, using (9.4.1), (9.2.8), and (9.4.3) in (9.2.7), the voltage equilibrium equation (9.2.7) becomes

$$V_G = -\frac{Q_i}{C_o} - \frac{Q_B}{C_o} + \psi_{si} = -\frac{Q_i}{C_o} + V_{th} + V \tag{9.4.4}$$

from which

$$Q_i = -C_o(V_G - V_{th} - V) = Q_i(y) \tag{9.4.5}$$

and so *the charge per unit interface surface stored in the inversion layer varies from point to point* along the length of the channel, from a maximum near the source to a minimum near the drain.

Considering the gate capacitor, one armature of which is made of a semiconductor material, it is interesting to realize that, because of the drain bias, *the semiconductor armature is not equipotential*, so that the voltage V_o across the oxide dielectric varies from point to point.

Drain Current

Having so visualized the potential and charge distributions along the channel, it is now possible to analyze the device behavior as a circuit element. In Fig. 9.4.1 it is assumed that the inversion mode is established, so that, under the voltage V_D, a current flows in the conducting, N-type channel.

Unfortunately, a direct computation of the channel resistance cannot be performed, because the carrier concentration in the inversion layer is not known *a priori*: from the previous considerations, it can only be predicted that this concentration depends on x (as symbolized in Fig. 9.4.1 by the variable concentration of the dots representing electrons). However, if z is the depth of the channel, considering an infinitesimal element $z\,dx$ of its cross section at an arbitrary coordinate y, as shown in the enlarged insert of the figure, the current through it can be computed as [cf. (4.2.7)]

$$dI\,(x,\,y) = -qn(x,\,y)\mu_e^* z\,dx\left(-\frac{dV}{dy}\right) \qquad (9.4.6)$$

where $n(x, y)$ is the unknown electron concentration distribution across the inversion layer, while the electric field, assumed to be directed along the length of the channel (long channel approximation), is the gradient of the channel potential distribution V. As for the mobility, the conditions under which the electrons move in the inversion layer are quite different from those prevailing in the bulk of a semiconductor. As a practical rough approximation, the inversion electron mobility, indicated by μ_e^* in (9.4.6), turns out to be about *one-half of the bulk mobility for the semiconductor material*: $\mu_e^* \approx \frac{1}{2}\mu_e$ [4, p. 49].

Integrating (9.4.6) with respect to x:

$$I(y) = \mu_e^* z \int_0^\infty -qn(x,\,y)\,dx\left(-\frac{dV}{dy}\right) = -\mu_e^* zQ_i(y)\frac{dV}{dy} \qquad (9.4.7)$$

in which it is recognized that the integral equals Q_i, the charge in the inversion layer per unit area of the oxide-semiconductor interface. Using (9.4.5) in (9.4.7) and separating the variables of the resulting differential equation:

$$I\,dy = \mu_e^* zC_o[V_G - V_{th}(V) - V]\,dV \qquad (9.4.8)$$

integrating the left side from 0 to L and the right side from 0 to V_D, remembering that, by Kirchoff's law, at steady state the current I is constant and equal to the drain current I_D:

$$I_D = \frac{\mu_e^* zC_o}{L}\left(V_G V_D - \frac{V_D^2}{2} - \int_0^{V_D} V_{th}(V)\,dV\right) \qquad (9.4.9)$$

expressing V_{th} according to (9.4.3), and performing the integration, finally

$$I_D = \frac{\mu_e^* z C_o}{L} \left\{ \left(V_G - 2V_{if} - \frac{V_D}{2} \right) V_D - \frac{2\sqrt{2\varepsilon q N_A}}{3C_o} [(V_D + 2V_{if})^{3/2} - (2V_{if})^{3/2}] \right\}$$

(9.4.10)

This expression yields the drain current under the assumption that a continuous inversion channel exists between the source and drain. This assumption requires that a nonzero charge Q_i be stored in the inversion channel at all points along the y axis. However, the potential V varies from 0 to V_D as the point y moves from the source to drain. If, at any point along this path, the voltage V becomes so large that Q_i in (9.4.5) becomes zero, then, at that point, *no electron charge is stored in the inversion layer* so that inversion is not achieved and *the channel is pinched off.*

Let the channel voltage at the point of pinchoff be V_{sat}; then pinchoff occurs whenever the MOSFET drain voltage satisfies the pinchoff (or saturation) condition:

$$V_D \geq V_{sat} = V_G - 2V_{if} + \frac{q\varepsilon N_A}{C_o^2} \left(1 - \sqrt{1 + \frac{2V_G C_o^2}{q\varepsilon N_A}} \right)$$

(9.4.11)

where V_{sat} has been computed from (9.4.5) by setting $Q_i = 0$, $V = V_{sat}$, and computing the threshold voltage $V_{th}(V_{sat})$ from (9.4.3) with $V = V_{sat}$.[4]

Considerations similar to those developed for the J-FET show that, *beyond pinchoff the drain current remains essentially constant* at the saturation value I_{Ds} obtained from (9.4.10) by setting $V_D = V_{sat}$. Operation beyond pinchoff constitutes the *saturation mode.*

Approximate Formulas

As usual, expression (9.4.10), together with condition (9.4.11), permit computation by points of the MOSFET I/V characteristics, as exemplified in Fig. 9.4.2. Unfortunately these expressions are rather complex and the computations are best performed with the aid of a computer. For pencil and paper calculations, approximate formulas can be used for order-of-magnitude results:

1. Below saturation. If the integration in (9.4.9) is performed under the assumption that V_{th} is a constant, the result is

---

[4] It is sometimes useful to remember that, at the pinchoff point, the charge in the inversion layer is negligible, so that the voltage equilibrium can be written

$$V_G - V_{th}(V_{sat}) = V_{sat}$$

$$I_D \approx \frac{\mu_e^* zC_o}{L}\left(V_G - V_{th} - \frac{V_D}{2}\right)V_D \qquad (9.4.12)$$

The question now arises: Eq. (9.4.3) shows that V_{th} is a function of V, yet, in (9.4.12), it is approximated by a constant: what is the most convenient value to adopt for the constant V_{th} and how does this choice affect the approximation?

For fastest results, letting $V_{th}(V) \approx V_{th}(0)$ is very convenient, but often yields unreliable results. In the following, the suggested values for V_{th} are those usually yielding best approximation. Some of these values, however, may still make for rather cumbersome calculations. For (9.4.12), acceptable approximations are often obtained using $V_{th} = V_{th}(V_D)$.

It should also be remembered that even (9.4.10) and (9.4.11) are theoretical formulas, derived under many implied simplifying assumptions, and, as such, only approximate reality. As an example, the theoretical curves of Fig. 9.4.2b, computed on the basis of Eqs. (9.4.10) and (9.4.11) for an N-channel enhancement MOSFET, should be compared with those of Fig. 9.4.2a, obtained by measurement.

2. Saturation current. Setting $V_D = V_{sat}$ in (9.4.12) and assuming $V_{th} = V_{th}(V_{sat})$:

$$I_{Ds} \approx \frac{\mu_e^* zC_0}{L}\frac{V_{sat}^2}{2} \qquad (9.4.13)$$

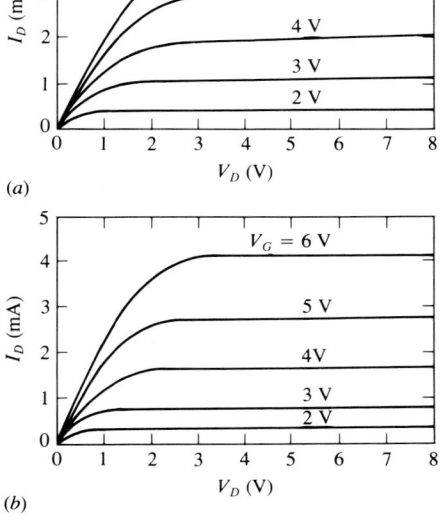

(a)

(b)

FIGURE 9.4.2
I/V characteristics of enhancement N-channel MOSFET (a) experimentally obtained from device measurement and (b) theoretically computed from Eqs. (9.4.10) and (9.4.11) [5].

with the equivalent expression:

$$I_{Ds} \approx \frac{\mu_e^* z C_o}{L} \frac{[V_G - V_{th}(V_{sat})]^2}{2} \qquad (9.4.13a)$$

where use has been made of the previously introduced identity $V_{sat} = V_G - V_{th}(V_{sat})$.

3. Very near the origin of the V/I graphs (*linear region*), the drain current can be approximated by introducing the assumption $V_D \ll V_G - V_{th}(0)$:

$$I_{DL} \approx \frac{\mu_e^* z C_o}{L} [V_G - V_{th}(0)] V_D \qquad (9.4.14)$$

Most of the approximate formulas may yield rather poor results, unless N_A and/or x_o are quite small.

Example 9.4.1. A MOSFET structure is characterized by the data of Example 9.2.1 and also has $L = 30$ μm, $z = 1.5$ mm. Compute the drain current for $V_G = 5$ V and drain voltages of (a) 4 V and (b) 1 V. Use the appropriate approximate formulas using both the suggested value of the threshold voltage and $V_{th}(0)$ for fastest computations. Compare the values obtained. Repeat if the impurity concentration is changed to $N_A = 10^{16}$ cm^{-3}.

Solution. $V_G = 5$. From (9.2.9), $V_{th}(0) = 0.77 < V_G$, so the MOSFET is not cut off.
From (9.4.11), $V_{sat} = 3.78$
(a) $V_D = 4$ V $> V_{sat}$. The MOSFET is *saturated*:

Setting $V_D = V_{sat}$ in (9.4.10) $\quad \rightarrow I_{Ds} = 8.87$ mA

From (9.4.3), $V_{th}(V_{sat}) = 1.22$ V:

From (9.4.13a), using $V_{th}(V_{sat}) \rightarrow I_{Ds} = 8.06$ mA (-9.1%)

using $V_{th}(0) \quad \rightarrow I_{Ds}'' = 10.1$ mA $(+13.8\%)$

(b) $V_D = 1 < V_{sat}$. The MOSFET is not saturated:

From (9.4.10), $\quad \rightarrow I_D = 4.1$ mA

From (9.4.3), $V_{th}(1) = 0.94$ V:

From (9.4.14), using $V_{th}(1) \rightarrow I_D' = 4.02$ mA (-1.9%)

using $V_{th}(0) \rightarrow I_D' = 4.2$ mA $(+2.4\%)$

For $N_A = 10^{16}$ cm^{-3}

$V_G = 5$. From (9.2.9), $V_{th}(0) = 2.1 < V_G$. The MOSFET is not cut off.
From (9.4.11), $V_{sat} = 1.71$.
(a) $V_D = 4$ V $> V_{sat}$. The MOSFET is *saturated*:

Setting $V_D = V_{sat}$ in (9.4.10) $\rightarrow I_{Ds} = 2.3$ mA

From (9.4.3), $V_{th}(V_{sat}) = 3.29$ V:

From (9.4.13a), using $V_{th}(V_{sat}) \to I'_{Ds} = 1.42$ mA (-38%)

using $V_{th}(0) \to I''_{Ds} = 4.1$ mA $(+78\%)$

(b) $V_D = 1 < V_{sat}$. The MOSFET is not saturated:

From (9.4.10) $\to I_D = 1.93$ mA

From (9.4.3), $V_{th}(1) = 2.87$ V:

From (9.4.14), using $V_{th}(1) \to I'_D = 1.58$ mA (-18%)

using $V_{th}(0) \to I''_D = 2.34$ mA $(+21\%)$

Comparison of the values obtained using the different approximations shows a large increment in variance for the larger doping. This has been predicted in the text. Notice that the more accurate value lies in between the predictions of the approximate formulas.

9.5 MOSFET SMALL-SIGNAL LINEAR EQUIVALENT CIRCUIT

From the formulation of Secs. 9.3 and 9.4 it is possible to deduce the *small-signal equivalent linear* behavior of the device, to express it in terms of a small-signal equivalent circuit, and to calculate its parameters.

A useful, approximate equivalent circuit is shown in Fig. 9.5.1. Appropriate formulas for the computation of the small-signal parameters appearing in this circuit usually depend on the region (or mode) of operation.

Linear Mode

In this region the approximation of (9.4.14) is usually quite good, and the corresponding small-signal parameters can be obtained by direct differentiation of this formula, as

$$g_{DL} = \frac{\partial I_{DL}}{\partial V_D} = \frac{\mu_e^* z C_o}{L} [V_G - V_{th}(0)] \qquad (9.5.1)$$

$$g_{mL} = \frac{\partial I_{DL}}{\partial V_G} = \frac{\mu_e^* z C_o}{L} V_D \qquad (9.5.2)$$

Other Modes

MUTUAL ADMITTANCE. From the approximate formula (9.4.12), taking the partial derivative of I_D with respect to V_G:

$$g_m = \frac{\mu_e^* z C_o}{L} V_D \qquad (9.5.3)$$

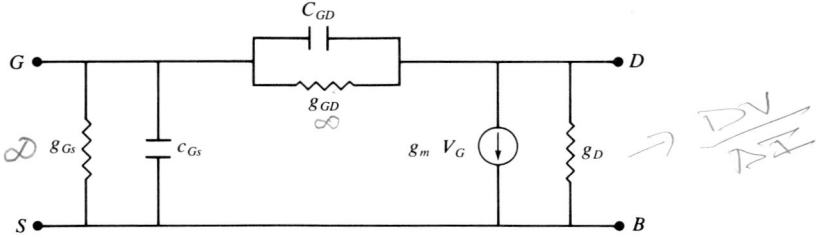

FIGURE 9.5.1
MOSFET small-signal linear equivalent circuit.

with the equivalent:

$$g_m = \frac{\mu_e^* z C_o}{L} [V_G - V_{th}(V_D)]$$

(9.5.3a)

The same result can be obtained by differentiating (9.4.10), so the formula is valid beyond the approximation of (9.4.12).

In the saturation mode, by direct differentiation of (9.4.10) (with $V_D = V_{sat}$),

$$g_{ms} = \frac{\partial I_{Ds}}{\partial V_G} = \frac{\mu_e^* z C_o}{L} V_{sat}$$

(9.5.4)

with the equivalent:

$$g_{ms} = \frac{\mu_e^* z C_o}{L} [V_G - V_{th}(V_{sat})]$$

(9.5.4a)

This formula, as also most of the others, is sometimes presented in textbooks in the approximate form, using $V_{th}(0)$ in place of $V_{th}(V_{sat})$. The approximation afforded by this form is often poor and the simple form (9.5.4) is recommended.

Formulas (9.5.4) and (9.5.4a) can also be obtained by simply substituting V_{sat} for V_D in (9.5.3).

DRAIN ADMITTANCE. From (9.4.10), differentiating with respect to V_D:

$$g_D = \frac{\mu_e^* z C_o}{L} \left[V_G - 2V_{if} - V_D - \frac{\sqrt{2\varepsilon q N_A(V_D + 2V_{if})}}{C_o} \right]$$

(9.5.5)

which, remembering (9.4.2), becomes

$$g_D = \frac{\mu_e^* z C_o}{L} [V_G - V_{th}(V_D) - V_D]$$

(9.5.6)

An equation derived from (9.5.2) is encountered in several books:

$$g_D \approx \frac{\mu_e^* z C_o}{L} [V_G - V_{th}(V_D)] \tag{9.5.6a}$$

in which, sometimes, $V_{th}(V_D)$ is even replaced by $V_{th}(0)$. The approximation of this equation is almost always quite poor, unless $V_D \ll V_{if}$. The form (9.5.6) is recommended (cf. Example 9.5.1).

In the saturation mode, as a first approximation, I_{Ds} can be assumed to be independent of the drain voltage, so that

$$\boxed{g_{Ds} = \frac{\partial I_{Ds}}{\partial V_D} \approx 0} \tag{9.5.7}$$

This infinite output resistance is, of course, a rough approximation. Using an approach similar to that applied to the evaluation of the J-FET drain admittance [cf. (8.5.17)], an order of magnitude value of the MOSFET output admittance can be obtained by using (8.5.18), in which δL can be computed as

$$\delta L = \frac{1}{2} \sqrt{\frac{2\varepsilon}{qN_A}} (V_D - V_{sat}) \, \alpha \tag{9.5.8}$$

where α is a constant depending mainly on the channel length L and on the characteristics of the drain n^+ region and can vary from 1 to as little as $\frac{1}{20}$.

The MOSFET *input conductances* approximate infinite resistances much more closely than in the case of the J-FET. The main source of possible gate leakage is generally traceable to fabrication imperfections introducing surface leakage at the oxide interface.

The *input capacitance* can be roughly approximated by the appropriate expressions developed in Sec. 9.3. Its influence on the frequency response of the device is identical with that analyzed for the J-FET. In particular, the same definition holds for a *cutoff frequency* parameter, which, using (9.5.2) and setting $C_G = C_o z L$, can be expressed as

$$f_{co} = \frac{g_m}{2\pi C_G} \approx \frac{\mu_e^* V_D}{2\pi L^2} \tag{9.5.9}$$

This is a worst case approximation, because the oxide capacitance represents the upper limit of the input capacitance, as discussed in Sec. 9.3. Notice that, as in the case of the J-FET, good high-frequency response mandates high mobility (low doping) and small channel length.

Ohmic drain and source resistances, of course, are present in MOSFETs, just as in J-FETs. Their effect is to modify the mutual and drain admittances, in the same way as described by (8.5.14).

Example 9.5.1. For the MOSFET device of Example 9.4.1, and under the same conditions, compute the mutual admittance, the drain admittance, and the cutoff frequency. In approximate formulas, use both the best approximation and the fastest computation value for threshold voltage V_{th}. Compare the results of best available theoretical formulas with approximate solutions.

Solution

(a) $V_G = 5$, $V_D = 4$. From Example 9.4.1 the MOSFET is in saturation:

From (9.5.4)	$\rightarrow g_{ms}$	$= 4.27$ mmho
From (9.5.4a), using $V_{th}(0)$	$\rightarrow g'_{ms}$	$= 4.78$ mmho $(+12\%)$
From (8.5.18), using $\alpha = 5$	$\rightarrow g_{ds}$	$= 38.5$ μmho
From (9.5.9)	$\rightarrow f_{co}$	$= 43.6$ MHz

(b) $V_D = 1$. From Example 9.4.1 the MOSFET is not in saturation:

From (9.5.3)	$\rightarrow g_m$	$= 1.13$ mmho
From (9.5.3a), using $V_{th}(0)$	$\rightarrow g'_m$	$= 4.78$ mmho $(+323\%)$

From (9.4.3), $V_{th}(1) = 0.94$ V:

From (9.5.6), using $V_{th}(1)$	$\rightarrow g_d$	$= 3.46$ mmho
using $V_{th}(0)$	$\rightarrow g_d^*$	$= 3.65$ mmho
From (9.5.6a), using $V_{th}(1)$	$\rightarrow g'_d$	$= 4.58$ mmho $(+32\%)$
using $V_{th}(0)$	$\rightarrow g''_d$	$= 4.78$ mmho $(+38\%)$
From (9.5.9)	$\rightarrow f_{co}$	$= 11.5$ MHz

For $N_A = 10^{16}$ cm^{-3}

(a) $V_G = 5$, $V_D = 4$, in saturation:

From (9.5.4)	$\rightarrow g_{ms}$	$= 1.66$ mmho
From (9.5.4a), using $V_{th}(0)$	$\rightarrow g'_{ms}$	$= 2.82$ mmho $(+70\%)$
From (8.5.18), using $\alpha = 5$	$\rightarrow g_{ds}$	$= 1.97$ μmho
From (9.5.9)	$\rightarrow f_{co}$	$= 17$ MHz

(b) $V_D = 1$, not in saturation:

From (9.5.3)	$\rightarrow g_m$	$= 0.968$ mmho
From (9.5.3a), using $V_{th}(0)$	$\rightarrow g'_m$	$= 2.8$ mmho $(+190\%)$

From (9.4.3), $V_{th}(1) = 2.87$ V:

From (9.5.6), using $V_{th}(1)$	$\rightarrow g_d$	$= 1.1$ mmho
using $V_{th}(0)$	$\rightarrow g_d^*$	$= 1.84$ mmho $(+67\%)$
From (9.5.6a), using $V_{th}(1)$	$\rightarrow g'_d$	$= 2.06$ mmho $(+87\%)$
using $V_{th}(0)$	$\rightarrow g''_d$	$= 2.82$ mmho $(+156\%)$
From (9.5.9)	$\rightarrow f_{co}$	$= 9.9$ MHz

The above computations clearly indicate that the approximate formulas yield values in excess of the best theoretical ones. Approximations improve when N_A and x_o are small. Comparison of the f_{co} values shows the influence of higher g_m and higher μ_e values (at lower doping μ_e is higher); the equivalence of (9.5.6a) and (9.6.4a) is evident by comparison of the values of g_d'' and g_{ms}'.

The extremely high percentage errors for the values of g_m computed with the fast approximation formula (9.5.3a) in the active, nonsaturated region, indicate that this formula is essentially useless in this mode of operation, especially as the higher accuracy formula (9.5.3) is so simple and fast to use.

All of the above considerations refer to a comparison of approximate formulas to the more sophisticated formulations of (9.4.10) and (9.4.11). These formulas, however, are themselves theoretical and represent only an approximation to the much more complex practical reality.

As usual the student is advised to develop some familiarity with the influence of different design parameter values on device performance by making appropriate parameter choices and predicting device performance using both the approximate and more accurate formulas. In the latter case, some simple programming on a personal computer can save much time and effort.

9.6 THE FLAT BAND VOLTAGE

The previous analysis of MOSFET behavior is based on the assumption that the metal and semiconductor work functions are equal, so that, in the band energy diagrams of Fig. 9.2.1, the Fermi levels are aligned. This is seldom the case, as implied by the data of Tables 7.1.1 and 7.3.1.

The band energy diagram of a MOS structure with $\phi_M \neq \phi_S$ is shown in Fig. 9.6.1. The figure assumes that the structure is open-circuited, so that the metal and the semiconductor are at the same potential, as shown by the horizontal vacuum level line. The metal and semiconductor Fermi levels are not aligned,

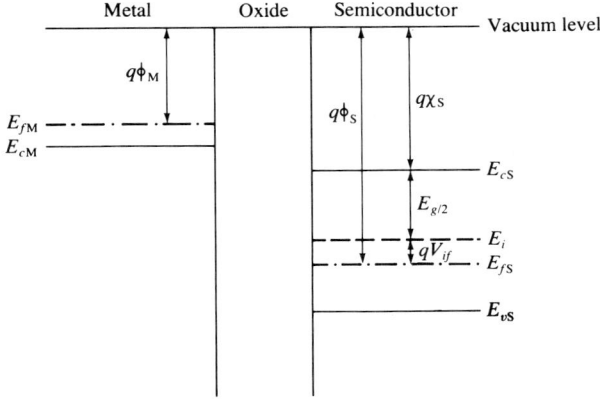

FIGURE 9.6.1
Energy band diagram of unbiased, open-circuited MOS structure with $\phi_M < \phi_S$.

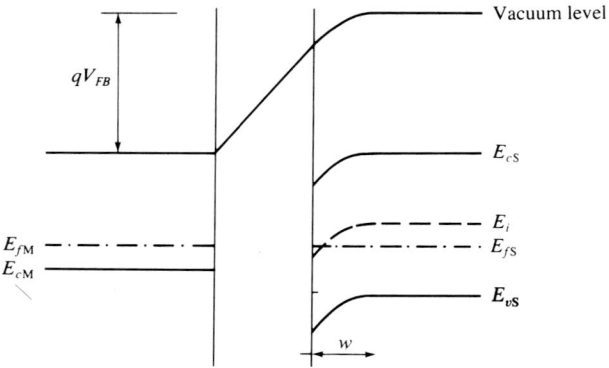

FIGURE 9.6.2
Structure of Fig. 9.6.1 under short-circuited condition, showing effect of contact potential.

although they are horizontal straight lines within the two conducting materials, as required by the no-current condition.

If, however, the metal and the semiconductor are short-circuited by an external galvanic connection, then, through the connecting conductor, there flows a current transient, because the Fermi levels are not aligned. In the case shown, electrons flow from the metal to the semiconductor, charging it negatively, so that, at equilibrium, a *contact potential* is established and the Fermi levels align.

The result, as usual, is the distorted band energy diagram of Fig. 9.6.2, which shows the formation of a depletion region w and, if the contact potential is large enough, even the possibility of inversion phenomena, as indicated by the relative position of E_i and E_f near the oxide-semiconductor interface.

The validity of all formulations of Secs. 9.2 through 9.5 is predicated on the assumption that at zero bias ($V_G = 0$) the band energy diagram is flat. This *flat band condition* can be simulated by an artifice: imagine that a voltage V_{FB} is added to the system between the metal and the semiconductor. If

$$V_{FB} = \phi_M - \phi_S \qquad (9.6.1)$$

this voltage balances the contact potential, so that the flat band condition is achieved. In mathematical terms this condition can be imposed by postulating that the bias voltage V_G starts from an initial value V_{FB}, so that the whole formulation remains valid *provided that, in each formula, V_G is substituted by $V_G - V_{FB}$.*

Applying this flat band rule, it is then seen that the effect of a nonzero V_{FB} is that the MOSFET I/V *characteristics are shifted* along the V_G axis by a quantity V_{FB}.

Charge Trapping in the Oxide

The initial bending of the band energy diagram can also result from other causes. Charges can be stored either within the oxide layer or at the oxide-semiconductor interface. Then they generate an electric field, consequently varying the contact potential. These charges can be incorporated in the structure during manufacture either on purpose or otherwise.

No matter what the cause, in any computation the initial bending of the bands can be compensated for by the same artifice: in all formulas subtract from V_G a flat band voltage V_{FB} equal to the total contact potential.

It is useful to analyze how the oxide charges are generated and how they affect V_{FB}. Wherever irregularities are introduced into the lattice, extra permissible states can be generated. When such states are filled, a net charge is locally stored. By this mechanism charges can be:

1. *Trapped at the interface* where the intimate contact between different crystal lattices can easily produce irregularities or
2. *Trapped in the oxide* where irregularities may be introduced during the oxidation process.
3. During oxidation *already ionized intrinsic atoms can be fixed in the oxide.*

All of the above phenomena can be controlled and minimized by low-temperature annealing, especially under appropriate crystal orientation. Charge per unit area can generally be limited to about one trapped charge per 10^5 atoms. This is equivalent to about 10^{10} charges per square centimeter.

4. Finally, *ionized atoms of foreign substances* (mostly alkali) can be introduced into the oxide during manufacture. Sometimes these charges are not bound to the lattice so that, under high electric field and/or high-temperature conditions, they may move around in the oxide, decreasing its insulation properties. Particularly clean and controlled manufacturing conditions are required to minimize this source of oxide charges.

In practice, under acceptable industrial clean conditions, it can be observed that the surface charge density Q_{SS} trapped at the semiconductor-oxide interface depends on the semiconductor crystal orientation, as indicated in Table 9.6.1 [4], which shows that, for low surface charge density, the 100 orientation is to be preferred.

On the other hand, charges may be purposely introduced into the oxide, usually by an ion implantation process. Ion implantation permits excellent

TABLE 9.6.1

Crystal orientation	Q_{ss}, C	Q_{ss}/q
(111)	$+8.0 \times 10^{-8}$ cm^{-2}	5×10^{11} cm^{-2}
(110)	$+3.2 \times 10^{-8}$ cm^{-2}	2×10^{11} cm^{-2}
(100)	$+1.4 \times 10^{-8}$ cm^{-2}	9×10^{10} cm^{-2}

Metal   Oxide   Semiconductor

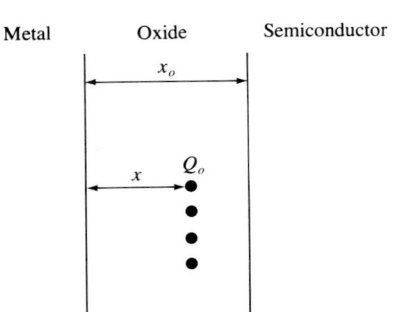

FIGURE 9.6.3
Schematic representation of a layer of charges embedded in the oxide at a distance x from the metal.

control of the flat band voltage and very refined techniques have been developed to achieve reliable and accurate control of this important characteristic.

It is useful to determine the relationship between the amount of charge introduced and V_{FB}. Let a layer of charge Q_o per unit interface area enter the oxide at a distance x from the metal, as shown in Fig. 9.6.3. This is equivalent to charging a capacitor per unit area ε/x, so that the flat band voltage is increased by

$$\Delta V_{FB} = -\frac{Q_o x}{\varepsilon_{ox}} = -\frac{Q_o}{C_o}\frac{x}{x_o} \tag{9.6.2}$$

The influence of the charge on V_{FB} is seen to be negligible if it is implanted near the oxide-metal interface and maximum if near the oxide-semiconductor interface.

If the charge is distributed at different depths in the oxide, then

$$\Delta V_{FB} = -\frac{\int_0^\infty \rho_v(x)x\,dx}{C_o x_o} \tag{9.6.3}$$

In general, the flat band voltage in a MOSFET characterized by $\phi_M \neq \phi_S$ and by several oxide charges can be computed by

$$V_{FB} = \phi_M - \phi_S - \frac{1}{C_o x_o}\sum Q_n x_n \tag{9.6.4}$$

where Q_n is the net charge per unit area in the oxide at a distance x_n from the metal. The computation of ϕ_S in (9.6.4) can be made with the help of (3.4.7) and (3.4.9), as indicated in Example 9.6.1.

Example 9.6.1. The MOSFET structure of Example 9.5.1 has a polysilicon gate electrode and, by ion implantation, 4×10^{11} B ions/cm^2 have been implanted in the oxide at a depth of 0.07 μm from the polysilicon. Compute the flat band voltage and the gate cutoff voltage.

Solution. From Table 7.1.1, for polysilicon, $\phi_M = 3.95$ V. For the P-type semiconductor of the problem, from Table 3.1.1, $\chi_S = 4.01$ V, $E_g = 1.12$ eV, and, from

Example 9.2.1, $V_{if} = 0.264$ V. By inspection of Fig. 9.6.1, $\phi_S = \chi_S + E_g/2q + V_{if} = 4.01 + 0.56 + 0.264 = 4.834$ V. From Example 9.3.1, $C_o = 3.45 \times 10^{-8}$. Then, from (9.6.4), remembering that B ions are negatively charged,

$$V_{FB} = 3.95 - 4.834 + \frac{1.6 \times 10^{-19} \times 4 \times 10^{11} \times 0.07}{3.45 \times 10^{-8} \times 0.1} = 0.415 \text{ V}$$

From Example 9.4.1, $V_{th}(0) = 0.77$ V. Then, from (9.4.5), setting $V = 0$, $Q_i = 0$, and applying the flat band rule, the gate cutoff voltage = gate voltage at the onset of inversion for $V_D = 0$ becomes

$$V_{G,co} = 0.77 + 0.415 = 1.185 \text{ V}$$

Effect of Flat Band Voltage on MOS Capacitance

Figure 9.3.2 displays the dependence of the gate capacitance C_G on the gate voltage in a MOS structure. The curve (reproduced in Fig. 9.6.4, solid line a, for ease of comparison) was computed under the assumption that $V_{FB} = 0$.

If $V_{FB} \neq 0$, then, in accordance with the rule, V_G must be substituted by $V_G - V_{FB}$, so that a nonzero V_{FB} results in a displacement of the C_G vs. V_G characteristics along the V_G axis by an amount V_{FB}.[5] This would result in the dashed line b of Fig. 9.6.4.

However, the phenomenon is a little more complex. Consider the extra states generated by lattice distortion at the oxide-semiconductor interface. These states become trapped charges (and so contribute to the oxide charge) only when they are filled by a current carrier. In turn, the probability that such a state be filled increases with the carrier concentration near the interface (increasing with Q_i). Therefore Q_o can be expected to increase with Q_i.

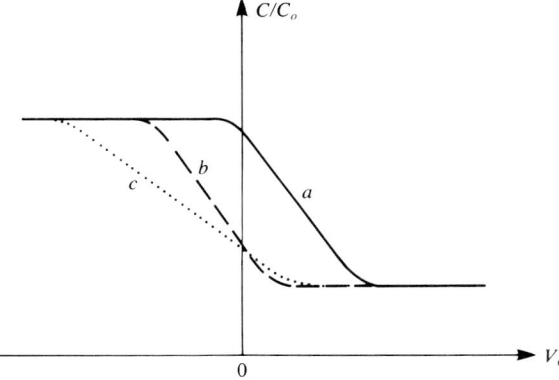

FIGURE 9.6.4
MOS capacitance vs. applied voltage showing displacement and distortion of the plot with increasing flat band voltage. (a) $V_{FB} = 0$; (b) $V_{FB} \neq 0$; (c) $V_{FB} \neq 0$ and significant lattice distortion.

---

[5] This displacement can be easily observed and measured and provides a simple method for determining V_{FB} experimentally.

If the number of interface states is small, their effect is negligible, but if it is large, then this dependence becomes significant. In this case, remembering that Q_i increases with V_G (9.4.5) and V_{FB} with the oxide charge (9.6.4), it can be concluded that V_{FB} increases with V_G.

When this happens, the displacement generated by V_{FB} is not constant but varies with V_G. This results not only in a shift but also in a distortion of the C_G/V_G characteristics, as shown in Fig. 9.6.4 (dotted line, c).

Trapping of ionized atoms of foreign substances gives rise to a technologically and historically important phenomenon: the *instability of the flat band voltage*. If the trapping occurs near the metal surface and the trapped charges are due to spurious ions that can migrate from point to point within the oxide by a slow *drift*, then originally their presence near the metal will only moderately influence the flat band voltage [x is small in Eqs. (9.6.2) through (9.6.4)]. If, however, the structure is subjected to a large bias for a long time at high temperature (so that the spurious ions can more easily drift in the oxide), then the flat band voltage is seen to increase because the ions are displaced toward the semiconductor and x increases in the above equations.

The phenomenon can be minimized by clean fabrication conditions, reducing the inclusion of unwanted impurities (especially Na atoms at the metal-oxide interface), and by a recovery process, by which the Na atoms are made to drift back toward the metal. Recovery is accomplished by aging at high temperature under short-circuit conditions [5].

Effect of Flat Band Voltage on Threshold

In accordance with the flat band rule, to account for a nonzero V_{FB}, Eq. (9.4.5) should be modified as

$$Q_i = -C_o(V_G - V_{FB} - V_{th} - V) \qquad (9.6.5)$$

from which V_{FB} can be interpreted to vary the threshold voltage, rather than V_G. The effect of a nonzero flat band voltage is then taken into account by computing V_{th} as

$$V_{th}^* = -\frac{Q_B}{C_o} + 2V_{if} + V_{FB} \qquad (9.6.6)$$

and using the modified V_{th}^* value in all formulas instead of changing V_G. This point of view is often used in discussing flat band voltage phenomena. It implies that it is possible to control the threshold voltage by charge implantation in the oxide.

9.7 DEPLETION MOSFETS

Setting $V = 0$ in (9.6.5) it is easily seen that the minimum gate voltage required to start inversion, i.e., to bring the device out of cutoff, is

$$V_{G,co} = V_{th} + V_{FB} = V_{th}^* \qquad (9.7.1)$$

In N-channel MOSFETs, if $V_G \leq V_{G,co}$, then the inversion charge $Q_i = 0$ inversion cannot occur, irrespective of the drain voltage ($V_{sat} = 0$), and the MOSFET is *cut off*.

The definition of enhancement MOSFET (no current flows at zero bias) implies that, in N-channel structures, the cutoff gate voltage $V_{G,co}$ is >0; however, if V_{FB} is a sufficiently large negative voltage, $V_{G,co}$ in (9.7.1) becomes negative, current can flow in the unbiased device, and the structure *becomes a depletion MOSFET*. Now the drain current can be varied by applying negative gate voltages, similarly to what was observed in J-FETs.

Depletion MOSFETs can be obtained using the enhancement MOSFET structure discussed so far, but forcing the modified threshold voltage V_{th}^* to be negative. This can be done either by using a metal semiconductor combination with a sufficiently negative value of $\phi_M - \phi_S$ or by adding to the fabrication sequence an ion implantation step, in which a sufficient positive charge per unit area is introduced. For P-channel MOSFETs a negative charge must be used, making V_{th}^* positive.

Example 9.7.1. An N-channel MOSFET has the following characteristics: Al metal electrode, P-type Si semiconductor with $N_A = 10^{15}$ cm^{-3}, oxide thickness $x_o = 0.1$ μm. Compute the threshold voltage and determine whether the MOSFET is of the enhancement or of the depletion type. Assume that (a) there is no net charge stored in the oxide or at the interfaces and (b) there is a charge surface charge Q_{SS}, as is to be expected, knowing that the semiconductor surface orientation is 111.

Solution

From Table 7.1.1, for Al: $\phi_M = 4.2$ V
From Table 3.1.1: $\chi_S = 4.01$ V
From (9.2.1): $C_o = 3.45 \times 10^{-8}$ F/cm^2
From (3.4.9): $V_{if} = 0.0259 \ln (10^{15}/1.5 \times 10^{10}) = 0.29$ V
From (9.2.10): $Q_B = 1.39 \times 10^{-8}$ C/cm^2
Then $\phi_S = 4.01 + 0.55 + 0.29 = 4.85$ V; $\phi_M - \phi_S = -0.65$ V
(a) From (9.6.1): $V_{FB} = -0.65$ V
   From (9.7.1) and (9.6.6): $V_{G,co} = 0.33$ V. As $V_{G,co} > 0$ this N-channel MOSFET is of the *enhancement type*.
(b) From Table 9.6.1, for 111 orientation: $Q_{SS} = 8 \times 10^{-8}$ C/cm^2 so that, applying the rule, $V_{FB} = -0.65 - 2.31 = -2.96$ and $V_{G,co} = 0.33 - 2.31 = -1.98$ V.
   Due to the trapped charges, the cutoff voltage has become <0 and the MOSFET is of the *depletion type*.

Depletion MOSFETs can also be obtained by diffusing a thin N-type layer under the gate oxide between the source and drain. Then, at zero bias, there is a continuous N semiconductor path from the source to drain. The ion implantation technique is almost universally preferred.

Depletion MOSFET characteristics are similar to those of enhancement MOSFETs, except that negative values of V_G appear in the active region.

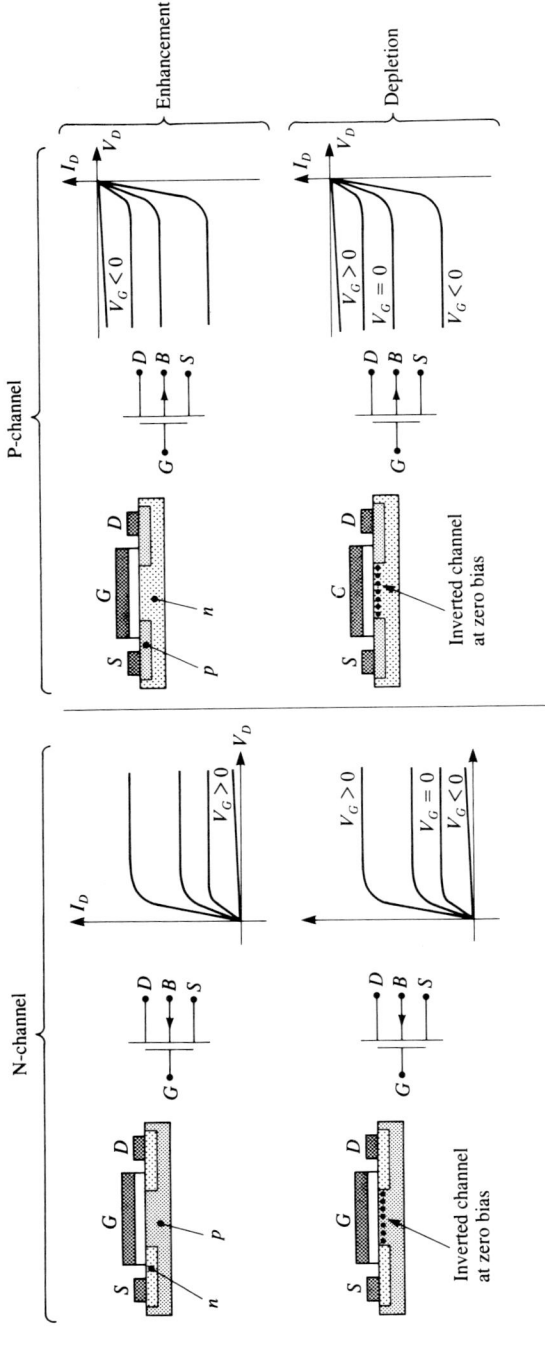

FIGURE 9.8.1
Structures, symbols, and I/V characteristics of P- and N-channel MOSFETs of the enhancement and depletion types.

In this section we have considered only N-channel structures. All the arguments presented still hold for P-channel MOSFETs, except that the sign of V_G and V_D and all inequality signs must be inverted.

9.8 COMPLEMENTARY MOSFETS— THE CMOS TECHNOLOGY

Section 9.1 describes the structure and fabrication of a P-channel enhancement MOSFET. This name refers to the fact that, due to the inversion phenomenon, in the active region, the current flowing in the device is mediated by holes (although the semiconductor material under the gate oxide, where conduction occurs, is N-type).

P-channel structures were chosen as the subject of our introductory description of MOSFET fabrication, because they present fewer problems and the sequence of fabrication steps is therefore somewhat simpler.

In analyzing MOSFET behavior, in Secs. 9.2 through 9.7, instead we discussed N-channel MOSFETS, in which conduction in the inverted channel is mediated by electrons, making the device much faster. The N-channel MOSFET structure was shown in Fig. 9.4.1, which differs from Fig. 9.1.2 only in that the type of all semiconductor materials is inverted.

P-MOS and N-MOS devices are therefore complementary both in structure and in circuit utilization, in the sense that, to obtain equivalent operation, the polarities of all voltages applied are opposite in complementary devices (e.g., a P-channel MOSFET is turned on by negative gate voltages).

The structures, circuit symbols, and I/V characteristics of P-MOS and N-MOS devices, of both the enhancement and depletion types, are summarized in Fig. 9.8.1.

Because of the technological difficulties encountered in the fabrication of N-channel MOSFETs, early MOSFETs were all of the P-channel type. In most modern applications, however, N-channel devices are usually preferred, because of their greater speed, resulting from the higher mobility of electrons.

Probably the most frequent and important application of complementary MOSFETs is the CMOS technology, based on the *CMOS inverter* of Fig. 9.8.2. The gates of a pair of complementary enhancement MOSFETs are connected together and to the input of the inverter; both their drains are connected to the output. The source of the P-channel MOSFET is connected to the positive terminal of the power supply V_{DD}, the source of the N-channel MOSFET to ground.

If V_{in} is positive ($V_{in} = V_{DD}$: high), then the N-channel MOSFET has a positive gate to source voltage and so is " on," and the P-channel MOSFET has zero gate to source bias and is therefore " off." The output terminal is essentially shorted to ground and isolated from the positive terminal of the power supply: $V_{out} = 0$ (low). If V_{in} is low ($V_{in} = 0$), the reverse is true, the N-channel MOSFET is off, the P-channel MOSFET is on, and $V_{out} = V_{DD}$ (high). The device operates as an inverter. To ensure stable operation, the magnitude of the cutoff voltage is

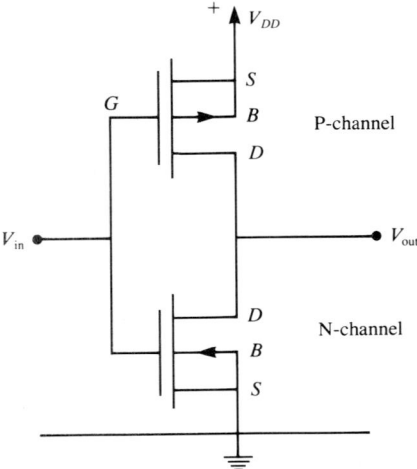

FIGURE 9.8.2
Circuit schematic of the CMOS inverter.

usually controlled by ion implantation and is set between 0.5 and 1 V (positive for the N-MOSFET, negative for the P-MOSFET).

The current through the inverter in each of the two states is determined by the inverse current in the off device, which, as a first approximation (valid for "long" channels), is of the order of the inverse saturation current in the reverse biased diode between the drain and the channel material.[6] This current is generally very low, of the order of picoamperes (cf. Chap. 6), so that large numbers of such inverters can be operated simultaneously with negligible total current drain from the battery. This property is extensively used, especially in hand-held calculators and in critical computer memories.

CMOS Fabrication

As mentioned in Sec. 9.1, the use of polysilicon for the gate contact material has been instrumental in permitting the industrial development of the CMOS technology.

One advantage of the polysilicon gate technology results from the high melting point of polysilicon. This permits the self-aligning polysilicon mask techniques described later in the section. Another advantage derives from the fact that the work function of the gate electrode can be made essentially equal to the semiconductor work function, as shown in the following example.

---

[6] More precise formulas for this current will be obtained when discussing the junction transistor (cf. Chap. 10).

Example 9.8.1. The N-channel MOSFET of Example 9.7.1 is fabricated with 100 crystal orientation. Taking the trapped charge into account, compute the gate cutoff voltage and determine whether the MOSFET is of the enhancement or depletion type (a) for Al gate electrode, (b) for P-type polysilicon electrode.

Solution

(a) From the results of Example 9.7.1, substituting a value $Q_{ss} = 1.4 \times 10^{-8}$ C/cm², obtained from Table 9.6.1 for 100 orientation:

$$V_{G,co} = \frac{1.39}{3.45} - \frac{1.4}{3.45} + 0.58 - 0.65 = -0.073 \text{ V}$$

The negative value of this quantity shows that the MOSFET is of the *depletion type*.

(b) The work function for a P-type polysilicon electrode, assuming the same doping as for the semiconductor, is

$$\phi_M = \chi_S + \frac{E_g}{2} + V_{if} = 4.01 + 0.55 + 0.29 = 4.85$$

so that, in this case $\phi_M - \phi_S = 0$ and, for 100 orientation,

$$V_{FB} = 0 - \frac{1.4}{3.45} = -0.4 \text{ V}$$

The gate cutoff voltage is therefore:

$$V_{G,co} = \frac{1.39}{3.45} + 0.58 - 0.4 = 0.58 \text{ V}$$

As this is a positive voltage, the device is of the enhancement type. Use of the polysilicon gate and choice of the 100 crystal orientation permits the fabrication of an N-channel enhancement MOSFET without requiring any additional fabrication step!

The following schematic description of a single-chip CMOS inverter fabrication sequence will also introduce the polysilicon gate technique. As already stated, polysilicon, thanks to its high melting point, can withstand most of the high temperatures required by other fabrication steps, so that polysilicon electrodes do not need to be put in place only after all other operations have been performed, as is the case for metal electrodes.

A typical fabrication sequence for a CMOS inverting gate is listed in the following outline.

The state of the device after each of the most important fabrication steps is shown in Fig. 9.8.3. Whenever required by clarity, the specimen is shown in cross section, better to observe the results of the current stage of fabrication. References to the pertinent figures are made at the appropriate stages in the following description of the fabrication sequence. The masks used in the process are shown in schematized form in Fig. 9.8.4. It is assumed that the photoresist used in the photostep is of the negative type.

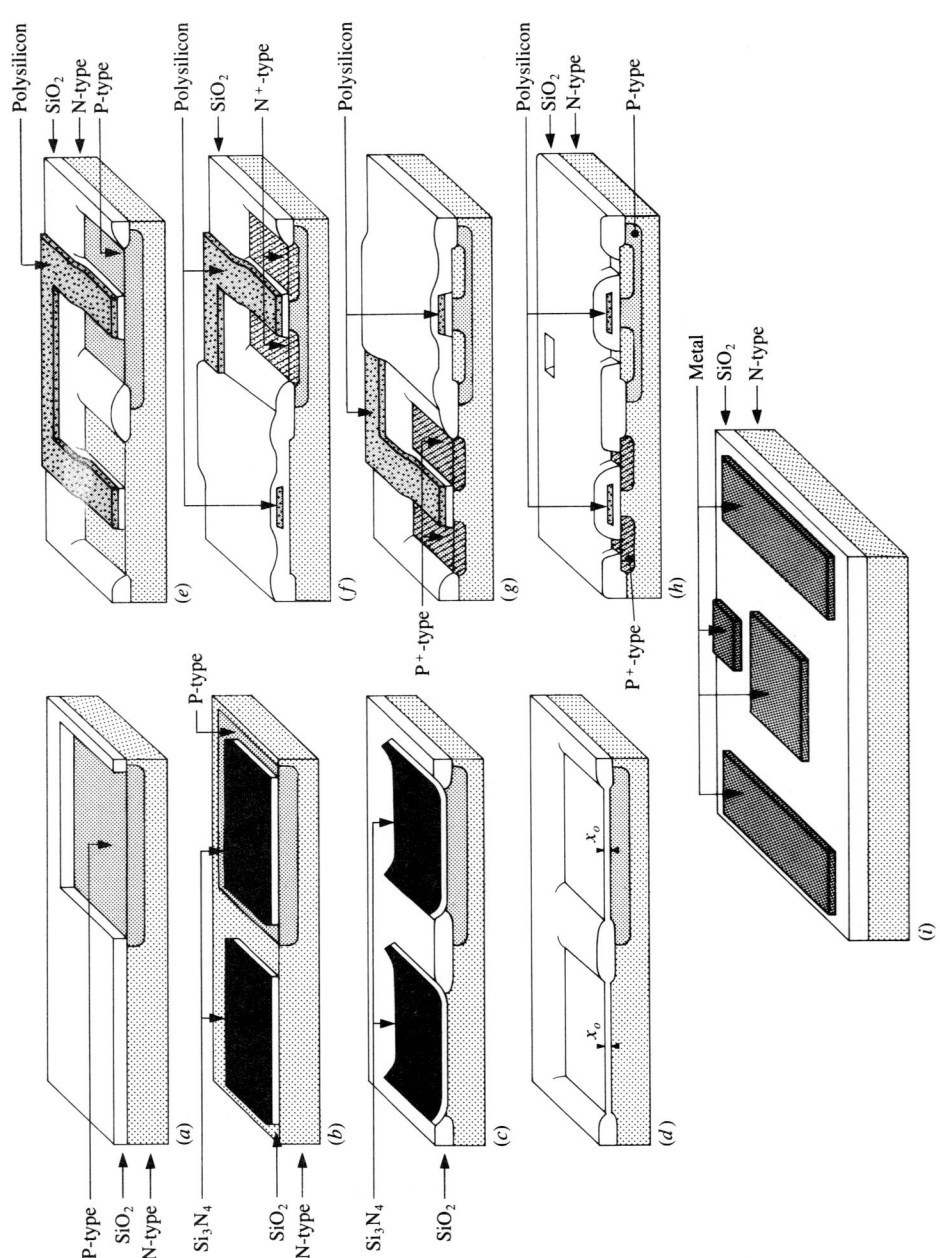

324

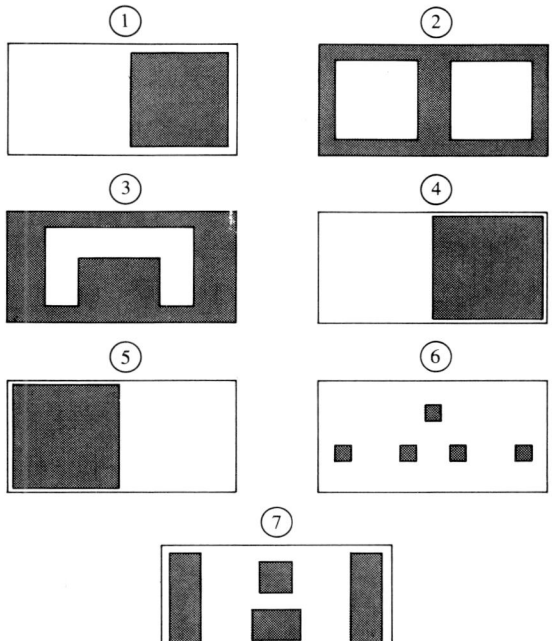

FIGURE 9.8.4
Simplified outline of the masks used in
the CMOS fabrication process.

1. An oxide layer is obtained on a lightly doped N-type substrate by the wet
   process.
2. Using a mask (#1), photostep, etching, and diffusion, a "well" of lightly
   doped P-type material is diffused in the substrate (cross section in Fig.
   9.8.3a).
3. The oxide is etched away and a thin oxide layer is grown on the whole face of
   the wafer.
4. A nitride film is deposited on the oxide and, using a mask (#2) and a photostep, it is etched, resulting in a nitride mask defining each device area. The
   result is shown in cross section in Fig. 9.8.3b.

FIGURE 9.8.3
CMOS inverter. Structure of the device at different stages of the fabrication sequence outlined in the
text. (a) Cross section after step 2. (b) Cross section after step 4. (c) Cross section after step 5 showing
field oxide extending under the Si_3N_4 mask. (d) Cross section after step 6 showing gate oxide of
accurately controlled width x_o. (e) Cross section after step 8. The two polysilicon gate electrodes are
connected together to form the CMOS inverter input terminal. (f) Cross section after step 10. Notice
the automatic registration of the gate oxide with respect to the channel, source, and drain layers. (g)
Cross section after step 12. (h) Cross section after step 13. (i) Complete device. Notice that only the
CMOS device terminals, rather than the individual MOSFET electrodes, are accessible.

5. Using this nitride mask, a field oxide layer is grown on the unprotected silicon surface. This constitutes the *recessed oxide* process, described in Sec. 5.3, and the resulting structure enjoys all the characteristics of this process (notice how the field oxide grows under the edges of the nitride mask, lifting it from the semiconductor: Fig. 9.8.3*c*).

6. The nitride and the underlying layer of thin oxide are stripped away and the whole wafer is covered with a thin layer of high-quality oxide of accurately controlled thickness x_o, the *gate oxide* (cross section in Fig. 9.8.3*d*).

7. The whole wafer is covered with a layer of doped polysilicon (with a sheet resistance from 30 to 1 $\Omega/\square$ depending on the gate dimensions).

8. Using a photostep with mask #3, the polysilicon and gate oxide layer are etched to obtain the gate electrodes, with their underlying channel oxide and uncovering the source and gate regions (cross section in Fig. 9.8.3*e*).

9. An oxide layer is deposited on the whole wafer and, by a photostep using mask #4, it is etched away, so that only the P-channel device is protected.

10. By a diffusion process, using the polysilicon and field oxide as a mask, two n^+ regions are diffused into the p well. They will become the source and drain of the N-channel MOSFET (cross section in Fig. 9.8.3*f*). The polysilicon can be used as a mask only thanks to its high melting point, as previously remarked. It should be noticed that this polysilicon mask is self-aligning. The diffused layers diffuse laterally under the gate oxide by just the right amount to insure proper alignment for correct device performance. Considering the critical registration problems of the corresponding two masks in the fabrication sequence of Sec. 9.1, the advantages of this self-registration technique are seen to be determinant in simplifying the whole process and decreasing fabrication costs.

11. A thick oxide layer is grown and, using mask #5, a photostep, and etching, the N-channel device is protected and the P-channel is exposed.

12. A self-aligning diffusion process, using the polysilicon gate as a mask, diffuses the p^+ source and drain regions of the P-channel device (cross section in Fig. 9.8.3*g*).

13. After passivating oxidation, a photostep using mask #6, and etching, appropriate wells are etched in the oxide to expose all source, drain, and gate regions (cross section in Fig. 9.8.3*h*).

14. The whole wafer is metallized and, using a final photostep with mask #7, it is etched to obtain the desired electrode pattern (three-dimensional view in Fig. 9.8.3*i*).

Figure 9.8.5 is a cross section of the completed device, where the reference numbers in circles refer to the above sequence of steps. Notice that only four terminals are available in this elementary integrated device: input, output, high voltage, and ground.

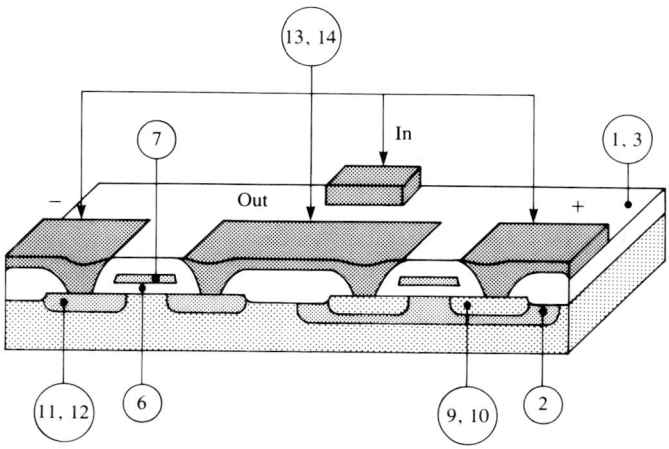

FIGURE 9.8.5
Cross section of the completed device referencing the pertinent steps of the fabrication outline by the numbers within the circles.

The several steps of Figs. 9.8.3 and 9.8.5 should be cross-referenced with the above process description.

A number of variations, improvements, and additional steps are often introduced to obtain specific characteristics required by the device and/or control device parameters. As one example several ion implantation procedures may be added to control the threshold values and to ensure proper device isolation (chanstop layers).

9.9 SUMMARY

In N-channel MOSFETs, source and drain regions are N-type, the body P-type. Semiconductor types are reversed in P-MOSFETs. The path between the source and drain therefore consists of two diodes back to back and, in enhancement MOSFETs, does not conduct at zero bias. Considering N-channel enhancement MOSFETs and assuming zero *flat band voltage* V_{FB} (cf. Sec. 9.6), three modes are possible when a gate bias V_G is applied:

1. Negative bias (gate negative): *accumulation* of majority carriers. No conduction.
2. Low positive bias $(0 < V_G < V_{th} + V_{FB})$: *depletion*. The surface potential is $\psi_s < 2V_{if}$. No conduction.
3. High positive bias $(V_G \geq V_{th} + V_{FB})$: *inversion*. The field induced in the P-type semiconductor distorts the energy bands; near the oxide-semiconductor interface E_i is bent below E_f, generating an inverted N-type channel. A continuous N-type channel exists from S to D and conduction is possible.

Beyond threshold the surface potential ψ_{si}, and with it the depletion region width w_{max} and bound charge per unit interface area Q_B, remain constant. Increasing V_G only increases the voltage V_o across the oxide and the inverted charge Q_i.

The threshold voltage V_{th} is the minimum gate bias required for inversion under zero drain voltage and zero flat band voltage. It is computed from Eqs. (9.2.9) and (9.2.10).

If a drain voltage V_D is applied, the channel voltage V varies from point to point, Q_B becomes a function of position and so does V_{th}, as indicated by Eq. (9.4.3).

Using the voltage equilibrium equation (9.4.4), the inversion charge per square centimeter of interface is computed from Eq. (9.4.5).

The gate capacitance per unit interface area depends on the mode of operation. At accumulation it equals C_o; at depletion it is given by Eq. (9.3.4), in inversion by Eq. (9.3.5) at high frequencies, or by C_o at low frequencies.

If $V_G \leq V_{G,co} = V_{th}(0) + V_{FB}$, the MOSFET is cut off. If $V_G \geq V_{G,co}$, the drain current is computed from Eq. (9.4.10), provided a continuous inversion channel exists between the source and drain.

If $V_D \geq V_{sat}$, with V_{sat} given by Eq. (9.4.11), somewhere along the channel the inversion condition is not satisfied and the channel is pinched off. In this case the drain current is given by (9.4.10) with V_{sat} substituting V_D. In this saturation mode the current is independent of V_D.

For fast paper and pencil computations several approximate formulas are available [(9.4.12) through (9.4.14)].

In the small-signal linear approximation, the device circuit behavior is described by the linear circuit of Fig. 9.5.1. The most important small-signal circuit parameters can be computed with the help of Eqs. (9.5.3), (9.5.6), and (9.5.7) [or (8.5.18)].

The bandwidth capabilities are usually represented by the cutoff frequency parameter f_{co} approximated by Eq. (9.5.9).

The flat band voltage V_{FB} results from unequal work functions and from oxide charges. It can be computed by Eq. (9.6.4).

The effects of V_{FB} can be taken into account by modifying all formulas obtained under the assumption $V_{FB} = 0$. Either substitute V_G with $V_G - V_{FB}$ or substitute V_{th} with $V_{th} - V_{FB}$.

By ion implantation techniques V_{FB} can be controlled during manufacture. This is reflected in the C_G vs. V_G diagram (Fig. 9.6.4) and in the MOSFET I/V characteristics that are displaced along the V_G axis. By this technique, an enhancement structure can be turned into a depletion structure, by forcing $V_{G,co}$ to become negative in N-MOSFETs, positive in P-MOSFETs.

Complementary MOSFET structures are summarized in Fig. 9.8.2.

Complementary MOSFET pairs are used in the CMOS technology. One of the most common CMOS structures is the CMOS inverter of Fig. 9.8.3.

Polysilicon gate technology is commonly used in MOSFET fabrication, particularly often in CMOS devices.

PROBLEMS

9.1. A MOS capacitor structure at 300 K has a layer of SiO_2 of thickness $x_o = 1$ μm between metal and an N-type semiconductor with $N_D = 10^{14}$ cm^{-3}.
(a) Does this correspond to an N- or a P-channel MOSFET structure? Why was this nomenclature chosen?
(b) What is the oxide capacitance per unit area of this structure?
   Qualitatively draw the energy band diagram and net charge distribution of the structure assuming that the metal is so chosen that, when metal and semiconductor are equipotential, the Fermi levels are aligned. Label all pertinent levels with their proper symbols, assuming that the energies are expressed in electronvolts. Indicate appropriate voltage polarities and qualitative relationships among voltages and energy levels. Do this for:
(c) No bias (compute V_{if})
(d) Enhancement conditions
(e) Depletion conditions
(f) Weak inversion (compute and show ψ_{si})
(g) Strong inversion
(h) Why does the Fermi level line not blend with the potential variations?
(i) What is the value of ψ_s at the onset of strong inversion?
(j) What does the value of ψ_s become if the bias is increased beyond the value generating strong inversion?
(k) How do Q_B, Q_i, and w vary as the bias increases beyond the threshold value for strong inversion?
(l) What charges constitute Q_B and Q_i?

9.2. A MOS structure between a metal and a P-type semiconductor with $N_A = 10^{15}$ cm^{-3} at 300 K has $x_o = 1$ μm, $\phi_M = \phi_S$. Compute the charge density in the depletion layer and in the inversion layer at the Si-oxide interface for:
(a) $\psi_s = |V_{if}| + 0.01$ V
(b) $\psi_s = \psi_{si}$
(c) $\psi_s = \psi_{si} + 0.01$ V
(d) Remembering the definitions of Q_B and Q_i as the charges per square centimeter of Si-oxide interface due to bound ionic charges and free inversion charges respectively, compute the order of magnitude of the percent variations of Q_B and Q_i due to a 0.01 V increment of ψ_s above ψ_{si}.

9.3. A MOS structure at 300 K is characterized by $\phi_M = \phi_S$, P-type Si semiconductor with $N_A = 10^{14}$ cm^{-3}, oxide thickness $x_o = 0.2$ μm. Compute V_o and ψ_s for:
(a) $V_G = 0.483$ V
(b) $V_G = 1$ V
(c) State any simplifying assumptions implied in the previous computations and estimate their influence on the results obtained.

9.4. Compute the value of w_{max} for Si with $N_A = 10^{16}$ cm^{-3} and compare with the value indicated by Fig. 9.2.7.

9.5. For a MOS structure at 300 K with a P-type semiconductor, assuming $\phi_M = \phi_S$, tabulate the threshold voltage vs. the oxide thickness x_o for $0.1 < x_o < 1$ μm in steps of 0.15 μm and for $N_A = 10^{15}$, 10^{16}, and 10^{17} cm^{-3}.

9.6. For the structure of Prob. 9.3 at $V_G = 1$ V:
(a) Compute Q_B and Q_i.
(b) Repeat with the same data, but assuming that the dielectric, instead of SiO_2, is Si_3N_4.

9.7. For the MOS structure of Prob. 9.3, assuming a Si–SiO_2 interface of 1 mm^2 surface, plot the structure's capacitance for $-2 < V_G < 2$ V.

9.8. Prove Eq. (9.3.5).

9.9. Prove Eq. (9.3.4).

9.10. An N-channel MOSFET at 300 K has a Si channel with an impurity concentration of 2×10^{15} cm^{-3}. The metal of the gate contact has $\phi_M = \phi_S$. The channel length is 20 μm and the thickness of the device is $z = 1$ mm. The thickness of the SiO_2 gate oxide is $x_o = 0.1$ μm.
(a) Is the semiconductor constituting the channel of N- or P-type?
(b) What is the minimum gate voltage required to obtain conduction?
(c) Assuming that inversion occurs in the channel, what is the mobility of the carriers?
(d) If the gate voltage is 2 V, what is the minimum drain voltage required to pinch off the device?
(e) Verify the result of part (d) by computing Q_i at the drain end of the channel.

9.11. Prove Eq. (9.4.11).

9.12. For the MOSFET of Prob. 9.10:
(a) Compute the saturation current for V_G varying in steps of 1 V from 2 to 6 V.
(b) For the above voltages and limiting the drawing to the pinchoff region, draw the field of V/I characteristics for $V_D < 10$ V.
(c) Repeat parts (a) and (b) using the appropriate approximate formulas and compare with the previous results.

9.13. Complete the set of I/V characteristics of Prob. 9.12, adding the region below pinchoff (see computer problem 9.27).

9.14. Prove Eq. (9.4.13).

9.15. (a) Compute g_m from (9.4.10) and compare with (9.5.3).
(b) Compute g_{ms} from (9.4.10) using (9.4.11) and compare with (9.5.4).

9.16. (a) Using the approximate expression of Prob. 9.15, compute g_{ms} at $V_G = 4$ V for the MOSFET of Prob. 9.10.
(b) Repeat the computation using the result of Prob. 9.15(b) and compare with the result of part (a).
(c) Compute g_{ms} around $V_G = 4$ V, using the results of Prob. 9.12.

9.17. The amplifier of Fig. P9.17 uses the MOSFET of Prob. 9.10. Compute the gain.

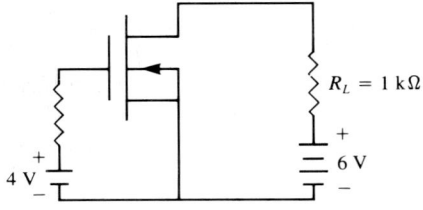

FIGURE P9.17
Circuit for Prob. 9.17.

9.18. The MOSFET of Prob. 9.10 is modified by changing the substrate doping to $N_A = 10^{14}$ cm^{-3}.

    (a) For $V_G = 3$ V compute V_{sat} and I_{DS}. Compare with the results of Prob. 9.12.

    (b) Compute the gain of the stage of Fig. P9.17 if the new MOSFET is used and the gate bias is changed to 3 V. Compare with the results of Prob. 9.17.

9.19. For the MOSFET of Prob. 9.18:

    (a) Compute the approximate cutoff frequency f_{co}.

    (b) Find a general formula expressing the high-frequency 3-dB point in terms of f_{co}, assuming the stage has an input impedance R_{in}, the input is provided by a small-signal current generator, and the input capacitance C_G all appears between the gate and the source.

    (c) Assuming $R_{in} = 10$ kΩ and using the experimental value $f_{co} = 70$ MHz published by the manufacturer, compute the high-frequency 3-dB point, supposing the MOSFET works in pinchoff with $V_G = 3$ V.

9.20. Verify the statement of Sec. 9.6 according to which one trapped charge per 10^5 atoms corresponds to $Q_{SS}/q = 10^{10}$ cm^{-3}, assuming that the atomic density in a semiconductor is of the order of 10^{22} cm^{-3}.

9.21. A MOSFET at 300 K is characterized by: P-type Si with $N_A = 10^{14}$ cm^{-3}, Al gate, $L = 10$ μm, $z = 1$ mm, $x_o = 0.4$ μm, and crystal orientation at the interface 100.

    (a) Compute the gate cutoff voltage.

    (b) Is this an enhancement or a depletion MOSFET?

    (c) Compute the gate cutoff voltage if the gate thickness is reduced to 0.1 μm.

    (d) Keeping $x_o = 0.1$ μm, change the substrate doping to $N_A = 10^{15}$ cm^{-3} and compute the new cutoff voltage.

9.22. A surface charge of 10 nC/cm^2 is deposited by ion implantation at the Si–SiO$_2$ interface of the MOSFET of Prob. 9.21(d). Compute the cutoff voltage and determine if the resulting device is of the enhancement or depletion type.

9.23. The CMOS inverter of Fig. 9.8.2 has $V_{DD} = 5$ V at 300 K. The inverse saturation currents of its component devices are both equal to 1 pA. Assuming the substrate dopings to be $N_A = N_D = 10^{14}$ cm^{-3}, the source and drain dopings $N_A = N_D = 10^{19}$ cm^{-3}, also $z/L = 100$, $x_o = 0.1$ μm, $|V_{th}| = 0.5$ V for both devices, and $V_G = 5$ V, determine:

    (a) The output voltage

    (b) Which device is on and which is off

9.24. The CMOS inverter of Prob. 9.23 is loaded with an identical inverter. At $t = 0$ the input of the first inverter is switched from 5 to 0 V. Assuming the Si–SiO$_2$ interface has a surface of 0.1 mm^2:

    (a) What is the total charge to be provided to the input of the second CMOS inverter?

    (b) Which device provides this charge?

    (c) Compute approximately the initial current flowing from the output of the first CMOS inverter.

Computer Problems

9.25. Write a computer program to solve Prob. 9.5. Interactively allow the operator to modify the data (channel doping and range of gate oxide thickness).

9.26. (*a*) Write a computer program to solve part (*d*) of Prob. 9.10. Interactively allow the operator to modify the data (channel impurity concentration, gate oxide thickness, and range of V_G).

    (*b*) Use this program to plot drain saturation voltage vs. V_G for $1 < V_G < 4$ V at 1 V intervals, gate oxide thickness 0.2 μm, and channel impurity concentration $N_A = 2 \times 10^{15}$ cm^{-3}.

9.27. Write a computer program to compute by points the V/I characteristics of the MOSFET of Prob. 9.10. Use $2 < V_G < 6$ V in steps of 1 V and $0 < V_D < 4$ V in steps of 0.25 V. Interactively allow the operator to modify the data.

9.28. Repeat Prob. 9.17 changing R_L to 2000 Ω. Solve the problem by both making use of the characteristics obtained in Prob. 9.27 and writing a suitable computer program to compute the work point by trial and error.

9.29. Solve Prob. 9.18, changing R_L to 3000 Ω.

REFERENCES

1. Sze, S. M.: *Semiconductor Devices, Physics and Technology*, John Wiley & Sons, Inc., New York, 1985.
2. Grove, A. S., B. E. Deal, E. H. Snow, and C. T. Sah: "Investigation of Thermally Oxidized Silicon Surfaces Using Metal-Oxide-Semiconductor Structures," *Solid State Elect.*, vol. 8, p. 145, 1965.
3. Grove, A. S., E. H. Snow, B. E. Deal, and C. T. Sah: "Simple Physical Model for the Space-Charge Capacitance of Metal-Oxide-Semiconductor Structures," *Appl. Phys.*, vol. 35, p. 2458, 1964.
4. Carr, William N., and J. P. Miza: *MOS/LSI Design and Application*, McGraw-Hill Book Company, Inc., New York, 1972.
5. Grove, A. S.: *Physics and Technology of Semiconductor Devices*, John Wiley & Sons, New York, 1967.

ADDITIONAL READING

Deal, B. E., E. H. Snow, and C. A. Mead: "Barrier Energies in Metal-Silicon Dioxide-Silicon Structures," *J. Phys. Chem. Solids*, vol. 27, p. 1873, 1966.
Frohman-Bentchowsky, D., and A. S. Grove: "Conductance of MOS Transistors in Saturation, *IEEE Trans. Elect. Devices*, vol. ED 16, p. 100, 1969.
Ihantola, H. K. S., and J. L. Moll: "Design Theory of a Surface Field Effect Transistor," *Solid State Elect.*, vol. 7, p. 423, 1964.
Nicollian, E. G., and J. R. Brews: *MOS Physics and Technology*, John Wiley & Sons, Inc., New York, 1982.
Sah, C. T.: "Characteristics of the Metal-Oxide-Semiconductor Transistor," *IEEE Trans. Elect. Devices*, vol. ED 11, p. 324, 1964.
Sze, S. M.: *Physics of Semiconductor Devices*, John Wiley & Sons, Inc., New York, 1969.

CHAPTER
10

THE BIPOLAR JUNCTION TRANSISTOR

10.1 BJT STRUCTURE

The bipolar junction transistor (BJT) consists of two PN junctions back to back. Two types of structure are used: the PNP in which an N-type semiconductor is sandwiched between two P semiconductors and the NPN structure in which the P-type is sandwiched between two N-types. Figure 10.1.1 represents the two structures in principle, also showing their circuit symbols. Three terminals are available, each connected to one of the semiconductor sections and labeled *emitter, base,* and *collector,* as shown in the figure. The junctions between emitter and base and between collector and base are designated as the *emitter junction* and the *collector junction* respectively.

The figure also shows the three terminal currents, I_E, I_B, and I_C. Each of these currents is conventionally assigned the positive sign if it is in the direction of the corresponding arrow (into the device). Currents in the opposite directions are negative, by convention. Kirchoff's current equilibrium results in the relationship:

$$I_E + I_B + I_C = 0 \tag{10.1.1}$$

If a voltage is applied between emitter and collector, essentially no current can flow through the device, because of the two bucking diodes. It will be shown that application of an appropriate bias between emitter and base permits conduction and that the collector current can be controlled by varying this bias.

333

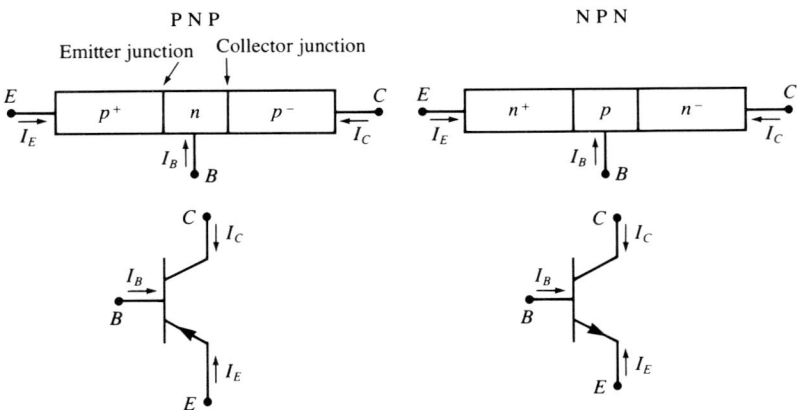

FIGURE 10.1.1
Structure and circuit symbols of PNP and NPN transistors.

For reasons that will become apparent in the following analysis, in order for the structure to function as a transistor, the base width must be very narrow. Also, for best results, the doping of the three sections should decrease from emitter to collector: large doping for the emitter, medium in the base, and low in the collector semiconductor: $N_E > N_B > N_C$.

In practice, the structure of Fig. 10.1.1 is implemented in a large variety of forms, to improve performance and/or optimize fabrication conditions. One possible sequence of fabrication steps, using planar technology, is described in the following for an NPN transistor.[1] A cross section of the device is shown in Fig. 10.1.2. The circled numbers correspond to the sequence of fabrication steps described later in this section. As usual, in following the outline of the sequence of fabrication steps, this figure should be systematically used for reference. Cross sections of the device at various fabrication stages are shown in Fig. 10.1.3 and the photomasks used in the corresponding photosteps in Fig. 10.1.4. The shapes of these photomasks are purely indicative and much oversimplified. Some characteristics of more complex patterns used in actual fabrication will be discussed later on. For clarity, the scale of the structure shown is distorted: the shapes and relative dimensions of the various parts are usually very different in real devices.

1. A lightly doped layer of N-type material (typical thickness about 10 μm) is epitaxially grown on a p^+ wafer. The n^- semiconductor will constitute the collector material. The p^+ support can be used to provide efficient device isolation when the device is used in integrated circuits. This isolation can be achieved by applying a strong negative bias to the p^+ semiconductor.

---

[1] The NPN configuration can be as much as 3 times faster than the PNP and is often preferred to it.

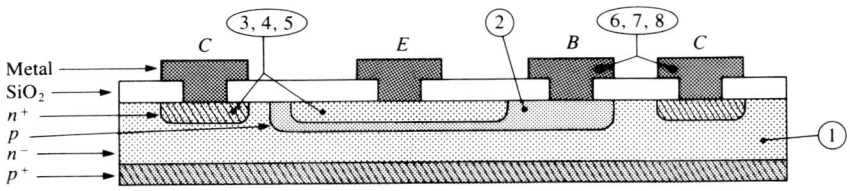

FIGURE 10.1.2
Planar structure for NPN transistor, cross section. Circled numbers refer to the list of fabrication steps outlined in the text.

2. The wafer surface is oxidized and, using mask #1 and a photostep, a *p* layer (the base region) is diffused in the *n⁻* material. The doping and geometry of this base region are carefully controlled at this stage. Figure 10.1.3*a* shows a cross section of the device at this stage of fabrication.

3. The oxide layer is stripped away and a new layer is grown, usually by the wet oxidation process.

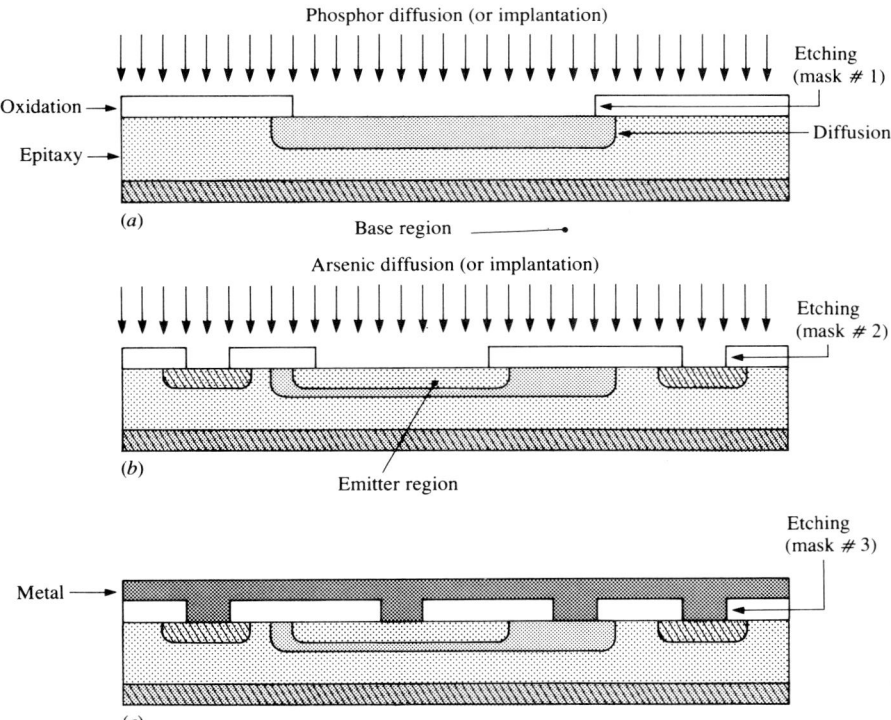

FIGURE 10.1.3
Cross sections of NPN structure at different stages during fabrication. (*a*) After base region diffusion. (*b*) After emitter region diffusion. (*c*) After metallization.

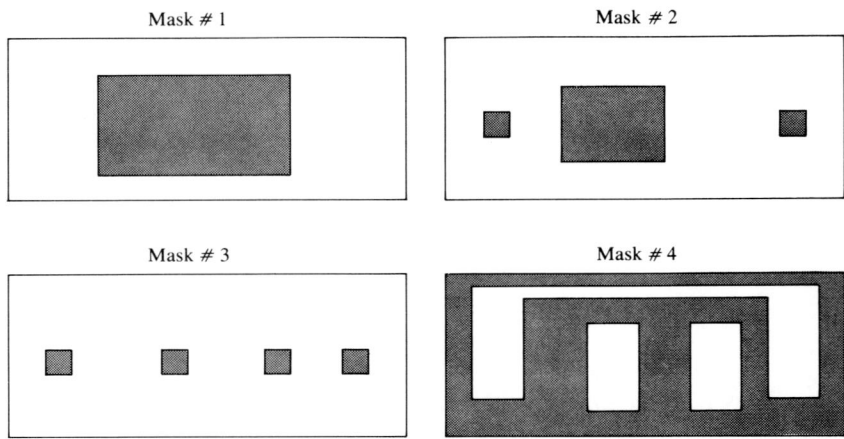

FIGURE 10.1.4

Masks corresponding to the fabrication sequence outlined in the text. Shapes and dimensions correspond to Figs. 10.1.2, 10.1.3, and 10.1.5 and are purely indicative and grossly simplified. Negative photoresist is assumed.

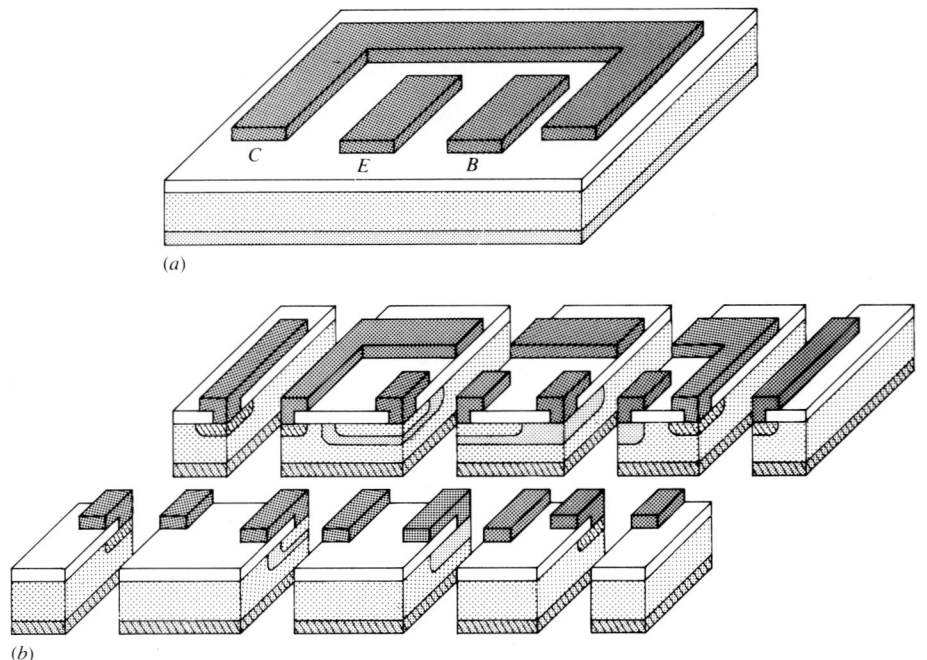

FIGURE 10.1.5

Completed transistor. Electrode shapes and dimensions are purely indicative and oversimplified. (a) External three-dimensional view. (b) Sectioned and exploded view showing internal structure.

4. Using mask #2 and a photostep, openings are etched to expose the semiconductor material at the proper positions for the diffusion of the emitter and of the collector contact regions.
5. By either diffusion or, in more modern devices, ion implantation, shallow heavily doped, n and n^+ regions are fabricated. They are the emitter and the ohmic contact regions of the collector (Fig. 10.1.3b).
6. The oxide is stripped away and a new oxide coating is grown.
7. Using mask #3 and a photostep, appropriate wells are opened in the oxide to permit contact with the various regions of the structure.
8. The whole face of the wafer is metallized (Fig. 10.1.3c) and, using mask #4, the metal is etched to yield the desired electrode configuration shown schematically in the cross section of Fig. 10.1.2.

An idea of the three-dimensional structure of the completed device can be obtained from Fig. 10.1.5a (external) and b ("exploded").

10.2   BJT BAND DIAGRAMS— THE TRANSISTOR EFFECT

Figure 10.2.1 shows the energy band diagrams of a PNP structure. In Fig. 10.2.1a the semiconductors are assumed to be insulated from each other, so that the junctions are not made and the effect of the different dopings on the Fermi levels of the three-component semiconductors is evident by inspection of the figure.

When the junctions are made, transients of current flow within the interior of the device, because the Fermi levels are not aligned. After the transients are over, the energy band diagram has the configuration shown in Fig. 10.2.1b, where the different contact potentials and depletion widths in the various regions correspond to the variations in doping (cf. Sec. 6.2).

In the majority of circuits, transistors operate in the *active mode*: a forward bias V_{EB} is applied to the emitter junction and the collector junction is reverse biased by a voltage V_{CB}, as shown in Fig. 10.2.1c. As discussed in Chap. 6, at the emitter junction the barrier voltage is decreased from V_{bi} to $V_{bi} - V_{EB}$ and the emitter depletion region width $x_{EE} + x_{BE}$ is correspondingly decreased, while the collector junction barrier is increased to $V_{bi} + V_{BC}$ (notice the sign of the voltages) and the collector depletion region width $x_{BC} + x_{CC}$ is correspondingly increased.

The behavior of the device can be analyzed with the help of the carrier flow lines representation introduced in Chap. 6 for the PN junction (cf. Fig. 6.6.2). The analysis is illustrated by Figs. 10.2.2 through 10.2.4.

To permit a gradual approach to the analysis, as a first step let the two junctions be separated, resulting in the two diode circuits of Fig. 10.2.2a. Figure 10.2.2b displays the carrier flow lines of the two diodes. The flow lines of the forward biased emitter junction are the same as those of Fig. 6.6.2 (in both cases $N_A > N_D$), while only the reverse saturation current flows in the reverse biased

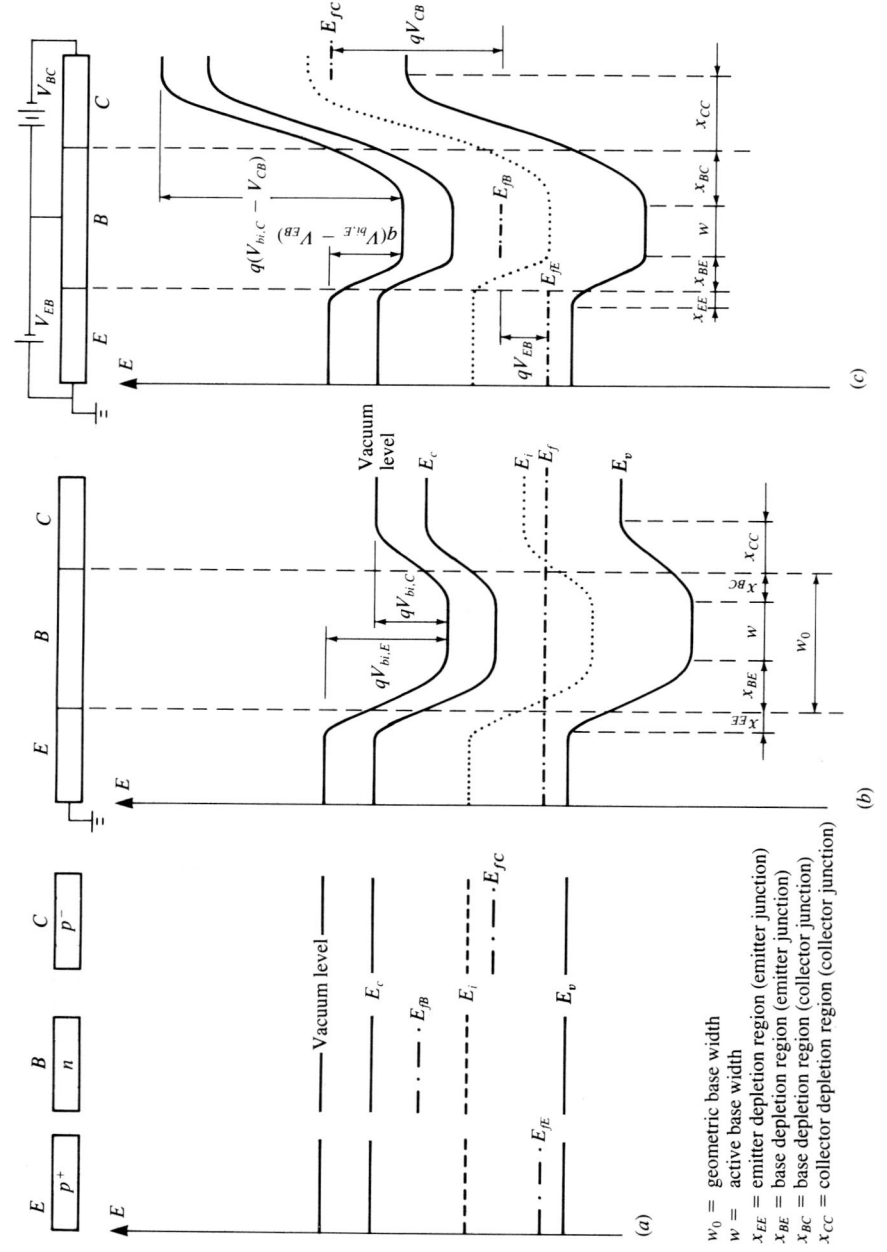

w_0 = geometric base width
w = active base width
x_{EE} = emitter depletion region (emitter junction)
x_{BE} = base depletion region (emitter junction)
x_{BC} = base depletion region (collector junction)
x_{CC} = collector depletion region (collector junction)

FIGURE 10.2.1
Energy band diagrams of the PNP structure. (a) Before junctions are made. (b) No applied bias voltages. (c) Forward biased emitter junction, reverse biased collector junction.

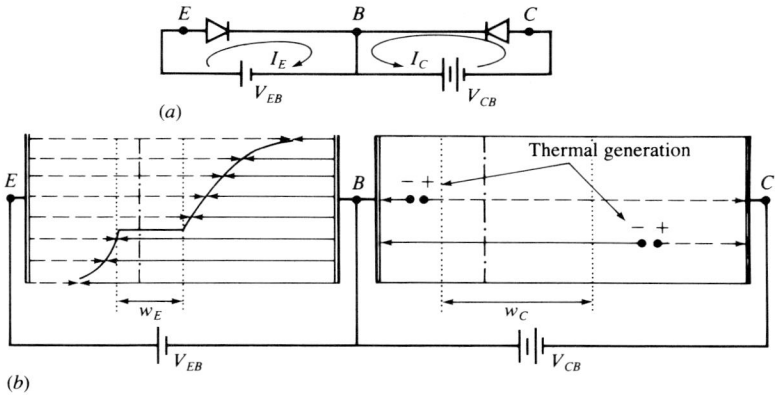

FIGURE 10.2.2
Circuit schematic and carrier flow lines of two separate diodes connected and biased as the emitter and the collector junctions in a PNP transistor. As in Fig. 6.6.2, dotted lines indicate hole flow, solid lines electron flow.

collector junction. The behavior of this structure is, of course, quite predictable and so not very interesting for the moment.

From the configuration of Fig. 10.2.2 it is possible to obtain the original structure of Fig. 10.1.1. To this end, the cathodes of the two diodes must be joined together to form a continuous base. To ensure continuity of the base material, the two metallic cathode electrodes must first be moved to the side of the base region, resulting in the structure shown in Fig. 10.2.3. Because of the new position of the base electrode, the flow lines of the carrier electrons in the

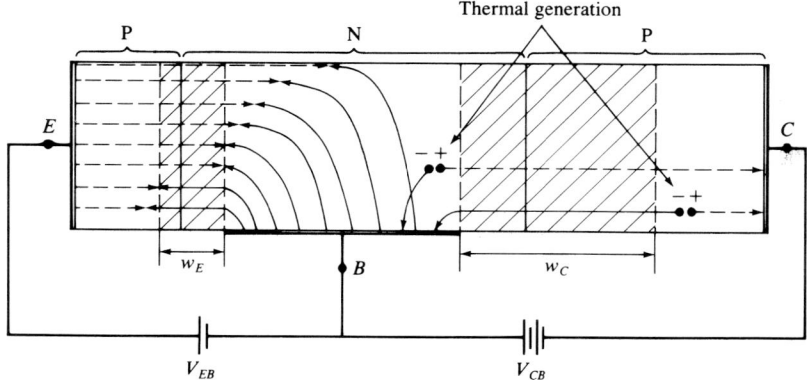

FIGURE 10.2.3
The separate diodes of Fig. 10.2.2 are joined into one PNP structure. The N semiconductors are merged into one base structure and the cathode electrodes are shifted to the side, becoming the base electrode. Notice the bending of the carrier flow lines.

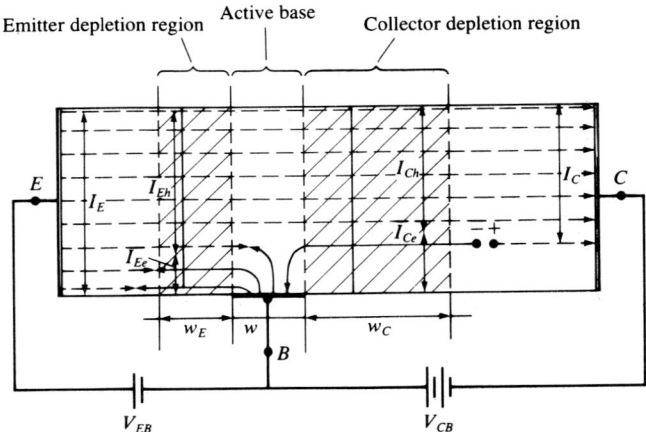

FIGURE 10.2.4
Same configuration as Fig. 10.2.3, but with a very thin base region. A large percentage of the injected hole flow lines cross the base into the collector without recombination.

base material bend to follow the new electric field configuration.[2] A similar bending is noticed for the flow lines of the inverse saturation current of the collector junction.

From the circuit point of view, this configuration is no more interesting than the previous one: the two PN junctions behave as two separate diodes. However, a new and important phenomenon, the *transistor effect*, occurs if the width of the base region becomes very small. This condition is represented in Fig. 10.2.4.

In accordance with the analysis of Sec. 6.6, the holes injected into the base across the emitter junction move by diffusion in the N-type semiconductor until they recombine with majority carrier electrons. A statistical measure of the distance travelled by these minority carriers before recombining is given by the diffusion length L_{hN} of the holes in the N semiconductor [cf. (6.7.13) and (6.7.15)].

Statistically, therefore, *if the width of the base material is smaller than the diffusion length*, then the probability that a minority carrier diffusing in the base will meet a majority carrier and recombine is smaller than the probability that it will diffuse right through the base into the collector junction depletion region. In other words, *it is probable that the diffusing minority carriers will enter the collector junction depletion region before they have a chance to recombine within the base.*

---

[2] As stated in Sec. 6.6, within the body of the N-type semiconductor, the current is mediated by *electron drift*; therefore the electrons move along the lines of the electric field established by the electrode configuration.

The large electric field in the collector depletion region is directed toward the collector, so each and all holes reaching the collector depletion region are swept across it into the collector, resulting in a collector current. If the base is narrow enough, this collector current represents a large percentage of the hole current injected into the base from the emitter.

In conclusion, analyzing the current distribution in the PNP device (Fig. 10.2.4), beginning at the emitter terminal:

1. Near the emitter, the current I_E is mediated by drift of the majority holes.
2. Not all of this hole current makes it across the emitter junction. Some is lost by recombination with the current of minority electrons injected from the base (cf. Sec. 6.6). Just before the edge of the emitter depletion region the total current I_E is mediated by a drift current of holes I_{Eh} and a diffusion current of electrons I_{Ee}. By current continuity,

$$I_{Eh} = I_E - I_{Ee} = \gamma I_E \qquad (10.2.1)$$

where an important parameter, the *emitter efficiency* γ, has been defined as

$$\gamma = \frac{I_{Eh}}{I_E} = \frac{I_{Eh}}{I_{Eh} + I_{Ee}} \qquad (10.2.2)$$

From the analysis of Sec. 6.7, γ can be expected to increase with the ratio N_E/N_B of the dopings in the emitter and in the base (cf. Prob. 10.4).

3. Within the emitter depletion region, as a first approximation, it can be assumed that no significant recombination occurs (because of the low carrier concentrations), so the current consists of two constant components: I_{Eh} (holes diffusing to the right) and I_{Ee} (electrons diffusing to the left).
4. Within the active base, the hole current I_{Eh}, in its diffusion motion, decreases by recombination with the majority base electrons. However, if the base is narrow, only a small portion of it has a chance to recombine; the rest of it, I_{Ch}, reaches the collector depletion region and is swept into the collector. From the above considerations, I_{Ch} is a fraction of I_{Eh}:

$$I_{Ch} = \alpha_T I_{Eh} \qquad (10.2.3)$$

where the *base transport factor* $\alpha_T < 1$ closely approaches 1 when the base width becomes very small.

5. The total current I_C across the collector junction consists of the hole current I_{Ch} injected into the collector from the emitter through the base, plus the small reverse saturation current I_{Ce} of minority electrons:

$$I_C = -(I_{Ch} + I_{Ce}) = -I_{Ch} + I_{Co} \qquad (10.2.4)$$

where the negative sign follows the convention of Sec. 10.1.1 and $I_{Co} = -I_{Ce}$ is the reverse bias leakage current from collector to base with open emitter.

Introducing a *common base current gain* coefficient α_0 equal to the product of the emitter efficiency times the base transport factor:

$$\alpha_0 = \frac{I_{Ch}}{I_E} = \frac{I_{Ch}}{I_{Eh}} \frac{I_{Eh}}{I_E} = \gamma \alpha_T \qquad (10.2.5)$$

then (10.2.4) can be rewritten in the familiar form:

$$I_C = -\alpha_0 I_E + I_{Co} \qquad (10.2.6)$$

It should be remembered that, by the assumptions made, the above relationships imply that no generation or recombination occurs in the depletion regions.

The reader is advised to review the above analysis following the current flow and clearly identify the various current components and their relationships.

Equation (10.2.6) shows that, if a variation ΔI_E is somehow produced in the emitter current, the transistor transfers it to the collector circuit as a current variation $\Delta I_C = -\alpha_0 \Delta I_E$. However, the emitter circuit, consisting of a forward biased diode, has a small series resistance r_f. On the other hand, the collector circuit, being reverse biased, can contain a very large load resistor $R_L \gg r_f$, through which the collector current variation ΔI_C can be made to flow.

It can then be concluded that the introduction of a signal into the emitter circuit of a transistor requires the expenditure of a signal power $\Delta I_E^2 r_f$. This signal is transferred to the collector circuit where it delivers to the load a power $\Delta I_C^2 R_L$, with $R_L \gg r_f$. This process yields a *power gain* of $(\Delta I_C^2 R_L)/(\Delta I_E^2 r_f) \approx \alpha_0^2 (R_L/r_f)$.

It is therefore seen that a transistor can achieve power gain by causing a current in a low resistance circuit to be

TRANS-ferred to a high res-*ISTOR*.

To ensure high gain, α_0 should be as large as possible, so that, by (10.2.5), both the emitter efficiency γ and the base transport factor α_T should be as close to unity as possible.

10.3 THE BJT IN THE ACTIVE MODE— QUANTITATIVE ANALYSIS

The currents flowing in the transistor under active bias conditions are shown in Fig. 10.2.4. In order to compute their magnitudes in terms of the structural parameters of the transistor (doping, geometrical dimensions, etc.) and of the applied bias voltages we shall be led to write and solve some differential equations. Although this will require no more than elementary calculus, yet the mathematics are sufficiently laborious to result in some confusion, so that in working out the computations one can easily lose sight of the simple logic of the investigation.

The analysis can be divided into four basic steps:

1. Determine the steady state minority carrier concentrations in the device.
2. Compute the various components of hole and electron currents at different locations. These internal currents I_{Eh}, I_{Ee}, I_{Ch}, I_{Ce}, etc., are shown in Fig. 10.2.4. As minority carriers move by diffusion, these currents can be computed from the carrier concentrations obtained in step 1.
3. From the currents computed in step 2 obtain the terminal currents of the device I_E, I_B, I_C.
4. Compute parameters γ and α_T defining the transistor effect.

Some of the expressions obtained will be uncomfortably complex and some approximations will be introduced. It is important that the limitations implicit in these approximations be clearly understood. The final results will be expressed in terms of a few newly introduced parameters.

The student is advised to become familiar with the logical sequence and with the results of each step and their physical meaning by first following the analysis in its general lines, without paying too much attention to the mathematical details. These can then be worked out once the concepts are clear. To help in this task, final formulas for carrier concentrations, currents, and parameter computations will be summarized at the end of steps 1, 2, 3, and 4. Each of these formulas will be enclosed in a frame for easy reference.

Minority Carrier Concentrations

As in the case of the PN junction in Sec. 6.7, the first step in the analysis will be to determine the minority carrier concentration distributions at steady state. Reference will be made to Fig. 10.3.1, where the x axis measures distances along the length of the device. The origin is chosen where the active base begins, at the edge of the emitter depletion region within the base ($x = 0$). The end of the active base, at the base edge of the collector depletion region, is at $x = w$. The other edges of the two depletion regions are at $x = -x_E$ for the emitter and $x = x_C$ for the collector.

In accordance with the PN junction analysis of Chap. 6, assuming low-level injection across the junctions so that the minority carriers move essentially by diffusion, the continuity equation (6.7.9) for holes in the base at steady state becomes

$$D_{hB} \frac{\partial^2}{\partial x^2} (p_B - p_{oB}) - \frac{p_B - p_{oB}}{\tau_{hB}} = 0 \qquad (10.3.1)$$

where the notation used is the same as in Eq. (6.7.9), except that the subscript B has been used to indicate characteristics valid in the base material [in (6.7.9) the subscript N, for N-type, was used]. Thus, for instance, D_{hB} designates the diffusion coefficient of holes in the base semiconductor material, τ_{hB} is the minority carrier (holes) lifetime in the base material, etc.

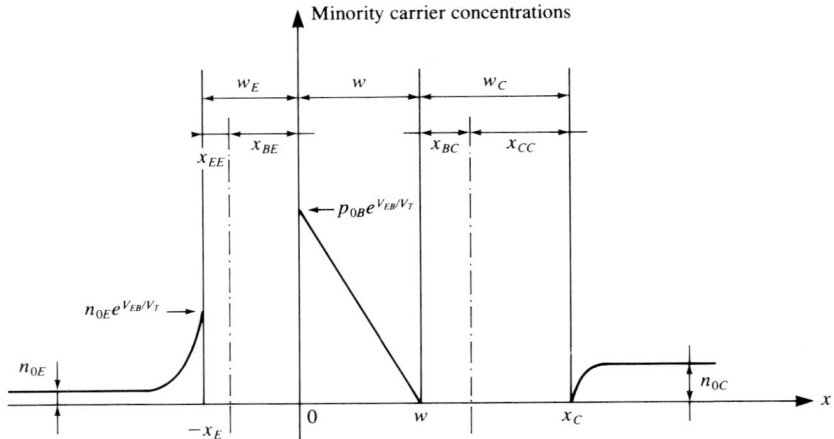

FIGURE 10.3.1
Minority carrier concentration distributions in a PNP structure biased in the active region. w_E and w_C are emitter and collector depletion region widths respectively, w is the active base width. Geometric junction locations are indicated by dash-dot lines.

The general solution of (10.3.1) is

$$p_B(x) = p_{oB} + C_1 e^{x/L_{hB}} + C_2 e^{-x/L_{hB}} \qquad (10.3.2)$$

which should be compared with (6.7.11), also remembering (6.7.15) and the above notation convention.

To compute the arbitrary constants C_1 and C_2, the boundary conditions must be determined. This can be done remembering that, under low-level injection, the minority carrier concentration at the edge of a biased depletion region is given by (6.7.8). Then, for the base at $x = 0$,

$$p_B(0) = p_{oB} e^{V_{EB}/V_T} \qquad (10.3.3)$$

because the emitter junction is forward biased with a voltage V_{EB}. At $x = w$,

$$p_B(w) = 0 \qquad (10.3.4)$$

because, for the reverse biased collector junction, the bias voltage is a large negative quantity.

Solution of (10.3.2), subject to these boundary conditions, yields (cf. Prob. 10.5):

Concentration in the base:

$$p_B(x) = p_{oB}\left\{(e^{V_{EB}/V_T} - 1)\frac{\sinh\left[(w - x)/L_{hB}\right]}{\sinh\left(w/L_{hB}\right)} + 1 - \frac{\sinh\left(x/L_{hB}\right)}{\sinh\left(w/L_{hB}\right)}\right\} \qquad (10.3.5)$$

Remembering the exponential expression of the hyperbolic sine, it is easy to prove (cf. Prob. 10.6) that, for large base widths ($w/L_{hB} \gg 1$),

$$p_B(x) \approx p_{oB} + p_{oB}(e^{V_{EB}/V_T} - 1)e^{-x/L_{hB}} \tag{10.3.6}$$

in accord with (6.7.13), and for narrow base widths, approximating to the first order of the small quantity $w/L_{hB} \ll 1$, Eq. (10.3.5) results in:

Approximate concentration in the base:

$$\boxed{p_B(x) \approx p_{oB}\, e^{V_{EB}/V_T}\left(1 - \frac{x}{w}\right) \qquad \text{for } 0 \le x \le w} \tag{10.3.7}$$

so that, for narrow base width, *a linear minority carrier concentration distribution is predicted*. This is usually the prevailing condition in transistor bases, as shown in Fig. 10.3.1.

The minority carrier distribution in the emitter is obtained by imposing on (10.3.2) the boundary conditions:

$$\text{for } x \to -\infty \quad \text{then} \quad n_E(-\infty) = n_{oE}$$

and

$$\text{for } x = -x_E \quad \text{then} \quad n_E(-x_E) = n_{oE}\, \exp\left(\frac{V_{BE}}{V_T}\right)$$

or:

Concentration in the emitter:

$$\boxed{n_E(x) = n_{oE} + n_{oE}(e^{V_{BE}/V_T} - 1)e^{(x+x_E)/L_{eE}} \qquad \text{for } x \le -x_E} \tag{10.3.8}$$

The minority carrier distribution in the collector is obtained from (10.3.2) under the boundary conditions:

$$\text{for } x = x_C \quad \text{then} \quad n_C(x_C) = 0$$

and

$$\text{for } x \to \infty \quad \text{then} \quad n_C(\infty) = n_{oC}$$

or:

Concentration in the collector:

$$\boxed{n_C(x) = n_{oC} - n_{oC}\, e^{-(x-x_C)/L_{eC}} \qquad \text{for } x \ge x_C} \tag{10.3.9}$$

In both emitter and collector, the minority carrier concentration *varies exponentially* from the edge of the depletion region, reaching the thermal equilibrium value deep in the body of the semiconductor.

The minority carrier distributions in the three regions of the transistor are indicated in Fig. 10.3.1, which should be compared with Fig. 6.7.1.

It is important to bear in mind that, for the transistor to reach steady state in a given condition of operation, the above minority carrier concentration distributions must have been established within the device. This implies that in the base there is a stored excess minority carrier charge:

$$Q_B = qA \int_0^w p_{oB}(e^{V_{EB}/V_T} - 1)\left(1 - \frac{x}{w}\right) dx = qA p_{oB}(e^{V_{EB}/V_T} - 1)\frac{w}{2} \quad (10.3.10)$$

where A is the area of the device cross section. This fact has important implications when the device is required to switch from one condition of operation to another, characterized by a different value of stored charge. Before steady state operation in the new condition can be achieved, the charge stored must be changed to the new value. This cannot be done instantaneously and imposes an upper limit on the speed of response of the device (cf. Secs. 10.4 and 10.9).

For NPN transistors, Eqs. (10.3.7) through (10.3.9) become

$$n_B(x) = n_{oB}\left[e^{V_{BE}/V_T}\left(1 - \frac{x}{w}\right)\right] \quad (10.3.11)$$

$$p_E(x) = p_{oE}[1 + (e^{V_{BE}/V_T} - 1)e^{(x + x_E)/L_{hE}}] \quad (10.3.12)$$

$$p_C(x) = p_{oC}[1 - e^{-(x - x_C)/L_{hC}}] \quad (10.3.13)$$

(see Prob. 10.7).

Electron and Hole Current Components (Fig. 10.2.4)

In Sec. 6.6 it was shown that the minority carriers move by diffusion. Under this assumption, the various current components can be found by applying (4.3.4) to the above concentration distributions. Therefore, from (10.3.7),

$$I_{Eh} = -qAD_{hB}\frac{dp_B}{dx}\bigg]_{x=0} \approx \frac{qAD_{hB}p_{oB}}{w}e^{V_{EB}/V_T} \quad (10.3.14)$$

where the $\approx$ sign indicates that the computation was based on the approximate formula (10.3.7). Similarly,

$$I_{Ch} = -qAD_{hB}\frac{dp_B}{dx}\bigg]_{x=w} \approx I_{Eh} \quad (10.3.15)$$

also limited to the first-order approximation based on the assumption $w \ll L_{hB}$. The implications of this approximation will be investigated later on. For the moment, bear in mind that (10.3.14) and (10.3.15) are first-order approximations.

For the electron current components, using (10.3.8),

$$I_{Ee} = qAD_{eE}\frac{dn_E}{dx}\bigg]_{x=-x_E} = \frac{qAD_{eE}n_{oE}}{L_{eE}}(e^{V_{EB}/V_T} - 1) \quad (10.3.16)$$

and, using (10.3.9),

$$I_{Ce} = qAD_{eC}\frac{dn_C}{dx}\bigg]_{x=x_C} = \frac{qAD_{eC}n_{oC}}{L_{eC}} = -I_{oC} \quad (10.3.17)$$

Terminal Currents (I_C, I_E, I_B)

From the above, using the current equilibrium conditions, the terminal currents can be obtained. Using (10.3.14) and (10.3.16),

$$I_E = I_{Eh} + I_{Ee} \approx qA\left(\frac{D_{hB}p_{oB}}{w} + \frac{D_{eE}n_{oE}}{L_{eE}}\right)e^{V_{EB}/V_T} - \frac{qAD_{eE}n_{oE}}{L_{eE}} \qquad (10.3.18)$$

Similarly, using (10.3.15) and (10.3.17),

$$-I_C = I_{Ch} + I_{Ce} \approx \frac{qAD_{hB}p_{oB}}{w}e^{V_{EB}/V_T} + \frac{qAD_{eC}n_{oC}}{L_{eC}} \qquad (10.3.19)$$

As will become apparent later on, it is useful to express (10.3.18) and (10.3.19) in a more symmetric form, which also better represents the physical meaning of the equations.

By simple algebraic manipulations, from (10.3.18),

$$I_E \approx qA\left[\left(\frac{D_{hB}p_{oB}}{w} + \frac{D_{eE}n_{oE}}{L_{eE}}\right)(e^{V_{EB}/V_T} - 1) + \frac{D_{hB}p_{oB}}{w}\right] \qquad (10.3.18a)$$

and, from (10.3.19), after some algebra,

$$I_C = -qA\left[\frac{D_{hB}p_{oB}}{w}(e^{V_{EB}/V_T} - 1) + \left(\frac{D_{eC}n_{oC}}{L_{eC}} + \frac{D_{hB}p_{oB}}{w}\right)\right] \qquad (10.3.19a)$$

and finally, from these, using current equilibrium (10.1.1),

$$I_B = -(I_E + I_C) \approx -qA\left[\frac{D_{eE}n_{oE}}{L_{eE}}(e^{V_{EB}/V_T} - 1) - \frac{D_{eC}n_{oC}}{L_{eC}}\right] \qquad (10.3.20a)$$

These equations are often, more concisely, written in terms of a set of coefficients as:

Device terminal currents:

$$\boxed{I_E = a_{11}(e^{V_{EB}/V_T} - 1) + a_{12}} \qquad (10.3.18b)$$

$$\boxed{I_C = -[a_{21}(e^{V_{EB}/V_T} - 1) + a_{22}]} \qquad (10.3.19b)$$

$$\boxed{I_B = -(I_E + I_C) = (a_{21} - a_{11})(e^{V_{EB}/V_T} - 1) + (a_{22} - a_{12})} \qquad (10.3.20b)$$

Computation of coefficients:

$$\boxed{a_{12} = a_{21} = qA\,\frac{D_{hB}p_{oB}}{w}} \qquad (10.3.21)$$

$$\boxed{a_{11} = qA\,\frac{D_{eE}n_{oE}}{L_{eE}} + a_{12}} \qquad (10.3.22)$$

$$\boxed{a_{22} = qA\,\frac{D_{eC}n_{oC}}{L_{eC}} + a_{21}} \qquad (10.3.23)$$

The formulas for the terminal currents should be compared with Eq. (6.7.20) for the forward biased diode. In both cases notice the exponential dependence on a junction voltage.

This formulation is limited to the active mode of operation (forward biased emitter, reverse biased collector) and to the first-order approximation of narrow base width [(10.3.7) and following]. A more general and accurate formulation will be developed in Sec. 10.7.

The above formulas have been obtained for a PNP structure. For NPN transistors the analysis is identical (see Prob. 10.8). It indicates that, in Eqs. (10.3.18b) through (10.3.20b), V_{EB} must be substituted with V_{BE} and that Eqs. (10.3.21) through (10.3.23) become

$$a_{12} = a_{21} = -qA \frac{D_{eB} n_{oB}}{w} \tag{10.3.21a}$$

$$a_{11} = -qA \frac{D_{hE} p_{oE}}{L_{hE}} + a_{12} \tag{10.3.22a}$$

$$a_{22} = -qA \frac{D_{hC} p_{oC}}{L_{hC}} + a_{21} \tag{10.3.23a}$$

Transistor Effect Parameters

Using the above formulation it is possible to compute the parameters defined in (10.2.2) and (10.2.3) to characterize the transistor effect. From (10.3.14) and (10.3.18), the emitter efficiency is

$$\gamma = \frac{I_{Eh}}{I_E} \approx \frac{1}{1 + \dfrac{D_{eE}}{D_{hB}} \dfrac{n_{oE}}{p_{oB}} \dfrac{w}{L_{eE}}} = \frac{1}{1 + \dfrac{D_{eE}}{D_{hB}} \dfrac{N_B}{N_E} \dfrac{w}{L_{eE}}} \tag{10.3.24}$$

where use was made of (3.3.13) and (3.3.14). Calculating the base transport factor from (10.3.14) and (10.3.15):

$$\alpha_T = \frac{I_{Ch}}{I_{Eh}} \approx 1 \tag{10.3.25}$$

This ideal value, corresponding to an infinitely narrow base, is obtained because (10.3.7) is a first-order approximation of the actual minority carrier concentration distribution in the base. If the narrow base assumption is dropped, differentiation of (10.3.5) yields rather complex formulas for I_{Eh} and I_{Ch}. The ratio of these two currents then yields a value that can be approximated by

$$\alpha_T \approx 1 - \frac{w^2}{2L_{hB}^2} \tag{10.3.26}$$

This second-order approximation is sufficient in most practical cases.

In Sec. 10.2 it was shown that gain optimization requires both γ and α_T to be close to unity. From (10.3.24) and (10.3.26) this suggests to the designer:

1. a narrow base (w much smaller than L_{hB} and L_{eE}) and
2. base doping much lower than emitter doping ($N_B \ll N_E$).

This condition also minimizes the mobility ratio D_{eE}/D_{hB}. For NPN structures, Eqs. (10.3.24) and (10.3.26) require self-evident modifications (see Prob. 10.10).

Equations (10.3.18) through (10.3.23) can be used directly to compute the BJT active mode currents for a given input voltage V_{EB}. However, the I/V characteristics of the BJT are usually displayed under constant input current conditions, so that the various currents must be computed for given I_E or I_B (rather than V_{BE}). This can be done using (10.2.6), and obtaining α_0 from (10.2.5), (10.3.24), and (10.3.26). The computation of I_{oC} appearing in (10.2.6) is often not necessary, but can be performed as shown in detail in Example 10.3.1.

Example 10.3.1. A Si PNP BJT at 300 K has device cross section $A = 1$ mm^2, base width 1 μm, $L_{eE} = L_{eC} = L_{hB} = 10$ μm, and dopings in the different semiconductors of $N_E = 2 \times 10^{18}$ cm^{-3}, $N_B = 10^{17}$ cm^{-3}, $N_C = 5 \times 10^{15}$ cm^{-3}. Compute I_C in the active mode for (a) $V_{EB} = 0.6$ V, (b) $I_E = 2.53$ mA, (c) $I_B = -9.15$ μA.

Solution. From (4.3.2) and (4.2.9) (or Table 4.2.1),

$$D_{hB} = 6.1 \text{ cm}^2/\text{s}; \qquad D_{eE} = 4.4 \text{ cm}^2/\text{s}; \qquad D_{eC} = 31 \text{ cm}^2/\text{s}$$

Also:

$$p_{oB} = 2.25 \times 10^{20}/10^{17} = 2250; \qquad n_{oE} = 112.5; \qquad n_{oC} = 45,000 \text{ cm}^{-3}$$

(a) From (10.3.21) through (10.3.23),

$$a_{12} = 1.6 \times 10^{-19} \times 10^{-2} \times 6.1 \times 2250/10^{-4} = 2.196 \times 10^{-13} \text{ A}$$

$$a_{11} = 1.6 \times 10^{-21} \times 4.4 \times 112.5/10^{-3} + a_{12} = 2.20392 \times 10^{-13} \text{ A}$$

$$a_{22} = 1.6 \times 10^{-21} \times 31 \times 45,000/10^{-3} + a_{21} = 2.4526 \times 10^{-12} \text{ A}$$

For $V_{EB} = 0.6$,

$$e^{0.6/0.0259} - 1 = 1.150471 \times 10^{10}$$

so that, from (10.3.19b),

$$I_C = -(2.196 \times 10^{-13} \times 1.15 \times 10^{10} + 2.4526 \times 10^{-12}) = -2.5254 \text{ mA}$$

For subsequent analysis of the results, compute I_E from (10.3.18b):

$$I_E = 2.196 \times 10^{-13} \times 1.15 \times 10^{10} + 2.20392 \times 10^{-13} = 2.534508 \text{ mA}$$

which implies $\alpha_0 = 2.5254/2.534508 = 0.9964064$.

(b) From (10.2.6), remembering (10.3.24) and (10.3.26),

$$\alpha_0 = \frac{1 - \dfrac{1}{2}\dfrac{10^{-8}}{10^{-6}}}{1 + \dfrac{4.4}{6.1}\dfrac{1}{20}\dfrac{1}{10}} = 0.99142$$

To compute I_{oC}, notice from (10.2.6) that it is equal to the value of I_C for $I_E = 0$. Introducing this value in (10.3.18b),

$$e^{V_{EB}/V_T} - 1 = -\frac{a_{12}}{a_{11}} = -0.9964064$$

indicating that, at open-circuited input, the transistor is slightly reverse biased. Indeed, $V_{EB} = 0.0259 \ \ln \ (1 - 0.9964064) = -0.1458$ V. Using this value in (10.3.19b), the value I_{oC} of the collector current with open-circuited emitter is

$$I_{oC} = -(2.196 \times 10^{-13} \times [-0.9964064] + 2.4516 \times 10^{-12})$$

$$= -2.2328 \times 10^{-12} \text{ A}$$

which is lower than the reverse saturation current of the collector junction, and so negligible in our computation. Finally, from (10.2.6),

$$I_C \approx -0.99142 \times 2.53 = -2.5083 \text{ mA}$$

This value is slightly less than the result obtained in part (a) for essentially identical conditions. As previously shown, the result of part (a) implies a value of $\alpha_0 \approx 0.9964064$, as compared with 0.99142. As will become apparent later, this difference is not negligible as it might appear at first sight. The discrepancy is due to the fact that in part (a) the use of (10.3.19b) implies the approximation of (10.3.15) and so a base transport factor $\alpha_T = 1$. The computation of part (b) employs the second-order approximation in computing α_T. In some cases, this factor can be the most important quantity in determining α_0.

Another, more direct, way of solving part (b) is to set $I_E = 2.53$ mA in (10.3.18b). Then:

$$e^{V_{EB}/V_T} - 1 = \frac{2.53 \times 10^{-3} - 2.196 \times 10^{-13}}{2.20392 \times 10^{-13}} = 1.148 \times 10^{10}$$

from which $V_{EB} \approx 0.599$ V, as expected. Now, from (10.3.19b),

$$I_C = -(2.196 \times 10^{-13} \times 1.148 \times 10^{10} + 2.4526 \times 10^{-12}) = -2.521 \text{ mA}$$

in good agreement with the previous value.

(c) Setting $I_B = -9.15 \times 10^{-6}$ in (10.3.20b):

$$e^{V_{EB}/V_T} - 1 = \frac{-9.15 \times 10^{-6} + (2.196 - 24.526) \times 10^{-13}}{(2.196 - 2.20392) \times 10^{-13}} = 1.15 \times 10^{10}$$

from which $V_{EB} = 0.6$. Now, from (10.3.19b),

$$I_C = -(2.196 \times 10^{-13} \times 1.15 \times 10^{10} + 2.4526 \times 10^{-12}) = -2.514 \text{ mA}$$

Repeated computations of this type can be used to draw by points the I/V characteristics of the BJT. This task is laborious enough to warrant the time and effort to generate computer programs for its implementation. The student with some computer experience will find that writing such a program may contribute to a deeper understanding of, and certainly to a greater familiarity with, the mechanics of transistor operation. It is suggested that the programs be made interactive, allowing choices of the most important design parameters for a possibly enlightening student experimentation and practice. This computer analysis can then be extended to somewhat more sophisticated equations, as described later on. This will permit comparison of the more accurate computations with those of rougher approximations, such as the computations of the previous example.

10.4 BJT CONFIGURATIONS AND SMALL-SIGNAL PARAMETERS

In many analog applications, the transistor operates under small-signal conditions. In this case, as usual, its behavior can be described in terms of small-signal equivalent linear circuits, characterized by small-signal parameters depending on the bias conditions. In transistor circuits it is customary to define a set of h parameters depending on the configuration in which the transistor is used in the circuit.

Three configurations are possible, depending on which electrode is set at reference signal potential (ground): common base, common emitter, and common collector (or emitter follower). These configurations are shown in Fig. 10.4.1. The small-signal parameters can be computed directly from their definitions using the set of equations of Sec. 10.3, which, of course, hold for all configurations.

The parameters for different configurations are related to each other, so that, once the parameters for one configuration are known, those for other configurations can be computed from them. For instance, from (10.2.6), using the current equilibrium equation (10.1.1) of the device,

$$I_C = \alpha_0(I_C + I_B) + I_{oC} \tag{10.4.1}$$

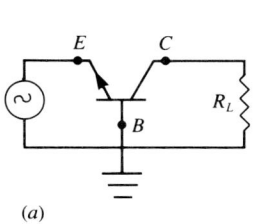 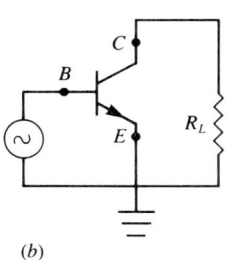 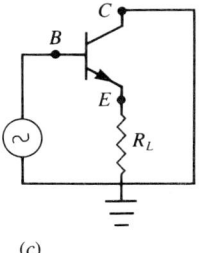

(a) (b) (c)

FIGURE 10.4.1
The three basic configurations of transistor amplifiers: (a) common base, (b) common emitter, (c) common collector (emitter follower).

from which

$$I_C = \frac{\alpha_0}{1 - \alpha_0} I_B + \frac{I_{oC}}{1 - \alpha_0} = \beta_0 I_B + I_{oCE} \tag{10.4.2}$$

where $\beta_0 = \alpha_0/(1 - \alpha_0)$ and $I_{oCE} = I_{oC}/(1 - \alpha_0)$ are the common emitter current gain and the common emitter inverse saturation current respectively.

A set of very useful approximate values for the magnitudes of the most important common base and common emitter parameters[3] is obtained using the approximate formulas and further approximating on the assumption that, at the chosen work point, the transistor is well out of cutoff (i.e., $V_{BE} > V_T$), so that the terms containing the exponential dependence on V_{BE} are much larger than the others.

For the *mutual conductance*, from (10.3.19),

$$g_m = \frac{\partial I_C}{\partial V_{EB}} \approx \frac{q A D_{hB} p_{oB}}{w V_T} e^{V_{EB}/V_T} \approx \frac{|I_C|}{V_T} \tag{10.4.3}$$

which, at 300 K, reduces to the familiar expression:

$$g_m \approx 40 |I_C| \tag{10.4.4}$$

The mutual conductance magnitude is essentially the same for the common base and common emitter configurations.

For the short-circuit forward *current gain*, in the common base configuration, from (10.3.18) and (10.3.19),

$$h_{fB} = \frac{\partial I_C}{\partial I_E} = \frac{\partial I_C}{\partial V_{BE}} \frac{\partial V_{BE}}{\partial I_E} \approx \frac{1}{1 + \dfrac{D_{eE}}{D_{hB}} \dfrac{n_{oE}}{p_{oB}} \dfrac{w}{L_{eE}}} \tag{10.4.5}$$

showing that, within these approximations, the *dynamic* current gain h_{fB} coincides with the *static* current gain α_0 [cf. (10.3.24) and (10.3.25)]. For the common emitter configuration, from (10.3.19) and (10.3.20a), analogously:

$$h_{fE} = \frac{\partial I_C}{\partial I_B} \approx \frac{D_{hB}}{D_{eE}} \frac{p_{oB}}{n_{oE}} \frac{L_{eE}}{w} \approx \frac{h_{fB}}{1 - h_{fB}} \tag{10.4.6}$$

in accord with (10.4.2). This equation shows that the design criteria 1 and 2 listed at the end of Sec. 10.3 after Eq. (10.3.26) also maximize h_{fE}.

For the *input impedance*, common base configuration, from (10.3.18),

$$h_{iB} = \frac{\partial V_{EB}}{\partial I_E} \approx \frac{V_T}{|I_E|} \tag{10.4.7}$$

---

[3] The sign of the parameters depends on the conventions adopted for the signs of transistor currents and of transistor bias voltages and on the transistor type (NPN or PNP). Here only the magnitude of these parameters is computed.

which should be compared with (6.8.1) and, at 300 K, reduces to the familiar formula

$$h_{iB} \approx \frac{1}{40|I_E|} \qquad (10.4.8)$$

In the common emitter configuration,

$$h_{iE} = \frac{\partial V_{BE}}{\partial I_B} = \frac{\partial I_C}{\partial I_B} \frac{\partial V_{BE}}{\partial I_C} = h_{fE} \frac{1}{g_m} \approx h_{fE} \frac{V_T}{|I_C|} \approx \frac{V_T}{|I_B|} \qquad (10.4.9)$$

For the *output admittance*, noticing that I_C in (10.3.19) is independent of the collector voltage V_C,

$$h_{oE} = \frac{\partial I_C}{\partial V_{CE}} \approx 0 \approx h_{oB} \qquad (10.4.10)$$

This is manifestly a very rough approximation, indicating that, in obtaining (10.3.19), some very important phenomena were overlooked. Finite values of the dynamic output resistance are obtained taking into account the base width modulation phenomenon (cf. Sec. 10.6). Experimental values, of the order of 10^4 to 10^6 Ω, are usually provided by the manufacturer.

In Sec. 10.3 it was noted that the charge stored in the base must be varied whenever a voltage is applied, so that the transistor represents a capacitive impedance to input signals. From (10.3.10),

$$C_{EB} = \frac{\partial Q_B}{\partial V_{EB}} = \frac{qAp_{oB}w}{2V_T} e^{V_{EB}/V_T} = g_m \frac{w^2}{2D_{hB}} \qquad (10.4.11)$$

where use was made of (10.4.3). Comparing (10.4.11) with (6.8.3), it becomes apparent that this is a diffusion capacitance, in which $w/2$ takes the place of the diffusion length because of the narrow base hypothesis. It can be shown (cf. App. 10A) that this diffusion capacitance is related to the finite *transit time* τ_w required for an injected minority carrier to cross the base.

The above diffusion capacitance is usually the most significant, but not the only transistor capacitance, so that C_{EB} is actually larger than predicted by (10.4.11). Other components are the depletion capacitance of the forward biased emitter junction and of the reverse biased collector junction [cf. (6.5.3)], the collector depletion region transit capacitance, etc. Experimental values for the transistor capacitances are sometimes furnished by the manufacturer, either directly or in the form of critical high-frequency parameters (cf. Sec. 10.5). For order-of-magnitude computations, (10.4.11) and (6.5.3) are usually sufficient for small-signal operation.

Although representing only first approximations, the above formulas yield acceptable values of transistor parameters for most order-of-magnitude computations, especially as they enjoy the advantage of being very simple and so prove to be very useful to the circuit designer whenever the use of published experimental data is impractical.

Example 10.4.1. For the BJT of Example 10.3.1 compute the small-signal parameters at a work point characterized by $I_C = -2.5$ mA.

Solution. From (10.4.3), $g_m \approx 40 \times 2.5 = 100$ mmho.
From (10.4.5), $h_{fB} \approx 0.9964064$, as seen in Example 10.3.1; however, a more reliable value is obtained from (10.3.24) and (10.3.26): $h_{fB} \approx 0.99142$. From this value, using (10.4.6),

$$h_{fE} \approx \frac{0.99142}{1 - 0.99142} = 115$$

Notice that use of the other, apparently closely approximate, value would yield $h_{fE} \approx 227$, almost double the previous one. From (10.4.8),

$$h_{iB} \approx \frac{1}{40 I_E} \approx 10 \ \Omega$$

where, because h_{fB} is so close to 1, $I_E \approx I_C$ has been assumed. From (10.4.9),

$$h_{iE} \approx \frac{115}{40 \times 2.5} = 1153 \ \Omega$$

Use of the other approximation for h_{fE} would yield $h_{iE} \approx 2768 \ \Omega$. For the moment h_{oB} and h_{oE} will be approximately assumed to be zero in accord with (10.4.10). A more sophisticated computation will be presented later. From (10.4.11),

$$C_{EB} \approx \frac{0.1 \times 10^{-8}}{2 \times 6.1} = 82 \text{ pF}$$

As stated in the text, this diffusion capacitance is usually the most significant; however, in this case, the collector depletion capacitance, computed from (6.5.3), can be comparatively quite large. Remembering that for the junction dopings indicated $V_{bi} = 0.0259 \ln (5 \times 10^{32}/2.25 \times 10^{20}) = 0.74$ V, then the unbiased capacitance becomes

$$C_{CB} = \sqrt{\frac{1.6 \times 10^{-19} \times 1.04 \times 10^{-12} \times 5 \times 10^{15}}{2 \times 0.74}} \times 10^{-2} = 237 \text{ pF}$$

For a collector bias of 5 V, $C_{CB} \approx 85$ pF—not negligible with respect to C_{EB}.

10.5 SMALL-SIGNAL EQUIVALENT CIRCUITS AND FREQUENCY RESPONSE

Figure 10.5.1 shows a simplified small-signal equivalent circuit for the transistor in the common base configuration. The parameters can be computed from the formulas of Sec. 10.4.

The input capacitance C_{in} is a combination of the transistor capacitances, in which C_{EB} of (10.4.11) is usually the most important component. Because of C_{in}, not all of the input current I_E contributes to the transistor effect; some of it goes to charge the capacitor, and only the current I'_E contributes to the transfer

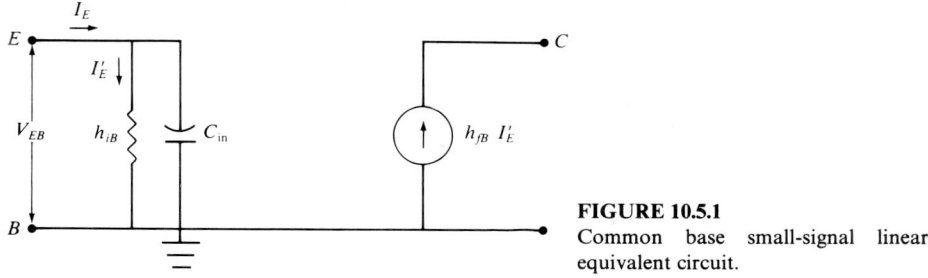

FIGURE 10.5.1
Common base small-signal linear equivalent circuit.

of current into the collector. From the current divider formula,

$$I'_E = I_E \frac{1/h_{iB}}{1/h_{iB} + j\omega C_{in}} = \frac{I_E}{1 + j\omega C_{in} h_{iB}} = \frac{I_E}{1 + j\omega/\omega_B} \qquad (10.5.1)$$

so that the current gain becomes

$$h_{fB}(\omega) = \frac{h_{fB}}{1 + j\omega/\omega_B} \qquad \text{with } \omega_B = \frac{1}{C_{in} h_{iB}} \qquad (10.5.2)$$

However, thanks to the relationship between h_{fE} and h_{fB},

$$h_{fE}(\omega) = \frac{h_{fB}(\omega)}{1 - h_{fB}(\omega)} \qquad (10.5.3)$$

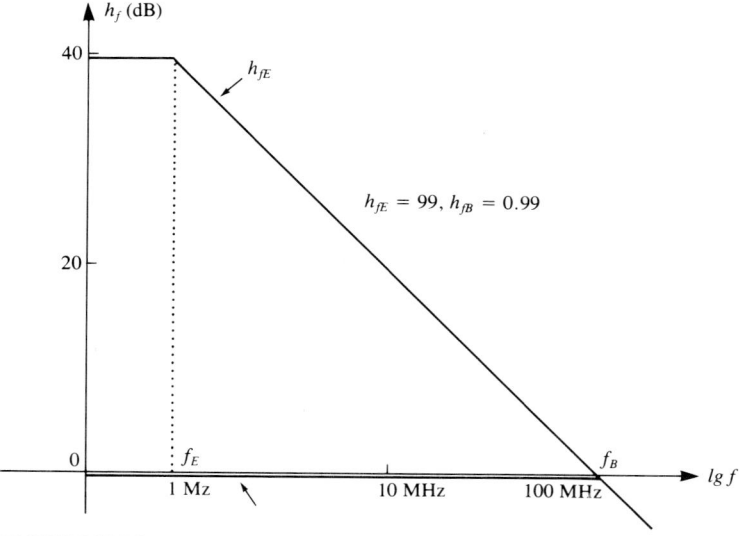

FIGURE 10.5.2
Frequency dependence of h_{fE} and h_{fB} for a transistor with low-frequency parameters $h_{fE} = 99$ and $h_{fB} = 0.99$. If $f_E = 1$ MHz, then $f_B = 100$ MHz.

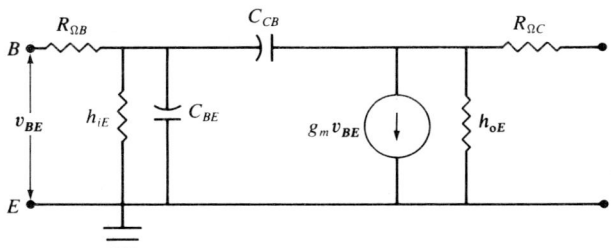

FIGURE 10.5.3
Common emitter small-signal linear equivalent circuit.

from which, using (10.5.2), after some algebra,

$$h_{fE}(\omega) = \frac{h_{fE}}{1 + j\omega/\omega_E} \qquad \text{with } \omega_E = \omega_B(1 - h_{fB}) \qquad (10.5.4)$$

Equations (10.5.2) and (10.5.4) show that the frequency response of the transistor has a high-frequency 3-dB point at ω_B for the common base and ω_E for the common emitter, with $\omega_E \ll \omega_B$, because $h_{fB} \approx 1$. The common emitter configuration therefore displays larger current gain but lower bandwidth than the common base. The gain-bandwidth products are essentially the same: $h_{fE}\omega_E = h_{fB}\omega_B$. The common emitter and common base current gains are compared at different frequencies in the semi-infinite asymptotic plots of Fig. 10.5.2.

Information about the high-frequency behavior of transistors is furnished by the manufacturer either in terms of the capacitances or of the 3-dB frequencies f_B and/or f_E, or in the form of a *cutoff frequency* f_{co} at which the common emitter current gain becomes unity. From (10.5.4), setting $h_{fE}(\omega_{co}) = 1$, after algebra,

$$f_{co} = f_E \sqrt{h_{fE}^2 - 1} \approx f_E h_{fE} \qquad (10.5.5)$$

A useful small-signal equivalent circuit for the transistor in the common emitter configuration is shown in Fig. 10.5.3. This circuit, which contains series ohmic base and collector resistances (cf. Sec. 10.6), lends itself well to computer-aided linear circuit analysis of the PCAP and ECAP types.

Example 10.5.1. For the BJT of Example 10.4.1 in the configuration of Fig. 10.5.4 (not to clutter the circuit it is assumed that the current generator i_E permits

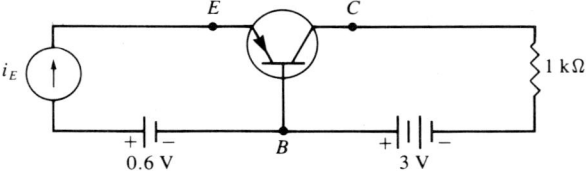

FIGURE 10.5.4
Common base amplifier schematic for Example 10.5.1. To avoid complicating the figure it is assumed that the small-signal current generator i_E permits unhindered flow of the dc component of the base current (notice that this is contrary to usual definitions).

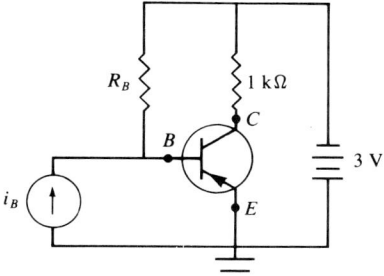

FIGURE 10.5.5
Common emitter amplifier schematic for Example 10.5.1.

unhindered flow of the dc bias current), draw the equivalent small-signal circuit, find the low-frequency voltage and current gain, and compute the high-frequency 3-dB point. Repeat for the configuration of Fig. 10.5.5, assuming the same work point as for Fig. 10.5.4 and $R_B \gg h_{iE}$.

Solution. The equivalent circuit is shown in Fig. 10.5.6, with the parameter values computed in Example 10.4.1. The effect of C_{EB} can be taken into account by assuming h_{fB} to be frequency dependent, in accordance with (10.5.2), with

$$\omega_B = \frac{1}{82 \times 10^{-12} \times 10} = 1.219 \times 10^{10} \text{ rad/s, or a 3-dB frequency}$$

$$f_B = \frac{\omega_B}{2\pi} = 194 \text{ MHz}$$

The voltage gain is

$$A = h_{fB}(\omega)1 \text{ K}\Omega/10 \text{ }\Omega = \frac{0.9914 \times 100}{1 + jf/194 \times 10^6} = \frac{99.44}{1 + jf/194 \times 10^6}$$

The effect of the output capacitance is to introduce a second break frequency at

$$f'_B = \frac{1}{2\pi C_{CB} R_L} = \frac{1}{2\pi \times 85 \times 10^{-12} \times 10^3} = 1.87 \text{ MHz}$$

The equivalent circuit for the common emitter configuration of Fig. 10.5.5 is shown in Fig. 10.5.7. Low-frequency gain is

$$A = -g_m R_L = -100$$

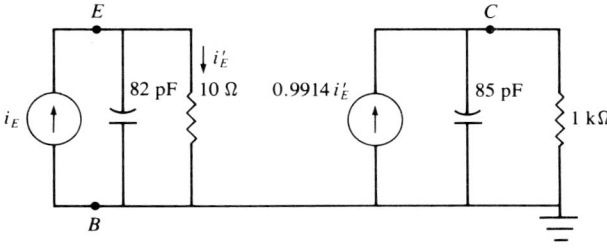

FIGURE 10.5.6
Small-signal equivalent circuit for Fig. 10.5.4.

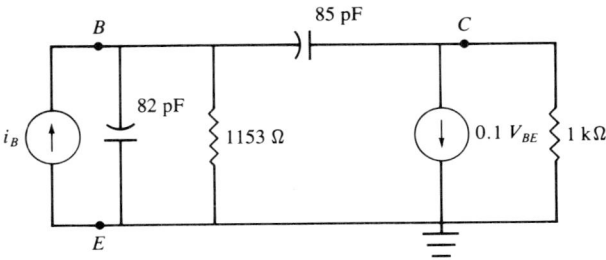

FIGURE 10.5.7
Small-signal equivalent circuit for Fig. 10.5.5.

High-frequency corner point, from (10.5.4), is

$$f_E = 194 \times (1 - 0.99142) = 1.67 \text{ MHz}$$

Another corner point exists due to the Miller effect capacitance.

10.6 ADDITIONAL PHENOMENA— HETEROJUNCTION BJT

In developing the elementary analysis of Secs. 10.3 through 10.5, only the most important phenomena occurring in the transistor structure were considered. The behavior of real transistors is influenced by several other phenomena. Following is a brief introduction to some of them and to their effects on transistor characteristics.

Base Width Modulation

Equation (10.3.19) indicates that the collector current is independent of the collector voltage. However, Fig. 10.3.1 shows that the active base width w equals the geometric base width w_0 minus the widths of the base depletion regions x_{BC} and x_{BE}. As stated in Sec. 6.5, x_{BC} varies with the collector reverse bias voltage; when V_C increases, w decreases (Early effect) and consequently I_C increases [cf. (10.3.19) and (10.3.21)].

As already mentioned, the Early effect results in a finite collector resistance.

Example 10.6.1. For the BJT of Example 10.5.1, but assuming a geometric base width $w_0 = 0.42 \ \mu\text{m}$, compute h_{oB} at an input current $I_E = 2.5$ mA and h_{oE} at an input $I_B = 5.63 \ \mu\text{A}$.

Solution. Neglecting the width of the forward biased emitter junction depletion region, the active base width is $w = w_0 - x_{CB}$ (cf. Fig. 10.3.1). From Example 10.4.1, $V_{bi} = 0.74$ V, so that, from (6.3.23),

at $V_{CB} = 0 \rightarrow x_{CB}(0) = 0.02 \ \mu\text{m}$ and $w(0) = 0.42 - 0.02 = 0.4 \ \mu\text{m}$

at $V_{CB} = 5 \rightarrow x_{CB}(5) = 0.06 \ \mu\text{m}$ and $w(5) = 0.42 - 0.06 = 0.36 \ \mu\text{m}$

Substituting these values in the expression for α [cf. (10.3.24) and (10.3.26)],

$$\alpha_0(0) = \frac{1 - \dfrac{1}{2} \dfrac{0.16 \times 10^{-8}}{10^{-6}}}{1 + \dfrac{4.4 \times 10^{17} \times 0.4 \times 10^{-4}}{6.1 \times 2.10^{18} \times 10^{-3}}} = 0.99775$$

and, analogously,

$$\alpha_0(5) = 0.99805$$

Approximating the collector current from (10.2.6) $(I_{oC} \ll)$ and using the above values,

$$I_C(0) \approx -0.99775 \times 2.5 = -2.494375 \text{ mA}$$

$$I_C(5) \approx -0.99805 \times 2.5 = -2.495125 \text{ mA}$$

The output admittance is approximated by $\Delta I_C / \Delta V_{CB}$, by definition, so that

$$h_{oB} \approx \frac{2.495125 - 2.494375}{5 - 0} = 1.5 \times 10^{-4} \text{ mmho (6.7 M}\Omega\text{)}$$

Analogously, from the above values of α_0, using (10.4.2),

$$\beta_0(0) \approx \frac{0.99775}{1 - 0.99775} = 443$$

$$\beta_0(5) \approx \frac{0.99805}{1 - 0.99805} = 512$$

Computing the approximate collector currents at the given I_B,

$$I_C(0) \approx -443 \times 5.63 \times 10^{-6} = -2.49409 \text{ mA}$$

$$I_C(5) \approx -512 \times 5.63 \times 10^{-6} = -2.88256 \text{ mA}$$

Thus, computing the common emitter output admittance:

$$h_{oE} \approx \frac{2.88256 - 2.49409}{5} = 0.0777 \text{ mmho (13 k}\Omega\text{)}$$

The common emitter output impedance is much smaller than in the common base configuration $[h_{oE} \approx h_{oB}/(1 - \alpha_0)]$.

Base Spreading Resistance and Current Crowding

Figure 10.6.1 shows the path of the current I_B for a PNP transistor having a geometric structure similar to the one shown in Fig. 10.1.2. The base region, being very thin, has a large ohmic resistance, so that the flow of base current produces voltage drops, and the potential in the base material decreases as the base terminal is approached. The emitter junction forward bias is therefore low at point A, larger at point B, and so on, with the result that the current density

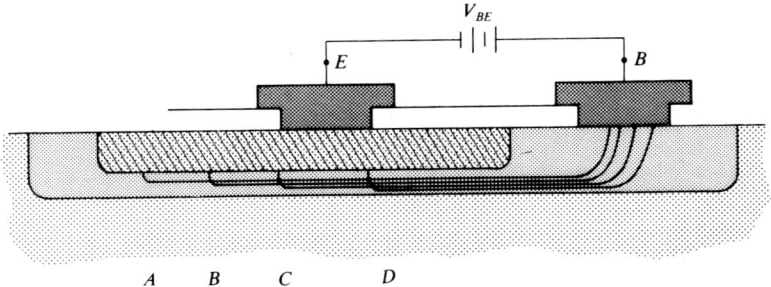

FIGURE 10.6.1
NPN transistor cross section showing paths of I_B current. Notice that voltage drops along the base vary the emitter junction voltage from a maximum near the base edge (D) to lower values further away (A).

increases toward the edge of the emitter region next to the base contact (*current crowding*).

This can give rise to local high-level injection, heating, failure, and other undesirable effects. A common artifice to make the current density distribution more uniform is the adoption of the interdigitated geometry shown in Figs. 10.6.2 and 5.4.1 in which the edge region between the base and emitter represents a large portion of the structure.

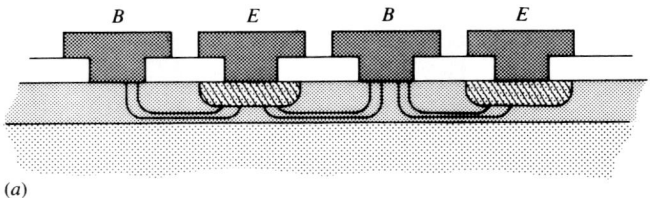

(a)

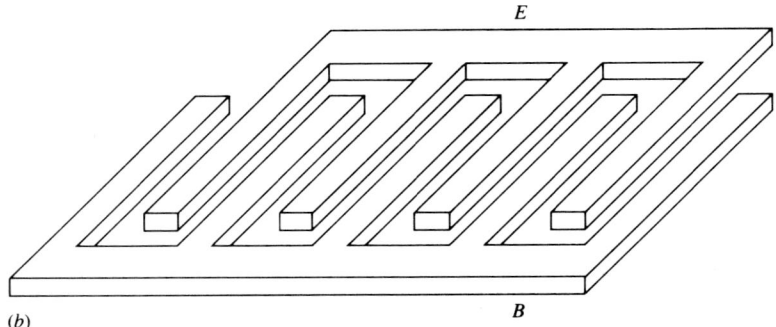

(b)

FIGURE 10.6.2
Interdigitated emitter-base geometry. (*a*) Schematic cross section showing I_B current paths (compare with Fig. 10.6.1). (*b*) Three-dimensional schematic (compare with Fig. 5.4.1).

High Injection Effects

High current operating conditions, current crowding, etc., can cause the rate of minority carrier injection to exceed the limits of low-level injection (cf. Sec. 4.4). This may invalidate the assumption that no generation and recombination occur in the depletion region and the other assumptions used in writing the continuity equation affecting the transistor characteristics. An analysis of this phenomenon is beyond the limits of this presentation.

Graded Junctions

In real transistors, the junctions, obtained by diffusion or implantation, are not abrupt but graded. As the base is very narrow, the nonuniform distribution of impurities extends to the whole base region and the resulting variable spatial charge density generates an electric field, which helps push the minority carriers across the base. This drift tends to decrease the base transit time, improving high-frequency performance and increasing base transport efficiency.

The effect of a graded distribution of impurities in the base can be roughly taken into account by considering the quantity p_{oB}/w appearing in the equations of Sec. 10.3. Remembering (3.3.13),

$$\frac{p_{oB}}{w} = \frac{n_i^2}{N_B w} \tag{10.6.1}$$

where N_B is the impurity concentration in the base and, in the case of uniform doping (abrupt junction), $N_B w$ is the number of impurity atoms in the base per unit interface area. If the dopant concentration in the base is variable, then this quantity becomes

$$\int_0^w N_B(x)\, dx \tag{10.6.2}$$

An often-acceptable description of the behavior of graded junction transistors can be obtained by using the formulas of Secs. 10.3, 10.4, and 10.5, but performing the substitution:

$$\frac{p_{oB}}{w} \to \frac{n_i^2}{\int_0^w N_B(x)\, dx} \tag{10.6.3}$$

Example 10.6.2. For the BJT of Example 10.3.1, but with graded base doping $N_B(x) = 10^{17}(1 - x/w)$, compute the collector current and h_{fE} at $I_E = 2.53$ mA and compare with the results of Examples 10.3.1 and 10.4.1.

Solution. From (10.6.3), using the given doping profile:

$$\int_0^{10^{-4}} 10^{17}(1 - 10^4 x)\, dx = 5 \times 10^{12}$$

operating the substitution (10.6.3) in the formula for α_0 [cf. (10.3.24) and (10.3.26)]:

$$\alpha_0 \approx 0.99695$$

yielding

$$I_C \approx 0.99695 \times 2.53 = 2.5223 \text{ mA}$$

This shows an increment above the corresponding value of 2.5083 mA computed in Example 10.3.1. The difference may appear negligible, but the common emitter current gain becomes

$$h_{fE} = \frac{0.99695}{1 - 0.99695} = 327$$

almost three times larger than the corresponding value of 115 computed in Example 10.4.1 for the abrupt junction case.

Heterojunction BJT

It was shown [cf. Eq. (10.3.24)] that, to maximize emitter efficiency, it is necessary to make the ratio N_B/N_E of the base to emitter doping as small as possible. This minimizes recombination in the emitter, making the emitter majority carrier component higher than the base majority carrier component ($I_{Eh} > I_{Ee}$ in a PNP transistor). Low doping in the base, however, results in high base resistivity, causing current crowding, and tends to increase the width of the base depletion region at the collector junction, increasing base width modulation and so decreasing the output impedance.

Improved operation can be obtained by using a heterojunction at the emitter-base interface. As shown in Sec. 6.10, appropriate choices of bandgaps can distort the relative heights of the conduction band vs. valence band barriers. In a PNP transistor, for instance, increasing the conduction band barrier while decreasing the valence barrier (as shown in Fig. 6.10.2) results in a higher hole current component and lower electron component, increasing the emitter efficiency $I_{Eh}/(I_{Eh} + I_{Ee})$. For an NPN transistor, the equivalent result would be obtained by reversing the barrier height variations, favoring electron over hole current component.

Using a heterojunction it is therefore possible to obtain excellent emitter efficiencies, even with comparatively high doping in the base. As previously noted this results in transistors with high β, low current crowding, high output impedance and improved high-frequency performance.

Real Characteristics

The above phenomena and others influence real transistor characteristics. A set of experimentally obtained BJT I/V characteristics is shown in Fig. 10.6.3 for a PNP and Fig. 10.6.4 for an NPN transistor. Comparing them with the ideal

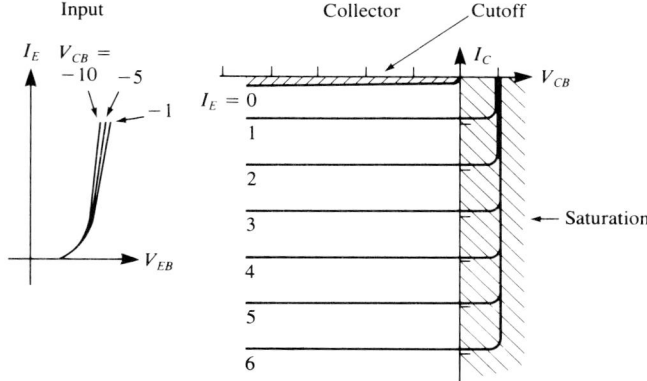

(a) Common base

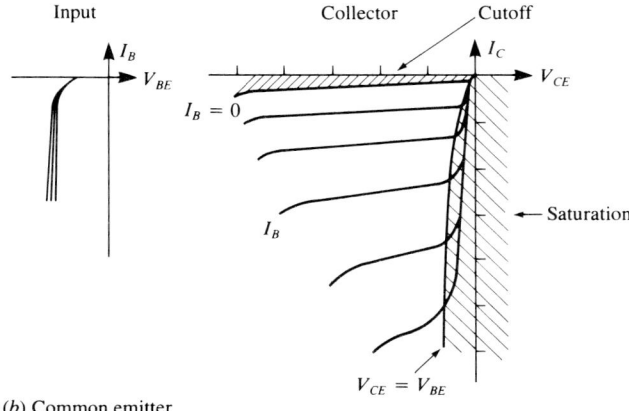

(b) Common emitter

FIGURE 10.6.3
PNP BJT characteristics showing active, cutoff, and saturation regions.

characteristics obtained in Secs. 10.3 through 10.5, a number of differences are apparent:

1. The real characteristics are not limited to the active mode (forward biased emitter junction and reverse biased collector junction), which, as stated in Sec. 10.2, is the only mode of operation described by the previous theory. Transistor behavior in the other modes is described later in Sec. 10.8. Only a limited region of the real characteristics corresponds to the active mode. As shown in the figures, the boundaries of the active region are:
   (a) For the common base configuration:
      (i) Curve I_{oC} ($I_E = 0$), where the emitter approximately ceases to be forward biased (for further details cf. Sec. 10.8 under "cutoff mode").

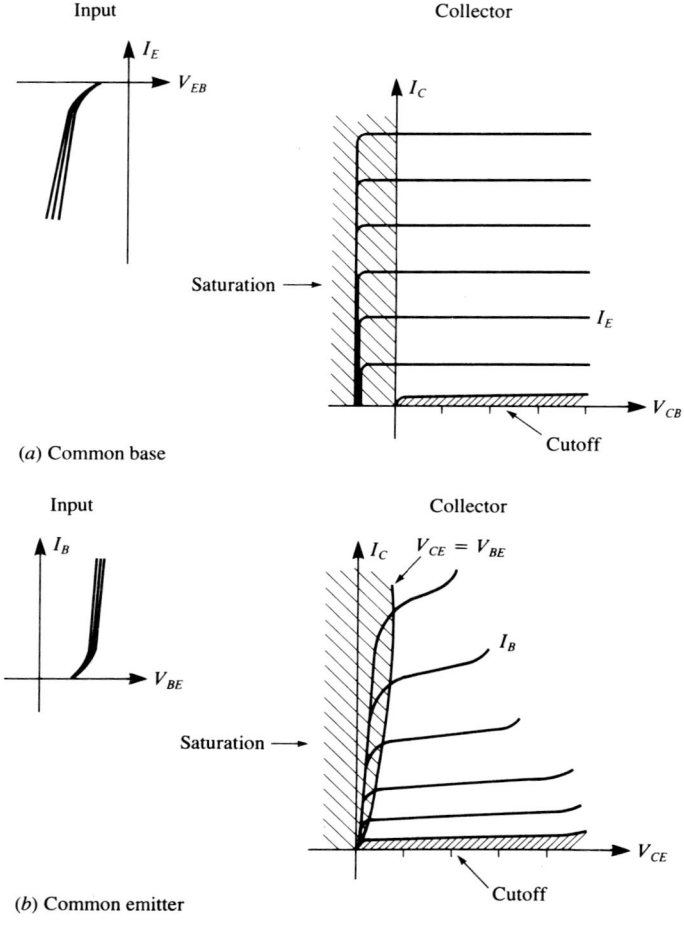

Input   Collector

(a) Common base

Input   Collector

(b) Common emitter

FIGURE 10.6.4
NPN BJT characteristics. Compare with Fig. 10.6.3.

   (ii) The I_C axis ($V_{CB} = 0$), where forward bias of the collector junction
    begins.
  (b) For the common emitter configuration:
   (i) Curve I_{oCE} ($I_B = 0$), approximately indicating the end of the emitter
    junction forward bias.
   (ii) The curve $V_{BE} = V_{CE}$, which is the locus of $V_{CB} = V_{CE} - V_{BE} = 0$
    where the collector junction bias becomes zero.
2. The input characteristics are seen to depend on the collector voltage. In
accordance with the Early effect, increasing the collector junction negative bias
results in a narrower active base width w. This in turn increases I_E [cf.
(10.3.18)] and α_0 [cf. (10.3.24) and (10.3.25)]. The common emitter character-

istics are even more strongly affected, because a small variation of α_0 results in a very large variation of β_0, as shown by (10.4.2).

3. The collector I/V characteristics have a slope, implying a finite output impedance. This slope is also a consequence of the Early effect and is small for the common base but significant in the common emitter configuration (cf. Example 10.6.1).

4. The collector current does not drop as soon as the collector bias goes to zero. Instead, it continues to follow the predictions of the active region theory even when the collector is forward biased, up to about 1 V. This phenomenon is discussed later in Sec. 10.8 under "saturation mode."

5. At high collector voltages the collector current suddenly tends to increase. This is the onset of the collector junction breakdown and will be discussed in Sec. 10.10.

10.7 THE EBERS-MOLL MODEL

The previous analysis has been limited to the assumption that the emitter is forward biased and the collector reverse biased. Additional assumptions (e.g., small signal) further limit the range of validity of the formulation and of the equivalent circuits.

A model of more general validity and of great conceptual value can be obtained by directly implementing in circuit form the mechanism of operation of the BJT. According to the description of Sec. 10.2, the BJT consists of two individually biased diodes connected back to back as shown in Fig. 10.7.1a, where

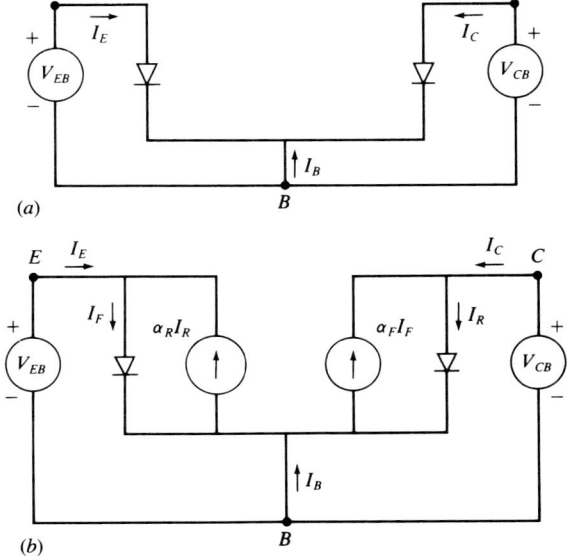

(a)

(b)

FIGURE 10.7.1
Equivalent circuits for a transistor structure. (a) Very wide base (no transistor action). Compare with Fig. 10.2.2 or 10.2.3. (b) Base width smaller than the base minority carrier diffusion length. Compare with Fig. 10.2.4.

the bias generator polarities and current directions are only indications of the positive sign conventions for these quantities, which, in practice, can all take on either positive or negative values. In any case, all currents can be computed from the theory developed for the PN junction [cf. (6.7.20)].

However, because in the BJT the base is very narrow, two modifications must be introduced:

1. As discussed in Sec. 10.3, the assumption of a narrow base modifies the boundary conditions under which the continuity equation must be solved. For the case considered (PNP), a first-order approximation (10.3.7) shows that I_{Eh}, the hole component of the emitter junction current, depends on the active base width w rather than on the diffusion length L_{hB}, as indicated in Eq. (10.3.14) [notice that, if the base were very wide, this current would be given by Equation (6.7.17) with $x = x_N$].

2. The transistor effect discussed in Sec. 10.2 requires that a portion α of each of the diode currents be transferred into the neighboring diode circuit.

Introducing these modifications in Fig. 10.7.1a leads to the circuit of Fig. 10.7.1b in which the current transfers due to the transistor effect are represented by the two dependent current generators $\alpha_F I_F$ and $\alpha_R I_R$. In Fig. 10.7.1b, currents are designated as "forward" (subscript F), if propagating from emitter to collector, and as "reverse" (subscript R), if propagating in the opposite direction.

The diode currents are computed from (6.7.20), modified as discussed above, under modification 1, resulting in

$$I_F = qA\left(\frac{D_{hB}\,p_{oB}}{w} + \frac{D_{eE}\,n_{oE}}{L_{eE}}\right)(e^{V_{EB}/V_T} - 1) = a_{11}(e^{V_{EB}/V_T} - 1) \qquad (10.7.1)$$

and
$$I_R = qA\left(\frac{D_{hB}\,p_{oB}}{w} + \frac{D_{eC}\,n_{oC}}{L_{eC}}\right)(e^{V_{CB}/V_T} - 1) = a_{22}(e^{V_{CB}/V_T} - 1) \qquad (10.7.2)$$

From Fig. 10.7.1b, the terminal currents are therefore

$$\boxed{I_E = I_F - \alpha_R I_R = a_{11}(e^{V_{EB}/V_T} - 1) - \alpha_R a_{22}(e^{V_{CB}/V_T} - 1)} \qquad (10.7.3)$$

$$\boxed{I_C = -\alpha_F I_F + I_R = -\alpha_F a_{11}(e^{V_{EB}/V_T} - 1) + a_{22}(e^{V_{CB}/V_T} - 1)} \qquad (10.7.4)$$

Notice that, if the collector junction is reverse biased, as is the case for formulas (10.3.18b) and (10.3.19b), then $V_{CB} \ll 0$ and, comparing (10.7.3) with (10.3.18b) and (10.7.4) with (10.3.19b),

$$\alpha_F a_{11} = a_{21} \qquad (10.7.5)$$

$$\alpha_R a_{22} = a_{12} \qquad (10.7.6)$$

so that

$$\alpha_F = \frac{a_{21}}{a_{11}} = \frac{1}{1 + \dfrac{D_{eE}}{D_{hB}} \dfrac{n_{oE}}{p_{oB}} \dfrac{w}{L_{eE}}} \tag{10.7.7}$$

$$\alpha_R = \frac{a_{12}}{a_{22}} = \frac{1}{1 + \dfrac{D_{eC}}{D_{hB}} \dfrac{n_{oC}}{p_{oB}} \dfrac{w}{L_{eC}}} \tag{10.7.8}$$

in perfect agreement with (10.3.24), so that the theory of Sec. 10.3 simply describes a special case (reverse biased collector) of the Ebers-Moll model.

As already pointed out in Sec. 10.3 and in Example 10.3.1, Eqs. (10.7.1) through (10.7.8), just as (10.3.7) through (10.3.23), are based on the first-order approximation of small base width w adopted in obtaining Eq. (10.3.7). As a consequence of this approximation, (10.7.7) yields $\alpha_F = \gamma$, implying $\alpha_T = 1$. In Example 10.4.1 it was shown that this can lead to gross miscalculations of the value of h_{fE}.

Whenever necessary, greater accuracy can be achieved at the expense of speed of calculation by abandoning the approximation of Eq. (10.3.7) and using (10.3.5) instead. This leads to the following new values of the Ebers-Moll parameters of Eq. (10.3.21) through (10.3.23):

$$\boxed{a_{12} = a_{21} = qA \frac{D_{hB} p_{oB}}{L_{hB}} \frac{1}{\sinh (w/L_{hB})}} \tag{10.7.9}$$

$$\boxed{a_{11} = qA \frac{D_{eE} n_{oE}}{L_{eE}} + a_{12} \cosh \left(\frac{w}{L_{hB}}\right)} \tag{10.7.10}$$

$$\boxed{a_{22} = qA \frac{D_{eC} n_{oC}}{L_{eC}} + a_{21} \cosh \left(\frac{w}{L_{hB}}\right)} \tag{10.7.11}$$

Equations (10.3.18b) through (10.3.20b) and (10.7.3) through (10.7.8) can still be used without modification to yield the desired greater accuracy, provided the Ebers-Moll parameters that appear in them are computed by means of Eqs. (10.7.9) through (10.7.11) instead of Eqs. (10.3.21) through (10.3.23). For instance, introducing the more accurate values of a_{21} and a_{11} in (10.7.7) and then approximating to the second order of small w values, it is possible to obtain for α_0 the value $\gamma\alpha_T$ predicted by Eqs. (10.3.24) and (10.3.26). The same holds true for the computation of α_F and α_R.

For NPN transistors, in Eqs. (10.7.3) and (10.7.4), V_{EB} is substituted by V_{BE} and V_{CB} by V_{BC}. Equations (10.7.9) through (10.7.11) become

$$a_{12} = a_{21} = -qA \frac{D_{eB} n_{oB}}{L_{eB}} \frac{1}{\sinh (w/L_{eB})} \tag{10.7.9a}$$

$$a_{11} = -qA \frac{D_{hE} P_{oE}}{L_{hE}} + a_{12} \cosh\left(\frac{w}{L_{eB}}\right)$$ (10.7.10a)

$$a_{22} = -qA \frac{D_{hC} P_{oC}}{L_{hC}} + a_{21} \cosh\left(\frac{w}{L_{eB}}\right)$$ (10.7.11a)

(see Prob. 10.16).

Unfortunately, use of (10.7.9) through (10.7.11) significantly slows down computer implementation, so that, for long calculations, (10.3.21) through (10.3.23) are often preferred.

For an example of the use of the Ebers-Moll model and equations, cf. Example 10.8.2 in the next section.

10.8   SATURATION, CUTOFF, AND INVERTED MODES

The active mode is of particular importance for analog operation, but for a proper analysis of many other transistor applications, such as digital circuits, the other modes should be investigated. The regions of the I/V characteristics corresponding to the active, saturation, and cutoff modes of operation are shown in Figs. 10.6.3 and 10.6.4, as discussed in Sec. 10.6.

Saturation Mode

This mode is characterized by forward bias of both the emitter and collector junctions (V_{EB} and V_{CB} both positive for PNP, negative for NPN). As shown in Figs. 10.6.3 and 10.6.4, for moderate forward bias of the collector junction, the current I_C retains values close to those of the active regions. It is only after the forward bias V_{CB} reaches values of about 1 V (for Si) that I_C decreases rapidly (and may go to zero or even change sign) under negligible variations of V_{CB}. This can be understood by comparing Figs. 10.3.1 and 10.8.1, displaying the plots of the minority carrier concentrations for a PNP transistor in the active and saturation modes respectively.

In both figures, near the edge of the forward biased emitter depletion region, the minority carrier concentration in the base (p_B) is high, in accordance with (10.3.3). In the active mode, Fig. 10.3.1 shows that, near the reverse biased collector depletion region, the minority carrier concentration in the base is zero, as predicted by (6.7.7). The resulting concentration gradient causes a diffusion current into the collector.

In the saturation mode, however, (6.7.7) shows that the minority carrier concentration in the base near the forward biased collector depletion region increases exponentially with the forward bias voltage V_{CB}. When V_{CB} approaches 1 V, the minority carrier concentration near the collector becomes significant and, as shown in Fig. 10.8.1 for three different values of V_{CB}, controls the concen-

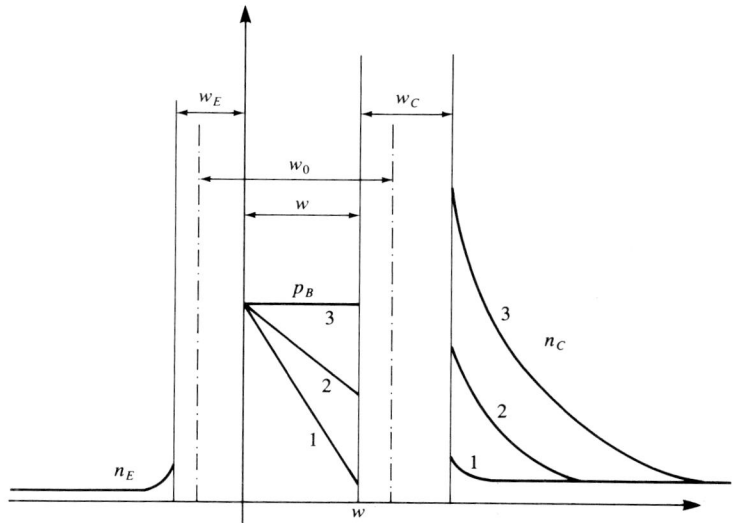

FIGURE 10.8.1
Minority carrier concentration distributions in PNP structure biased in saturation region.

tration gradient in the base (remember that the concentration near the emitter remains constant because it depends exclusively on the input conditions). Notice that in case 3 of the figure the concentration is uniform throughout the base, there is no minority carrier diffusion in the base, and I_C drops to zero. Still higher values of V_{CB} can even reverse the sign of I_C.

As already noticed, the concentration gradient, and so I_C, varies exponentially with V_{CB}, so that practically any value of current can be sustained by an almost constant small collector voltage. Notice that this is a pretty good description of a *short circuit* between the collector and base. An even better approximation to short-circuit conditions exists between the collector and emitter, where the voltage is even smaller because $V_{CE} = V_{CB} - V_{EB}$ is the difference between two almost identical voltages (cf. Example 10.8.2). This property of the saturation mode is often used in digital circuits (cf. Section 10.9).

Example 10.8.1. Under what conditions does the BJT of Fig. 10.8.2 operate in the saturation mode?

Solution. Assuming $V_{BE} \approx -0.6$ V (for Si), the base current is

$$I_B = \frac{V_{CC} + 0.6}{R_B}$$

If the transistor is in saturation, the load line for R_L in Fig. 10.8.2b determines the work point, indicating, in this case, a saturation collector-emitter voltage of -0.2 V,

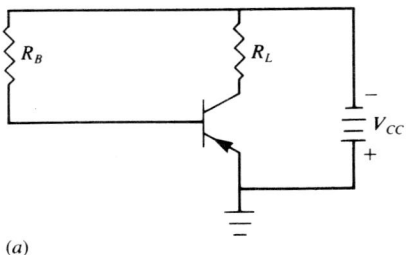

(a)

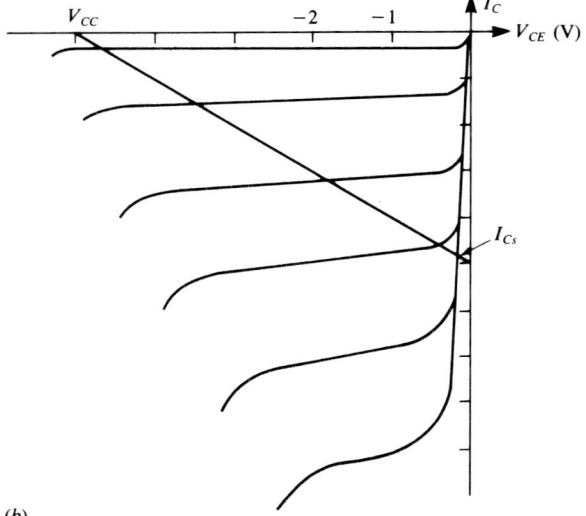

(b)

FIGURE 10.8.2
Example 10.8.1. (a) Circuit schematic. (b) V/I characteristics with load line.

so that the collector current is

$$I_C = \frac{V_{CC} + 0.2}{R_L} = I_{Cs}$$

As shown in Fig. 10.8.2b, the saturation point is independent of the value of I_B as long as I_B is large enough. The condition for saturation is, therefore,

$$\beta_0 \left| \frac{V_{CC} + 0.6}{R_B} \right| \geq \left| \frac{V_{CC} + 0.2}{R_L} \right|$$

The mechanism by which the transistor reaches current equilibrium at saturation under the circuit constraints should be understood in the light of Fig. 10.8.1. As indicated, if the base is narrow, the minority carrier concentration distribution in the base remains essentially linear. Remembering the expression for the diffusion current [implied in (4.3.4)] and computing the slope of the concentration graph, using (6.7.7), the current equilibrium at saturation becomes

$$|I_C| \approx |I_{Ch}| = qA \frac{D_{hB} P_{oB}}{w} (e^{V_{EB}/V_T} - e^{V_{CB}/V_T}) = \left| \frac{V_{CC} - V_{Cs}}{R_L} \right| = |I_{Cs}| \quad (10.8.1)$$

where V_{Cs} is the value of V_{CE} at saturation. Using (10.3.21) and noticing that, as already remarked, under usual conditions, the saturation value of the collector voltage is very low:

$$e^{V_{CB}/V_T} \approx e^{V_{EB}/V_T} - \frac{|I_{Cs}|}{a_{12}} \quad (10.8.2)$$

from which it is possible to compute V_{Cs}. In conclusion, when the input current increases, the output current also increases and the collector voltage drops by an amount equal to the voltage drop in the load. If the input current is so high that $\beta I_B R_L > V_{CC}$, then the collector voltage tends to reverse its sign. This being manifestly impossible, the system seeks the equilibrium condition indicated by (10.8.2). The base current ceases to be the quantity controlling the collector current flow, $I_C \neq \beta I_B$, and I_C is controlled by V_{CB}.

Because the equilibrium of the load circuit keeps (or *clamps*) the collector voltage at the low value V_{Cs} imposed by (10.8.2), this mechanism is often referred to as *bottoming*.

Example 10.8.2. The BJT of Example 10.3.1 operates in the circuit of Fig. 10.8.2 under an input current $I_B = -100 \; \mu A$. If $R_L = 1000 \; \Omega$ and $V_{CC} = -5$ V, compute the collector current I_C and the collector voltage V_{CE}.

Solution. For the circuit shown, from (10.8.1) the saturation current $I_{Cs} \approx -5/10^3 = -5$ mA, where the small voltage V_{Cs} has been neglected. From Example 10.4.1, $\beta_0 \approx 115$, so the condition for saturation, $115 \times 0.1 \times 10^{-3} = 11.5$ mA $> |I_{Cs}| = 5$ mA, is satisfied (notice that the absolute values of the currents have been used in the inequality). Computing the coefficients of the Ebers-Moll equation from (10.7.9) through (10.7.11),

$$a_{12} = \frac{1.6 \times 10^{-19} \times 10^{-2} \times 6.1 \times 2250}{10^{-3} \sinh (0.1)} = 2.192 \times 10^{-13}$$

$$a_{11} = \frac{1.6 \times 10^{-21} \times 4.4 \times 112.5}{10^{-3}} + 2.192 \times 10^{-13} \cosh (0.1) = 2.211 \times 10^{-13}$$

$$a_{22} = \frac{1.6 \times 10^{-21} \times 31 \times 45{,}000}{10^{-3}} + 2.192 \times 10^{-13} \cosh (0.1) = 2.452 \times 10^{-12}$$

It should be noted that, although only three decimal figures are shown in the above, the actual computation was performed in double precision and extended to the twelfth significant figure. Such precision is required for this type of computation.

Using the current equilibrium equation (10.1.1) on (10.7.3) and (10.7.4) and setting $I_C = I_{Cs} = -5$ mA and $I_B = -0.1$ mA, in accordance with the data, from the Ebers-Moll equations,

$$-10^{-4} = (a_{12} - a_{11})(e^{V_{EB}/V_T} - 1) + (a_{12} - a_{22})(e^{V_{CB}/V_T} - 1)$$

$$-5 \times 10^{-3} = -a_{12}(e^{V_{EB}/V_T} - 1) + a_{22}(e^{V_{CB}/V_T} - 1)$$

The system can be solved for the exponential expressions using the computed values of the a parameters. After some algebra, this yields

$$V_{CB} = 0.44159 \text{ V}$$

$$V_{EB} = 0.61804 \text{ V}$$

showing that the transistor is indeed in the saturation condition and yielding for the saturation clamp voltage,

$$V_{Cs} = V_{CE} = V_{CB} - V_{EB} = -0.17645 \text{ V}$$

which confirms the initial hypothesis that V_{Cs} is negligible with respect to 5 V.

Cutoff Mode

This mode is characterized by reverse bias on both emitter and collector junctions. As can be observed in Figs. 10.6.3 and 10.6.4, the current is extremely low. Indeed, as shown in Fig. 10.8.3, the minority carrier concentration throughout the base is essentially zero because no accumulation of minority carriers exists next to the reverse biased junctions (6.7.7). There is no minority carrier diffusion in the base and so essentially no collector current, irrespective of the magnitude of the applied reverse bias voltage. This condition closely approximates an *open circuit* and is often used in digital circuits.

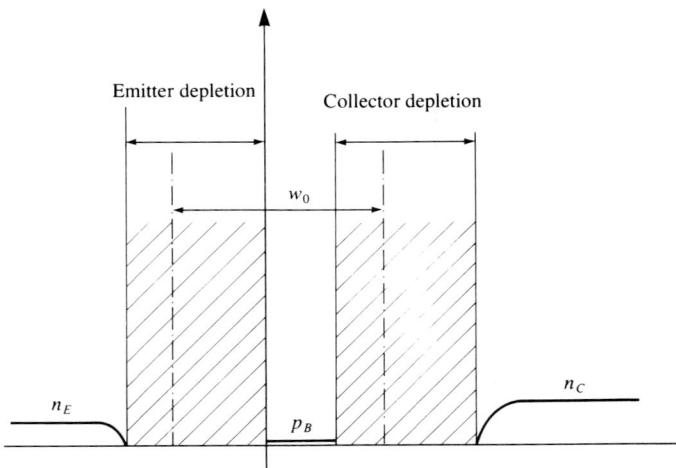

FIGURE 10.8.3
Minority carrier concentration distributions in PNP structure biased in cutoff region.

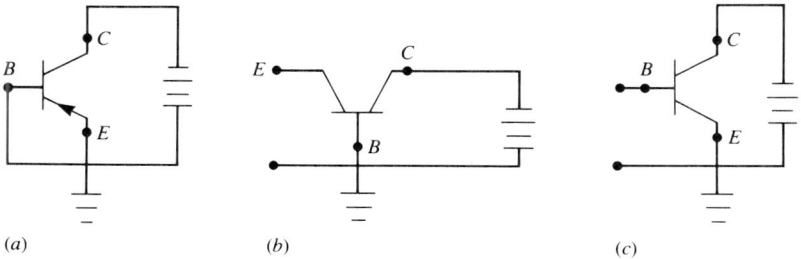

(a)   (b)   (c)

FIGURE 10.8.4
Circuit schematics of transistor in three typical cutoff conditions: (a) shorted input, (b) open emitter, (c) open base.

Three cases of considerable practical importance are often referred to in the literature under the designation of "cutoff conditions." As shown in Fig. 10.8.4, displaying their PNP implementation, they all have reverse biased collector junctions. The three cases are:

1. *Short-circuited input*, with collector current I_0. Imposing the shorted input condition $V_{EB} = 0$ in (10.3.19), the PNP collector current becomes

$$I_0 = -a_{22} = -qA\left(\frac{D_{hB}\, p_{oB}}{w} + \frac{D_{eC}\, n_{oC}}{L_{eC}}\right) \qquad (10.8.3)$$

which, subject to the sign convention of Sec. 10.1, is the inverse saturation current of the narrow-cathode PN junction at the collector, as previously remarked (Sec. 10.7).

The minority carrier concentration distribution is shown in Fig. 10.8.5, curve A, for $V_{EB} = 0$. Near the emitter junction, (10.3.3) yields a concentration p_{oB}, as shown. Near the reverse biased collector junction the concentration is 0, as usual. Equation (10.8.3) can be verified from this graph by computing the slope of the concentration distribution and applying the diffusion equation.

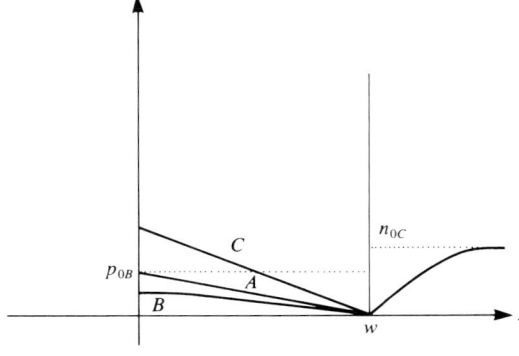

FIGURE 10.8.5
Minority carrier concentration distributions in the active base for the three conditions of Fig. 10.8.4. (a) Shorted input: $V_{BE} = 0$ ($I_C = I_0$). (b) Open emitter: $I_E = 0$ ($I_C = I_{oC}$). (c) Open base: $I_B = 0$ ($I_C = I_{oCE}$).

2. *Open emitter* (common base open input) with collector current I_{oC} of Eq. (10.2.6). Introducing the condition $I_E = 0$ into Eq. (10.3.18b) yields

$$e^{V_{EB}/V_T} - 1 = -\frac{a_{12}}{a_{11}} = -\alpha_F \qquad (10.8.4)$$

where use was made of (10.7.5). This negative value implies $V_{EB} < 0$ and so some reverse bias of the emitter junction. Using this value in (10.3.19b),

$$I_{oC} = \alpha_F a_{21} - \alpha_{22} \qquad (10.8.5)$$

Remembering (10.7.6) and (10.8.3),

$$I_{oC} = I_0(1 - \alpha_F \alpha_R) \ll I_0 \qquad (10.8.6)$$

showing that the common base leakage current with open input is much smaller than the inverse saturation current of the corresponding PN junction.

This conclusion is supported by the graphs of Fig. 10.8.5. For $I_E = 0$, there must be no diffusion across the emitter junction. By (10.3.14), at the edge of the emitter junction depletion region the slope of the minority carrier concentration graph for $I_E = 0$ must be $dp_B/dx = 0$, as shown by curve (B). It is easy to see that the slope of this graph near the collector junction is smaller than for the shorted input case ($V_{EB} = 0$), so that $I_{oC} < I_0$.

3. *Open base* (common emitter open input). From (10.3.20b), imposing the condition $I_B = 0$,

$$(e^{V_{EB}/V_T} - 1) = \frac{a_{12} - a_{22}}{a_{21} - a_{11}} = \frac{\alpha_F}{\alpha_R} \frac{1 - \alpha_R}{1 - \alpha_F} \qquad (10.8.7)$$

in which use was made of (10.7.5) and (10.7.6). Notice that this positive value implies that $V_{EB} > 0$, i.e., that the emitter junction is forward biased, so that, rigorously, the device is not in the cutoff mode!

Using (10.8.7) in (10.3.19b) and subsequently factoring a_{22} :

$$I_{oCE} = -a_{12} \frac{\alpha_F}{\alpha_R} \frac{1 - \alpha_R}{1 - \alpha_F} - a_{22} = I_0 \frac{1 - \alpha_F \alpha_R}{1 - \alpha_F} = \frac{I_{oC}}{1 - \alpha_F} \qquad (10.8.8)$$

where use was made of (10.7.6), (10.8.3), and (10.8.6).

This large value for I_{oCE} can be justified physically. As previously remarked, for the common emitter configuration, open-circuiting the base results in forward biasing the emitter junction. This is displayed in curve C for $I_B = 0$ in Fig. 10.8.5. It is easy to observe the increment of slope at $x = w$ corresponding to the much larger value of I_{oCE} predicted by (10.4.2).

Example 10.8.3. For the BJT of Example 10.3.1, compute I_0, I_{oC} and I_{oCE}.

Solution. From (10.7.5) and (10.7.6), using the a parameter values computed in Example 10.8.2:

$$\alpha_F = \frac{2.192}{2.211} = 0.991412$$

(which should be compared with the result of Example 10.3.1)

$$\alpha_R = \frac{2.192}{24.52} = 0.089398$$

so that the currents requested are:

From (10.8.3): $I_0 = -2.452$ pA

From (10.8.6): $I_{oC} = -2.452(1 - 0.0894 \times 0.9914) = -2.235$ pA

smaller than I_0, as expected.

$$\text{From (10.8.8): } I_{oCE} = \frac{-2.235}{1 - 0.9914} = -259.88 \text{ pA}$$

much larger than I_0, as expected.

Inverted Mode

This mode of operation is characterized by the forward biased collector junction and the reverse biased emitter junction. It is analogous to the active mode, except that the roles of the collector and emitter are interchanged. If the BJT were a symmetrical structure, transistor performance would be identical to that in the active mode; however, due to differences in doping and in geometry, the symmetry is destroyed and both γ and α_T (and consequently α_0) are usually much lower than in the active mode.

Introducing these conditions ($V_{EB} \ll 0$ and $V_{CB} > 0$) into the Ebers-Moll equations:

$$I_E = -a_{11} - a_{12}(e^{V_{CB}/V_T} - 1) \tag{10.8.9}$$

$$I_C = a_{21} + a_{22}(e^{V_{CB}/V_T} - 1) \tag{10.8.10}$$

where the a parameters are computed from (10.3.21) through (10.3.23) or, more rigorously, from (10.7.9) through (10.7.11).

Notice that now I_C is the input and I_E the output current, so that the forward current gain is $\alpha_R < \alpha_F$ [cf. (10.7.7), (10.7.8), the assumptions about the dopings in the different regions, and the results of Example 10.8.3]; therefore the deviation from symmetry results in poor performance in this mode. If the structure is asymmetric, not only in doping but also geometrically (as might be the case in interdigitated emitter junctions), the inverted mode operation may even result in damage to the BJT (e.g., by excessive current crowding).

10.9 THE BJT AS A SWITCH

In digital circuits the transistor is often employed as a switch, i.e., as a device capable of producing either an open circuit (*off* state) or a short circuit (*on* state) between two terminals. This operation can be performed by switching the BJT mode of operation back and forth between the cutoff and saturation modes. As

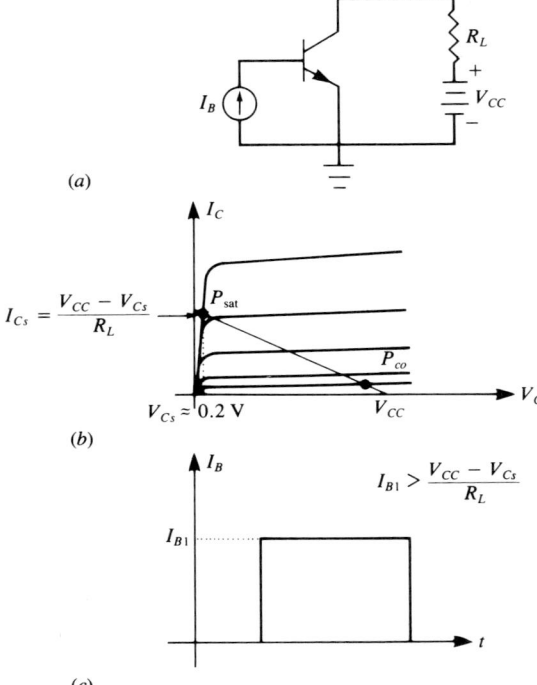

(a)

(b)

(c)

FIGURE 10.9.1
Saturation inverter. (a) Circuit schematic. (b) On and off work points. (c) Input current waveshape.

discussed in Sec. 10.8, in the cutoff mode practically no current can flow between the collector and emitter terminals, irrespective of the voltage between them, i.e., an open-circuit condition exists between these two terminals. In the saturation mode the voltage between collector and emitter is clamped to almost zero and, for a given supply voltage, the current between the two terminals is essentially determined by the load resistance: a good approximation to a short-circuit condition.

A simple example of this type of operation is provided by the saturation inverter circuit of Fig. 10.9.1, which also displays the waveshape of the input current and the field of I/V characteristics with the load line and the work points corresponding to the two possible levels of input current.

When the input current is zero, the transistor is cut off and its work point is P_{co} in the figure.

Under the load conditions indicated, at saturation, the collector voltage is V_{Cs}, as shown, and the collector current can easily be computed as $I_{Cs} = (V_{CC} - V_{Cs})/R_L$, so that the saturation work point P_{sat} depends on the load line. However, as long as the base current $I_B > I_{Bs} = I_{Cs}/\beta = (V_{CC} - V_{Cs})/\beta R_L$, saturation is assured and the work point is essentially independent of the actual value of I_B.

With the square wave input current shown in the figure, one might expect the output to jump from one state to the other instantaneously, but the speed of the BJT response is limited by the phenomena described in Secs. 10.4 and 10.5, when discussing the junction capacitances and frequency response. All of these phenomena affect the switching speed, but usually the most important influence is exerted by charge storage in the base. We shall discuss this phenomenon for an NPN transistor. This will require modification of some of the formulas in the text. The reader should consider the signs (e.g., the minority carrier charge in the P-type base is $Q_B < 0$).

It is easy to see that the collector current I_C is related to the charge Q_B stored in the base by minority carrier accumulation. Indeed, comparing (10.3.10), (10.3.15), and (10.3.14), properly modified to fit NPN transistors,

$$I_C \approx -I_{Ce} = -\frac{2D_{eB}}{w^2} Q_B = -\frac{Q_B}{\tau_w} \tag{10.9.1}$$

where τ_w is the base transit time of the minority carriers discussed in Sec. 10.4, page 353, and App. 10A. Equation (10.9.1) mathematically expresses the fact that, as repeatedly noted in previous sections, in order to change the transistor currents, it is necessary to change the charge stored in the base. However, this requires as much time as it takes to move the corresponding amount of charge into, or out of, the base region. To determine the speed of response it is therefore important to compute how long it takes to perform this operation.

If the minority carrier concentration in the base is $n_B(x)$, then the excess minority carrier charge accumulated in the base region is

$$Q_B = -qA \int_0^w [n_B(x) - n_{oB}] \, dx \tag{10.9.2}$$

From the equation of continuity, (4.5.13), multiplying by qA, integrating from 0 to w, and using (10.9.2),

$$\frac{dQ_B}{dt} = -A[J_{eB}(w) - J_{eB}(0)] - \frac{Q_B}{\tau_{eB}} \tag{10.9.3}$$

where use was made of the identity $dn_B/dt = d(n_B - n_{oB})/dt$. Remembering that $AJ_{eB}(w) = I_{Ce} = -I_C$ and $AJ_{eB}(0) = I_{Ee} \approx I_E$ and using the current equilibrium equation (10.1.1),

$$I_B = -\left[\frac{dQ_B}{dt} + \frac{Q_B}{\tau_{eB}}\right] \tag{10.9.4}$$

known as the *charge control equation*. Considering now the leading edge of the input current pulse of Fig. 10.9.1, it is possible to compute the turn-on and turn-off times.

Turn-On Time

Supposing the step occurs at $t = 0$; then integrating (10.9.4) under the boundary condition:

$$\text{for } t = 0 \text{ then} \rightarrow Q_B = 0$$

results in

$$Q_B = -I_B \tau_{eB}(1 - e^{-t/\tau_{eB}}) \tag{10.9.5}$$

so the charge stored in the base increases exponentially with time toward an asymptotic value $I_B \tau_{eB}$, depending on the base current. The collector current correspondingly varies as required by (10.9.1):

$$I_C = I_B \frac{\tau_{eB}}{\tau_w} (1 - e^{-t/\tau_{eB}}) \tag{10.9.6}$$

However, while the stored charge eventually reaches the asymptotic value, the current is constrained by the equilibrium of the collector circuit and cannot rise beyond the saturation value I_{Cs} because the collector "bottoms," as mentioned in Sec. 10.8 under "saturation mode."

The time required to reach the saturation condition is designated as the turn-on time τ_{on} and can be computed from (10.9.6), setting $I_C = I_{Cs}$:

$$\tau_{on} = \tau_{eB} \ln \left(\frac{1}{1 - \dfrac{V_{CC} - V_{Cs}}{I_B R_L} \dfrac{\tau_w}{\tau_{eB}}} \right) \tag{10.9.7}$$

Notice that, everything else being equal, this turn-on time can be minimized by increasing the on level I_B of the input current. This, however, may affect the turn-off time, as discussed in the following. Also notice that, when the turn-on time has elapsed and the output current reaches its final value, the charge stored in the base has reached the corresponding value:

$$Q_{B,\text{sat}} = -\frac{V_{CC} - V_{Cs}}{R_L} \frac{w^2}{2D_{eB}} = -I_{Cs} \tau_w \tag{10.9.8}$$

but it does not remain at this value. Instead, the high-input current I_B forces it to keep increasing until it reaches the asymptotic value $-I_B \tau_{eB}$.

Summarizing, at turn-on the output current increases exponentially with a time constant τ_{eB} toward an asymptotic value $I_B \tau_{eB}/\tau_w$, but never reaches it; instead, it stops rising after a turn-on time τ_{on} and then remains at the saturation value I_{Cs}. At this point, however, steady state has not yet been achieved and the minority carrier charge stored in the base keeps increasing exponentially with a time constant τ_{eB} until it asymptotically reaches the value $-I_B \tau_{eB}$, proportional to the magnitude of the input current.

The turn-on waveshapes are displayed in Fig. 10.9.2, assuming the input current step to occur at $t = 0$.

Turn-Off Time

Suppose that the inverter has been in the on condition a long time, with input current $|I_B| > (V_{CC} - V_{Cs})/\beta R_L$ sufficient to ensure saturation. The initial value of

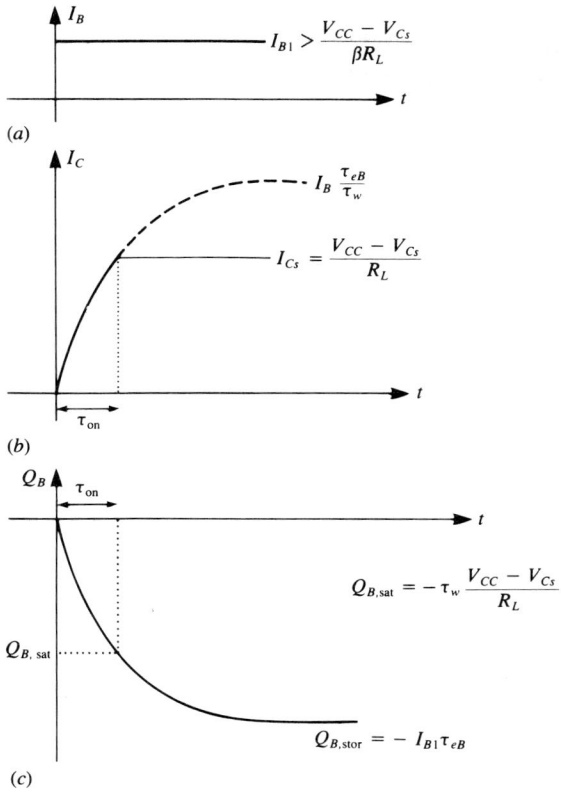

FIGURE 10.9.2
Turn-on waveshapes for the satura-tion inverter of Fig. 10.9.1. (*a*) In-put step of amplitude $I_{B1}(V_{CC} - V_{Cs})/\beta R_L$. (*b*) Collector current: for $0 < t < \tau_{on}$, I_C rises exponen-tially toward $I_{B1}\tau_{eB}/\tau_w$ with time constant τ_{eB}; for $t > \tau_{on}$, $I_C = I_{Cs} = (V_{CC} - V_{Cs})/R_L$. (*c*) The minority charge stored in the base rises expo-nentially with a time constant τ_{eB} toward an asymptotic value $I_{B1}\tau_{eB}$.

Q_B is therefore $Q_{B,\text{init}} = -I_B \tau_{eB}$ and depends on I_B. Assuming that the input current is turned off at $t = 0$, then, from (10.9.5):

$$Q_B = -I_B \tau_{eB} e^{-t/\tau_{eB}} \tag{10.9.9}$$

The charge stored in the base decreases exponentially with a time constant τ_{eB}. However, because of the collector circuit constraints, the collector current I_C does not change and remains at the value I_{Cs} until Q_B reaches the value $Q_{B,\text{sat}}$ of Eq. (10.9.8). The time required for Q_B to reach this value is the delay time τ_Δ and can be computed from (10.9.9) by setting $Q_B = Q_{B,\text{sat}}$:

$$\tau_\Delta = \tau_{eB} \ln \left(\frac{Q_{B,\text{init}}}{Q_{B,\text{sat}}} \right) = \tau_{eB} \ln \left(\frac{I_B \tau_{eB}}{I_{Cs} \tau_w} \right) \tag{10.9.10}$$

Consequently, while the time dependence of the stored charge during turnoff is described by (10.9.9), the collector current follows the equation:

$$I_C = \frac{V_{CC} - V_{Cs}}{R_L} [1 - (1 - e^{(t - \tau_\Delta)/\tau_{eB}}) u(t - \tau_\Delta)] \tag{10.9.11}$$

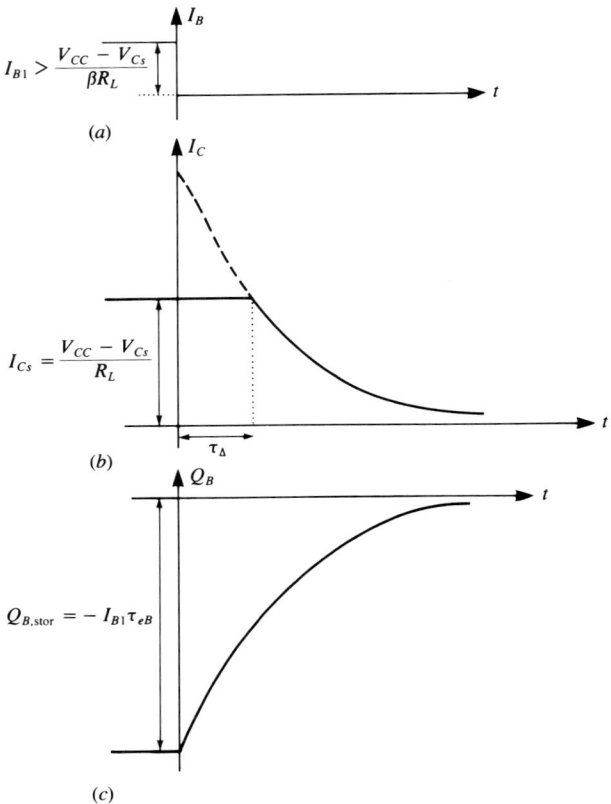

(a)

(b)

(c)

FIGURE 10.9.3
Turn-off waveshapes for the saturation inverter of Fig. 10.9.1. (a) Downward input step of amplitude I_{B1}. (b) Collector current: for $t < \tau_\Delta$, $I_C = I_{Cs}$; for $t > \tau_\Delta$, I_C decays exponentially toward zero with a time constant τ_{eB}. (c) The minority charge stored in the base decays exponentially to zero with a time constant τ_{eB}.

Turn-off is therefore characterized by a delay time τ_Δ, followed by an exponential decay of the output current with time constant τ_{eB}, as depicted in Fig. 10.9.3. If the dominant recombination mechanism is indirect (cf. Sec. 4.4), then τ_{eB} in the above formulas should be substituted by τ_r.

Example 10.9.1. An NPN BJT, characterized by $\tau_{eB} = 0.2$ μs, active base width $w = 1$ μm, and minority carrier diffusivity in the base $D_{eB} = 5$ cm^2/s, operates in the circuit of Fig. 10.9.1 with $V_{CC} = 5.2$ V, $R_L = 1000$ Ω, under an input current pulse from $I_B = 0$ to $I_B = 100$ μA. The pulse starts at $t = 0$ and ends at $t = 100$ ns, as shown in Fig. 10.9.4. At saturation, the clamping voltage magnitude is assumed to be 0.2 V. Compute and draw the output waveshape.

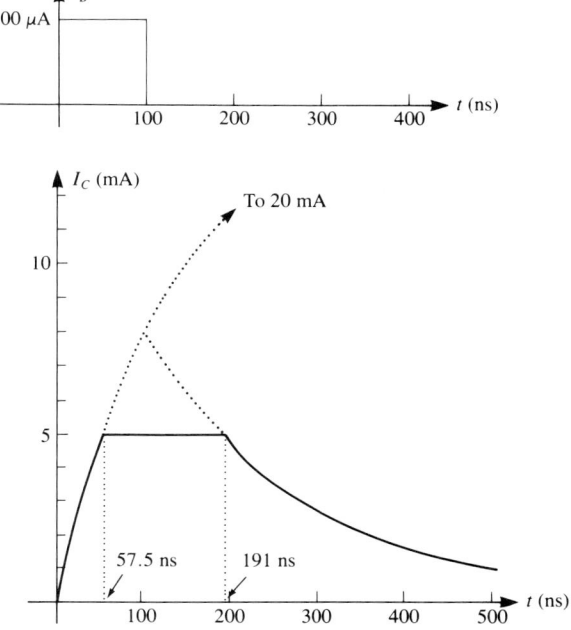

FIGURE 10.9.4
Response of the saturation inverter of Example 10.9.1 to an input pulse. (a) Input pulse of
amplitude 100 μA and duration 100 ns. (b) Collector current. For calculation of values of
critical quantities refer to Example 10.9.1.

Solution. The saturation current is $I_{Cs} = (5.2 - 0.2)/1000 = 5$ mA. The transit time
from (10.9.1) is $\tau_w = 10^{-8}/(2 \times 5) = 10^{-9}$ s. From (10.9.7), the turn-on time is

$$\tau_{on} = 2 \times 10^{-7} \ln \frac{1}{1 - 5 \times 10^{-12}/2 \times 10^{-11}} = 57.5 \text{ ns}$$

From (10.9.8), $Q_{B,sat} = -5 \times 10^{-3} \times 10^{-9} = -5$ pC. From (10.9.9), using the pulse
duration,

$$Q_{init} = -10^{-4} \times 2 \times 1^{-7}\left[1 - \exp\left(\frac{-10^{-7}}{2 \times 10^{-7}}\right)\right] = -7.87 \text{ pC}$$

instead of the asymptotic value of $-10^{-4} \times 10^{-7} = -10$ pC, so that, from
(10.9.10),

$$\tau_\Delta = 2 \times 10^{-7} \ln\left(\frac{7.87}{5}\right) = 91 \text{ ns}$$

The output current pulse is shown in Fig. 10.9.4, solid line.

10.10 COLLECTOR BREAKDOWN

In the active mode, breakdown may occur at the collector junction, where the
electric field is largest. Both Zener and avalanche breakdown phenomena are

observed and their analysis follows the principles and data outlined in Sec. 6.6 and in App. 6A.

In the BJT, breakdown may also occur because of *punch through*. This occurs when the collector depletion region extends so far into the base that it reaches the emitter depletion region, reducing the active base width to zero. In this case, every emitter majority carrier entering the depletion region is accelerated right through to the collector and the current is limited only by the external circuit constraints and the resistances of the emitter and collector bulk material.

For abrupt collector junctions of geometric base width w, remembering (6.3.19), the punch-through voltage can be approximated by

$$|V_{pt}| \approx \frac{qN_B w^2}{2\varepsilon} \tag{10.10.1}$$

For the special case of common emitter configuration with open-circuited base, still another mechanism is often responsible for breakdown: the avalanche growing of the current I_{oCE} discussed in Sec. 10.8.

In Sec. 10.8 it was remarked that, when the base is open-circuited, the current I_{oCE} is comparatively large, because the transistor is slightly forward biased. This forward bias is produced by the voltage that must exist across the emitter junction to allow the current to flow, and so it is an increasing function of I_{oCE}.

When avalanche multiplication begins, even before avalanche breakdown occurs, the current I_{oCE} is multiplied by the factor M, as noted in Sec. 6.6. This increased current, in turn, increases the bias, and M increases with it, in accordance with (6A.9), further increasing I_{oCE} and so on. If the avalanche multiplication phenomenon is large enough, this *positive feedback* condition may cause I_{oCE} to grow without bounds, resulting in breakdown.

In accordance with the definition of M, under these conditions, in which the emitter and collector currents have the same amplitude, I_{oCE}, the current in the transistor, is given by

$$I_{oCE} = M(\alpha_0 I_{oCE} + I_{oC}) \tag{10.10.2}$$

from which

$$I_{oCE} = \frac{I_{oC}}{1 - \alpha_0 M} \tag{10.10.3}$$

showing that breakdown occurs (i.e., I_{oCE} grows without bounds) when

$$\alpha_0 M = 1 \tag{10.10.4}$$

Using this condition in (6A.9), after some algebra, it is easy to prove that, if V_{bdB} is the common base avalanche breakdown voltage, then, in the common emitter, open base configuration, breakdown occurs at a collector voltage:

$$V_{bdE} \approx V_{bdB}(\beta)^{-(1/\eta)} \tag{10.10.5}$$

with η between 2 and 4 for Si. Notice that, for high β transistors, the common emitter breakdown voltage can be much lower than for the common base configuration.

10.11 THE SILICON CONTROLLED RECTIFIER

In the conventional rectifier of Fig. 10.11.1a, the diode acts as a voltage-operated switch connecting the load to the source when current flows in the positive direction, as in Fig. 10.11.1b, and isolating the load whenever the current flow would be in the negative direction, as in Fig. 10.11.1c.

This means of controlling the direction of the current in the load by selecting, through the switching device, the portion of the waveform to be applied to the load is very economical, because almost no power is dissipated in the diode. Notice that this economy is achieved only if the diode acts as a device operating either in cutoff or in the saturation state. Indeed, in the cutoff state, the voltage across the device is considerable but essentially no current flows through it, while, in saturation, the current may be high but the device voltage drop is small, so the power dissipated in the device can be very small in both states.

It would be desirable to use a similar, economic principle of operation to control the average value of the load current. This could be achieved by connecting in series between the generator and load a device that could be switched at will from cutoff to saturation by an appropriate signal. The circuit of Fig. 10.11.2 can perform this task.

The control device consists of a PNP and an NPN transistor in such a configuration that, with the usual current polarity conventions of Sec. 10.1 and

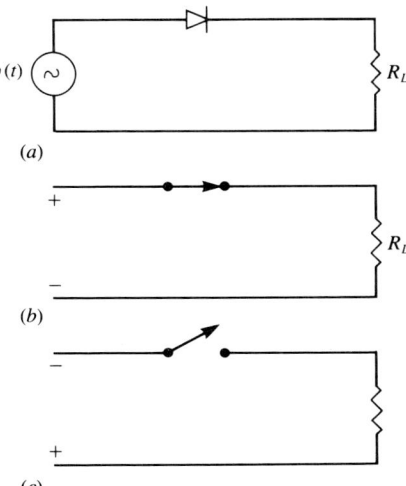

(a)

(b)

(c)

FIGURE 10.11.1
Diode rectifier. (a) Circuit schematic. (b) Equivalent piecewise linear circuit for forward bias. (c) Equivalent piecewise linear circuit for reverse bias.

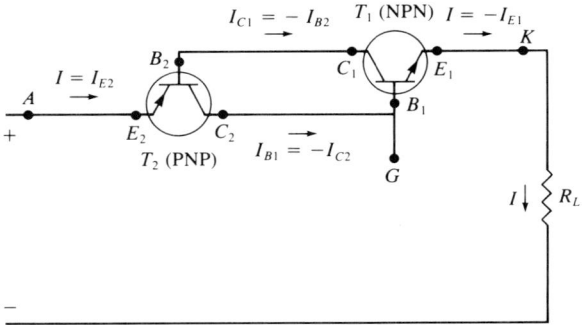

FIGURE 10.11.2
Control circuit with two-transistor control.

Fig. 10.1.1,

$$I_{B1} = -I_{C2}$$

$$I_{B2} = -I_{C1}$$

(10.11.1)

while, in accordance with Kirchoff's first law, the load current flowing in the main loop is

$$I = I_{E2} = -I_{E1}$$

(10.11.2)

Originally, if the base current of transistor T_1 is zero, then $I_{C1} = 0$, $I_{B2} = 0$, $I_{C2} = 0 = I_{B1}$, and both transistors are cut off.

If, however, a pulse of current is injected at G, this momentary base current takes T_1 into the active region, and the collector current I_{C1} then provides an input to transistor T_2, which also comes out of cutoff. The resulting I_{C2} flows into the base of T_1 keeping it on even if the outside control pulse is now off. The student should observe that, in Fig. 10.11.2, the actual currents so generated are in the proper directions to bring the transistors out of cutoff.

If I_{C2} is large enough, it will drive T_1 further into conduction, increasing I_{C1}, which in turn will increase I_{C2}, and so on, in a *regenerative* process, which would tend to increase the currents without bounds. In reality, the equilibrium of the circuit and the limitations of the devices will simply drive both transistors into saturation.

Mathematically, using Eq. (10.2.6),

$$I_{C1} = -\alpha_1 I_{E1} + I_{oC1}$$

$$I_{C2} = -\alpha_2 I_{E2} + I_{oC2}$$

(10.11.3)

However, from the equilibrium of transistor T_2,

$$I + I_{C2} - I_{C1} = 0$$

(10.11.4)

or $I - \alpha_2 I_{Ee} + I_{oC2} + \alpha_1 I_{E1} - I_{oC1} = I(1 - \alpha_1 - \alpha_2) + (I_{oC2} - I_{oC1}) = 0$

(10.11.5)

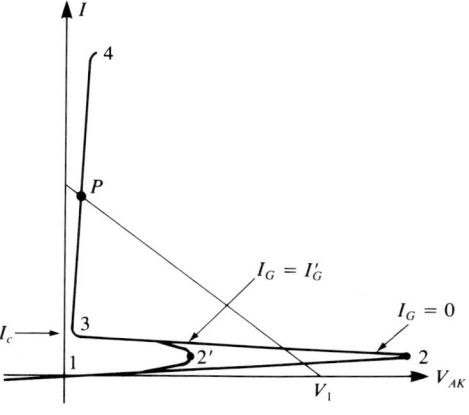

FIGURE 10.11.3
Voltampere characteristics of two-transistor control of Fig. 10.11.2. Notice the dependence on the gate current I_G.

in which use was made of (10.11.2). In conclusion, letting $I_{oC2} - I_{oC1} = -I_0$,

$$I = \frac{I_0}{1 - (\alpha_1 + \alpha_2)} \tag{10.11.6}$$

showing that, as $\alpha_1 + \alpha_2$ tends toward 1, then I tends to increase without bounds and is limited only by the load resistance R_L.

The voltampere characteristics of the control device, therefore, are as shown in Fig. 10.11.3. Starting with the device in cutoff at point 1, as the voltage V_{AK} increases, the current remains negligible and the work point moves along line 1–2. As the work point moves, the transistors get closer and closer to the active state. In terms of their parameters, α_1 and α_2 increase from zero. When $\alpha_1 + \alpha_2$ reaches 1, suddenly the regenerative mechanism described above comes into action, the transistors come out of cutoff all the way into saturation, the work point jumps to point 3, and the current increases along branch 3–4 of the characteristics. The current can now take on any value from I_c (the *maintaining current*) upwards, under an essentially negligible voltage (saturation). The final equilibrium position of the work point depends on the R_L load line, as shown in Fig. 10.11.3.

Notice that the condition $\alpha_1 + \alpha_2 = 1$ for the start of the regenerative process depends not only on the voltage V_{AK} but also on the current I_G injected at point G. This is shown in the characteristics of Fig. 10.11.3, where the critical point is 2 if $I_G = 0$, or 2′ if $I_G = I'_G$. As an example, let the voltage $V_{AK} = V_1$ and $I_G = 0$; then the device is in cutoff, but if a current pulse of amplitude I'_G is now applied at G, then the transition is triggered and the device goes into saturation.

In practice, the two-transistor control device of Fig. 10.11.2 is implemented in the silicon controlled rectifier (SCR) or thyristor. Figure 10.11.4 shows the circuit symbol of this device together with its basic structure. Comparing Fig. 10.11.4 with Fig. 10.11.2 it is seen that in the new device, the base-collector connections of the two-transistor implementation are obtained by fabricating each

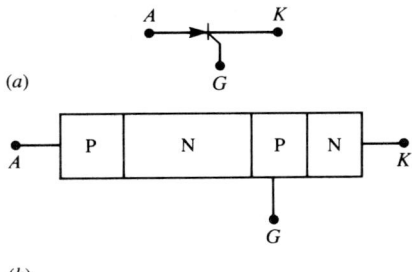

(a)

(b)

FIGURE 10.11.4
Silicon controlled rectifier (thyristor). (a) Circuit symbol. (b) Schematic representation of internal structure.

corresponding collector-base pair out of a single semiconductor crystal. The device terminals A, G, and K correspond to the analogous points in Fig. 10.11.2 and are designated as anode, gate, and collector.

In the typical control circuit of Fig. 10.11.5a, starting at $t = 0$ with $I_G = 0$, as shown by the waveshapes of Fig. 10.11.5b, the device remains in cutoff as $v(t)$ increases, so that the current $i(t)$ is zero until time $t = t_1$.

At this instant, a current I_G is momentarily injected, triggering the regenerative transition to the saturation state *and then the device remains saturated, irrespective of the value of I_G.* The load current becomes $i(t) = v(t)/R_L$ and current flow is maintained until, at $t = t_2$, the current decreases below the maintaining level I_c shown in Fig. 10.11.3. Then the device reverts to the cutoff state. The device now remains in cutoff until time t_4, when the gate pulse is applied again during the next positive cycle of the anode voltage.

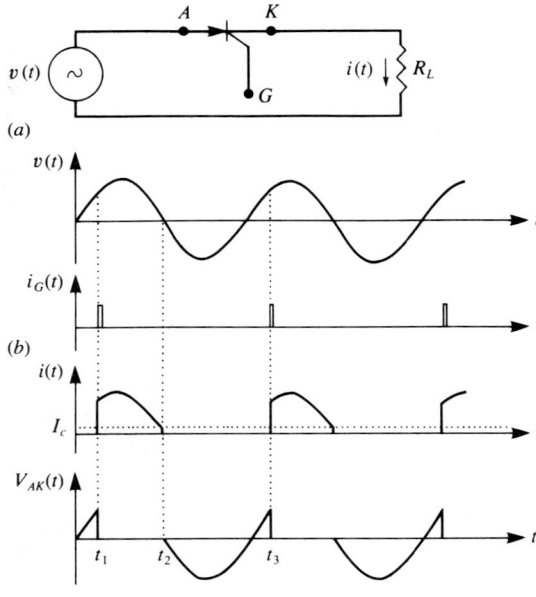

(a)

(b)

FIGURE 10.11.5
Typical SCR current control circuit. (a) Circuit schematic. (b) Voltage and current waveshapes.

It should be noticed that, once triggered, the SCR remains in saturation until the current is decreased below the critical maintaining value I_c and that conduction can occur only during the positive cycle of the anode voltage (rectifying action).

Several different thyristor devices are also available, in which conduction can be arrested by gate action and/or conduction can occur in both directions.

It is evident that, by delaying the trigger pulse, the SCR duty cycle can be varied, thereby varying the average value of the current provided to the load (current control).

Figure 10.11.5 also shows the voltage waveshape $V_{AK}(t)$ across the SCR.

10.12 SUMMARY

The BJT consists of two PN junctions back to back. The terminals are emitter, base, and collector. NPN and PNP structures are possible.

In the active mode of operation (forward biased emitter and reverse biased collector junctions), a large percentage of the minority carriers injected into the base from the emitter never reach the base terminal, but cross right through into the collector. This *transistor effect* occurs because the base is very narrow.

Using the PNP structure for the analysis, a portion γ of the emitter current is injected into the base in the form of holes. Of this, a portion α_T makes it across the base into the collector, so that the current transferred across the base is

$$I_{Ch} = \gamma \alpha_T I_E = \alpha_0 I_E$$

The collector current is the superposition of this current and of the minority carrier current leaking across the collector junction, given by Eq. (10.2.6), where the signs result from the conventions of Sec. 10.1.

In the active mode, the PNP BJT currents are given by Eqs. (10.3.18*b*) through (10.3.20*b*), the parameters of which are defined by Eqs. (10.7.9) through (10.7.11), or, in the first-order approximation of small active base widths w, by Eqs. (10.3.21) through (10.3.23). Analogous expressions hold for the NPN structure.

These equations show that the transistor currents are proportional to the slopes of the minority carrier concentration distribution in the base at the edges of the depletion regions (cf. Fig. 10.3.1). They permit construction of the BJT I/V characteristics in the active region by points.

Three circuit configurations are defined: common base, common emitter, and common collector. The relationship between common emitter and common base parameters is implied in Eq. (10.4.2).

The small-signal behavior of the transistor is described by the small-signal equivalent circuits of Figs. 10.5.1 and 10.5.2 in the common base and common emitter configurations respectively.

The pertinent parameters are traditionally presented in the "h" or hybrid form of linear circuit analysis. They can be computed from Eqs. (10.4.3) through (10.4.11).

The frequency response of the transistor depends on the capacitances associated with the two junctions. Of these, usually, the most important is the diffusion capacitance of the forward biased junction, related to the base transit time by Eq. (10.4.11). High-frequency behavior parameters are usually provided in terms of the capacitances themselves, of the 3-dB frequencies f_B and f_E, or of the cutoff frequency f_{co}, as discussed in Sec. 10.5.

Real transistor I/V characteristics, as shown in Figs. 10.6.3 and 10.6.4, differ from the theoretical curves because of the influence of several phenomena, the most important of which, such as base width modulation, base spreading resistance, current crowding, and effects of high-level injection and of graded base doping, are discussed in Sec. 10.6.

The Ebers-Moll model of the BJT, shown in Fig. 10.7.1, is a direct schematic implementation of the fundamental mechanisms of transistor behavior (PN junction characteristics and transistor effect). It is characterized by Eqs. (10.7.3) through (10.7.6), again using the parameters of Eqs. (10.7.9) through (10.7.11) [or the rougher approximations of Eqs. (10.3.21) through (10.3.23)].

The Ebers-Moll equations are a generalization of (10.3.18b) through (10.3.20b), and are valid in all modes of operation.

In the saturation mode both junctions are forward biased, the collector bottoms and the output acts essentially as a short circuit supporting a practically unlimited current under a very small voltage. This is particularly true in the common emitter mode. In accordance with the conditions depicted in Fig. 10.8.1, the system seeks the collector voltage that satisfies the equilibrium conditions of both the carrier diffusion in the base and of the external circuit, as discussed in Sec. 10.8 and Example 10.8.2.

In the cutoff mode both junctions are reverse biased, the minority carrier concentration distribution in the base is shown in Fig. 10.8.3, and the output acts as an open circuit, with essentially no current for unlimited voltages. Three important cases, usually classified as cutoff conditions, correspond to $V_{EB} = 0$, $I_B = 0$, and $I_E = 0$ respectively. Their behaviors correspond to the minority carrier concentration diagrams of Fig. 10.8.5 and are analyzed in Section 10.8 under "cutoff mode." In reality, the open base condition is shown to fall in the active mode.

The BJT is often used as a switch, taking advantage of the almost short-circuit behavior of the output at saturation and almost open-circuit behavior at cutoff.

The speed of response of the BJT is limited by the several capacitances previously described and by charge accumulation and decay times. Other phenomena related to the properties of the saturation and cutoff modes may complicate the response. The square pulse response is characterized by an exponential time constant τ_{eB}, which is illustrated in Figs. 10.9.2 through 10.9.4. This example is characteristic of the use of transistors in digital circuits.

Both Zener and avalanche phenomena may cause collector junction breakdown. Other possible mechanisms include punch through (when the base width is decreased to zero by the Early phenomenon) and unbounded growing of I_{oCE} under common emitter conditions (a positive feedback avalanche phenomenon).

A PNPN device, the thyristor, can be switched from cutoff to saturation by a pulse applied to the gate. Once saturated the device remains in this condition until the current decreases below a maintaining level I_c; then the device reverts to the cutoff condition. Varying the position of the gate pulse within the positive cycle of the anode voltage permits control of the average current delivered to the load. The power dissipated in the controlling device can be a very small fraction of the power delivered to the load.

APPENDIX 10A
TRANSIT TIME

The time it takes the minority carriers injected into the base to diffuse across it and reach the collector depletion region is known as the *transit time* τ_w and determines several important characteristics of transistor behavior.

As the time required to cover the infinitesimal distance dx is dx/v, then

$$\tau_w = \int_0^w \frac{1}{v} \, dx \tag{10A.1}$$

As described in Sec. 10.2, the minority carriers move by diffusion, so that, for PNP structures, remembering the definition of diffusion density rate and (4.3.1),

$$F = p_B v = -D_{hB} \frac{\partial p_B}{\partial x} \tag{10A.2}$$

Solving for the diffusion velocity v and introducing into (10A.1),

$$\tau_w = \int_0^w \frac{p_B}{-D_{hB} \, \partial p_B / \partial x} \, dx \tag{10A.3}$$

Obtaining p_B and its derivative from (10.3.7) and substituting into the integral, after some algebra,

$$\tau_w = \int_0^w \frac{w - x}{D_{hB}} \, dx = \frac{w^2}{2D_{hB}} \tag{10A.4}$$

which should be compared with (10.4.11). A similar expression, with D_{hB} substituted by D_{eB}, holds for NPN structures.

PROBLEMS

10.1. (a) What is the purpose of growing an NPN transistor on a p^+ substrate?

(b) In step 6 of the fabrication sequence shown in the text, much care is taken to control the depth of penetration of the n^+ layer. What important fabrication parameter is determined by this depth of penetration? Which method of fabrication best controls this quantity?

(c) Why are n^+ regions diffused into the n^- semiconductor?

(d) During fabrication what determines the emitter junction cross section?

10.2. (a) From inspection of Fig. 10.2.1a comment on the relative positions of the Fermi levels and their relationship to the doping of the various regions.

(b) Draw graphs equivalent to Fig. 10.2.1a and b for an NPN BJT characterized by geometric base width 1 μm and dopings $N_E = 10^{17}$ cm^{-3}, $N_B = 10^{16}$ cm^{-3}, $N_C = 10^{14}$ cm^{-3}. Compute the contact potentials and V_{ifE}, V_{ifB}, and V_{ifC}.

(c) For the transistor of Fig. 10.2.1, describe the current transients occurring when the junctions are made and how they result in the band diagram of Fig. 10.2.1b. Which of these currents flow in any possible outside electrical connections to the transistor terminals?

(d) Comment on the relative width of the four depletion regions shown in Fig. 10.2.1b.

(e) Compute the depletion region widths for the transistor of part (b) above. What is the active base width of the unbiased transistor?

(f) Describe the current transients occurring when the transistor is biased and resulting in the transition from the band diagram of Fig. 10.2.1b to that of Fig. 10.2.1c. Which of these transient currents flows in the outside connections to the transistor terminals?

(g) Draw a graph equivalent to the band diagram of Fig. 10.2.1c for the transistor of part (b) biased with 0.5 V across the emitter junction and 5 V across the collector junction.

(h) For the conditions of part (g) compute the active base width. Change the collector bias from 5 to 10 V and compute the new active base width.

10.3. Qualitatively discuss the role played by the base width in determining the base transport factor.

For the transistor of Prob. 10.2(b), assuming the minority carrier lifetime in the base to be 0.1 μs, indicate the order of magnitude of the base width below which the transistor effect becomes important.

10.4. Logically justify the statement of Sec. 10.2 that the emitter efficiency should increase with the ratio N_E/N_B of the emitter-base doping.

10.5. Prove Eq. (10.3.5).

10.6. Prove Eqs. (10.3.6) and (10.3.7).

10.7. Write the continuity equation for the minority carriers in the emitter, base, and collector of an NPN BJT with appropriate boundary conditions and prove Eqs. (10.3.11) to (10.3.13).

10.8. Prove that, for NPN transistors, in Eqs. (10.3.18b) through (10.3.20b), V_{EB} must be substituted by V_{BE} and the coefficients must be computed by Eqs. (10.3.21a) through (10.3.23a).

10.9. A Si NPN transistor at 300 K has device cross section $A = 1$ mm^2, base width 2 μm, dopings $N_E = 2 \times 10^{18}$ cm^{-3}, $N_B = 10^{17}$ cm^{-3}, $N_C = 5 \times 10^{15}$ cm^{-3}, and minority carrier lifetimes $\tau_E = 10^{-7}$ s, $\tau_B = 2.5 \times 10^{-7}$ s, $\tau_C = 10^{-6}$ s. Compute I_C in the active mode for (a) $V_{BE} = 0.6$ V, (b) $I_E = -2.53$ mA, and (c) $I_B = 6.8$ μA.

10.10. For the BJT of Prob. 10.9, compute α_0 and β_0 for conditions (a), (b), and (c) and compare them with the values obtained using (10.3.24) together with the first-order approximation (10.3.25) and the second-order approximation (10.3.26). Comment.

10.11. Compute the reverse saturation current I_{oC} of the transistor of Prob. 10.9. Compare with that of the collector junction acting alone. Comment.

10.12. The circuit of Fig. P10.12 employs the transistor of Prob. 10.9. Compute (in magnitude *and* sign): (*a*) the base-emitter voltage V_{BE}, (*b*) the collector-base voltage V_{CB}, (*c*) the current and voltage small-signal gain, and (*d*) the input impedance.

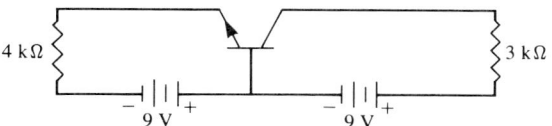

FIGURE P10.12
Schematic for Prob. 10.12.

10.13. The circuit of Fig. P10.13 employs the transistor of Prob. 10.9. Compute: (*a*) the base-emitter voltage V_{BE}, (*b*) the collector-base voltage V_{CB}, (*c*) the small-signal voltage gain, and (*d*) the input impedance.

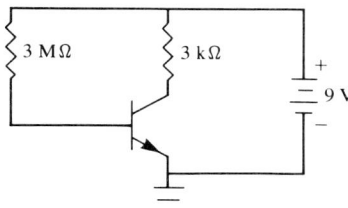

FIGURE P10.13
Schematic for Prob. 10.13.

10.14. The specifications of a problem entailing an NPN transistor indicate that, at $V_{BE} = 0.6$ V and $V_{CE} = 5$ V, the collector current is either (*a*) $I_C = -5$ mA or (*b*) $I_C = 3.42$ mA. Which of the two specifications is admissible?

10.15. Compute a new value for the emitter resistor in Fig. P10.12 to make the collector to base voltage become $V_{CB} = 6$ V.

10.16. Compute a new value for the base resistor in Fig. 10.13 to make the collector to emitter voltage become $V_{CE} = 3$ V.

10.17. Prove Eqs. (10.3.24) and (10.3.26) by using the higher precision coefficients (10.7.9) through (10.7.11) in (10.7.7) and approximating to the first and the second order respectively of small quantities.

10.18. Repeat Prob. 10.15 using the higher precision Ebers-Moll equation parameters of Eqs. (10.7.9a) through (10.7.11a).
    Hint: Assume the work point to be well into the active region, but verify this assumption.

10.19. Repeat Prob. 10.16 using the higher precision Ebers-Moll equation parameters of Eqs. (10.7.9a) through (10.7.11a). Compare with the results of Prob. 10.16 and comment.
    Hint: Assume the work point to be well into the active region, but verify this assumption.

10.20. Repeat Prob. 10.9 using the higher precision Ebers-Moll equation parameters of Eqs. (10.7.9a) through (10.7.11a).
    Hint: Assume the work point to be well into the active region, but verify this assumption. If the assumption is incorrect, then use the Ebers-Moll equations, rather than the approximate expressions.

10.21. Repeat Prob. 10.15 using the higher precision coefficients (10.7.9a) through (10.7.11a). Compare results and comment.

10.22. In Fig. P10.13 the base resistor is changed to yield $V_{CE} = 0.2$ V. Use higher accuracy coefficients.
(a) What is V_{BE}?
(b) Compute the new value of R_B.

10.23. Taking into account base width modulation compute the active base width of the NPN transistor of Prob. 10.9, assuming $V_{BE} = 0.6$ V and (a) $V_{CB} = 0$ V, (b) $V_{CB} = 10$ V.

10.24. In the schematic of Fig. P10.13 the base resistor and collector resistor are changed, so that the BJT is at the onset of saturation with $I_B = 22$ μA and $I_C = 2.93$ mA.
(a) Compute the new resistor values.
(b) If a pulse of current of -100 μA and 2 μs duration is applied to the base, compute and sketch the output voltage.

10.25. Compute the work point of the transistor of Prob. 10.9 if:
(a) $V_{CB} = 2$ V, $I_E = -3$ mA
(b) $I_E = -3$ mA, $I_C = 2$ mA
(c) $V_{BE} = 0.55$ V, $V_{CB} = -0.4$ V
(d) $V_{BE} = 0.55$ V, $V_{CE} = 3$ V
(e) $V_{BE} = -0.55$ V, $V_{CE} = 3$ V

Computer Problems

10.26. Write a computer program to solve Prob. 10.20, interactively permitting the user to change the values of base width and of dopings and minority carrier lifetimes in emitter, base, and collector.

10.27. Write a computer program to compute α_0 and β_0 for the transistor of Prob. 10.9 to the second order of small quantities, interactively permitting the user to change the values of the base width and of dopings and minority carrier lifetimes in emitter, base, and collector.

10.28. Write a computer program to solve Prob. 10.12, interactively permitting the user to change the values of the base width, of the dopings and minority carrier lifetimes in emitter, base, and collector, and of the external resistors and voltages.

10.29. Write a computer program to solve Prob. 10.13, interactively permitting the user to change the values of the external resistors and voltages.

10.30. Write a computer program to solve Prob. 10.22. Interactively permit choice of supply voltage and collector load resistance. Solve using the values of Fig. P10.13.

10.31. Write a computer program to solve Prob. 10.24a.

10.32. Write a computer program to solve Prob. 10.25.

ADDITIONAL READING

Burns, S., and P. Bond: *Principles of Electrical Circuits*, West Publishing Co., St. Paul, Minn., 1987.
Fortino, A.: *Fundamentals of Integrated Circuits*, Reston Publishing Co., Inc., Reston, Va., 1984.
Gray, P., and R. Meyer: Analysis of Integrated Circuits, 2d ed., John Wiley & Sons, Inc., New York, 1984.
Grove, W.: *Physics and Technology of Semiconductor Devices*, John Wiley & Sons, Inc., New York, 1967.

Hanavati, R.: *Semiconductor Devices*, Intex Educational Publishers, New York, 1975.

Hayt, E.: *Engineering Electronics*, 4th ed., McGraw-Hill Book Co., Inc., New York, 1981.

Horowitz, P., and W. Hill: *The Art of Electronics*, Cambridge University Press, 1980.

Millman, J., and C. Halkias: *Integrated Electronics, Analog and Digital Circuits and Systems*, McGraw-Hill Book Co., Inc., New York, 1972.

Mitchell, F., Sr., and F. Mitchell, Jr.: *Introduction to Electronic Design*, Prentice Hall, Inc., Englewood Cliffs, N.J., 1988.

Streetman, B.: *Solid State Electronic Devices*, Prentice Hall, Inc., Englewood Cliffs, N.J., 1972.

Sze, S. M.: *Semiconductor Devices, Physics and Technology*, John Wiley & Sons, Inc., New York, 1985.

Sze, S. M.: *Physics of Semiconductor Devices*, 2d ed., John Wiley & Sons, Inc., New York, 1981.

Wang, S.: *Introduction to Solid State Electronics*, North Holland, New York, 1980.

Yang, E.: *Microelectronic Devices*, McGraw-Hill Book Co., Inc., New York, 1988.

INDEX